"Devant le Deluge" and Other Essays on Early Modern Scientific Communication

David A. Kronick

The Scarecrow Press, Inc.
Lanham, Maryland, and Oxford
2004

SCARECROW PRESS, INC.

Published in the United States of America
by Scarecrow Press, Inc.
A wholly owned subsidiary of
The Rowman & Littlefield Publishing Group, Inc.
4501 Forbes Boulevard, Suite 200, Lanham, Maryland 20706
www.scarecrowpress.com

PO Box 317
Oxford
OX2 9RU, UK

British Library Cataloguing in Publication Information Available

Library of Congress Cataloging-in-Publication Data

Kronick, David A. (David Abraham), 1917–
"Devant le deluge" and other essays on early modern scientific communication / David A. Kronick.
p. cm.
Includes bibliographical references and index.
ISBN 978-0-8108-5003-3
1. Journalism, Scientific—History. I. Title.
PN4784.T3 K67 2004
070.4'495—dc22

2003024607

∞™ The paper used in this publication meets the minimum requirements of American National Standard for Information Sciences—Permanence of Paper for Printed Library Materials, ANSI/NISO Z39.48-1992.
Manufactured in the United States of America.

Contents

Preface

Dissertation subjects are sometimes the result of a marriage of expedience and desperation engendered by an anxious search for a research subject. It is also hoped that the subject is manageable, will satisfy the high standards of a dissertation committee, and can be completed before the candidate has grown old and gray. As a graduate student at the University of Chicago in 1952, I had the happy inspiration to choose the subject of the history of the early scientific journal for my research topic. The subject not only earned me a Ph.D., but also resulted in two editions of a book: *A History of Scientific and Technical Periodicals: The Origins and Development of the Scientific and Technical Press* (2d ed., Metuchen, N.J.: Scarecrow Press, 1976). It seems to have filled a void in the literature. I take pleasure in finding it cited from time to time, probably not because of its merits, but because there has been so little written on this important and interesting subject. I learned a number of facts in the process of compiling a list of relevant titles for the project. One was that the periodicals I wished to examine were widely scattered in libraries all over Europe and North America.

During the intervening period between my inspiration for graduate work and my retirement from a long career as a medical librarian, I have spent many pleasant hours in libraries all over this country and abroad looking for these sometimes elusive records. I was able to spend extensive time in Great Britain, once for six months on a Ford Foundation Grant and again a year after retirement. I used other travel opportunities to search libraries in Göttingen, Munich, Paris, and Strasburg. My choice of the subject

---

Drawn in part from "The Scientific Paper Devant le Deluge" published in *Current Contents. Agriculture, Biology and Environmental Science* 23 (1992): 6–10.

also gave me a wonderful opportunity to become acquainted with one of the most interesting and influential periods in both our social and intellectual histories and to meet some luminous and fascinating characters. The seventeenth and eighteenth centuries were a period not only of a revolution in science in which new ways of putting the question to nature were explored, but also a period of intellectual awakening to problems of science and society and to an enhanced awareness that science could be used for improving the quality of human life.

Over the years, I have found occasion to draw upon this research to write several papers on some aspects of the scientific communication process. They were published in a wide variety of journals, few of which had any relation to the history of science or technology. On my retirement, I had hoped to put together a comprehensive picture of the intellectual, social, and economic aspects involved in the initiation and publication of scientific and technical periodicals in the early modern period. A number of competitive opportunities and projects intervened. With the support of the National Endowment for the Humanities, I was able to put together a guide to the periodicals of this period that I hope will be useful to historians to facilitate the study of early periodicals. The bibliography brings together a large number of titles garnered from many sources, and provides access to them by subject, editor, and associated learned or scientific society (Kronick 1991).

I was delighted, therefore, when Scarecrow Press accepted my proposal to publish a collection of my earlier published papers, ranging from 1960 to 1994, along with some unpublished papers that I was in process of completing. The published papers appear as they did in print with minor editorial changes. I am most grateful to Scarecrow Press not only for this project but also for their support over the years. It started when I met Ralph Shaw, the former distinguished librarian of the U.S. Department of Agriculture and then dean of the Library School at Rutgers University, who had founded the press. I had a chance meeting with him on a railroad platform in New Jersey when he asked me about the fate of my dissertation and I sadly told him it was safely stored on a shelf in my apartment. He graciously offered to take it under his wing and see it through to publication. I told him I was at the time involved with the forthcoming birth of my second child, and he asked me to tell him whether it was a boy or a girl. The first edition of my *History* consequently came out in a pink cloth cover.

As science is a cumulative form of knowledge, the need to access scientific records becomes less and less important as they grow older. There are a number of reasons, however, for studying these sometimes quaint and archaic-looking periodicals. They are important, not only as a resource for learning about the diffusion of scientific and technical information in this early period, but also for studying the characteristics of the scientific information dissemination and storage systems as they evolved. We can witness many of the same concerns and activities that exist in contemporary scientific and technical journalism, such as the problems of authority, documentation, priority, and quality control. All these factors can be studied in embryo in this period.

Science and technology were difficult to separate in this early period, when utilitarian objectives were clearly embodied in the charter of most scientific societies, such as the Académie des Sciences in Paris, and the Royal Society of London. Many of the other scientific and learned societies that sprang up all over Europe also engaged in the improvement of agriculture and industry. Above all, they were formed to generate and communicate information, not only for their own members, but also for wider audiences. In many instances, the preparation and reading of papers were requirements of membership. The societies kept close control over their publications through mechanisms that foreshadow the peer review that is characteristic of our best scientific periodicals today. The specialized journals in chemistry, mining, agriculture, and other subjects dealt primarily with practical problems, although theoretical and scientific papers were included when they were regarded as relevant.

The seventeenth- and eighteenth-century scientific community differed greatly in many aspects from the scientific community at the end of the twentieth century. The differences are not only those of size, complexity, structure, and social significance. They were bigger in the sense that a larger part of the literate population could read and understand much of the literature being produced, but smaller in the number of disciplines that had established their identity. However, there is ample evidence that many of the behavioral and psychological aspects of publishing in science have not changed very much over the centuries. Editors, particularly those of publications issued by scientific societies, were concerning themselves with the problems of peer review. Authors showed as much interest in establishing priorities and were as jealous of their property rights in ideas as they are today. Editors in their prefaces to new journals then showed as much need to justify starting a new journal.

Many of the independent journals (those not associated with the learned societies of the period) were largely derivative in that they often used material borrowed from other journals or included abstracts or reviews of work published elsewhere. In scanning some of these early periodicals, one is sometimes reminded of the legendary village in which all the inhabitants made their living by taking in each other's laundry. In that sense, these periodicals resembled the newspaper that had been established as a format almost a century before 1665 when *Philosophical Transactions*, the first scientific journal, appeared. The newspaper also provided the scientific journal with a format and mode of distribution. As in the newspaper, which may carry the same news no matter where it is published, dissemination of news was the object, and redundancy was not considered a factor.

To understand the development of the standard format of the scientific paper, we need to look at the early scientific journals, such as *Philosophical Transactions*. It owed its format largely to the form of the erudite letter, a primary form of communication for scientists preceding the introduction of the journal. The origins of the scientific paper are also likely to be found in the memoirs presented at scientific societies and selected for insertion into their proceedings. Another likely source for these formats may be found in the "prize essays" that were composed in competition for awards offered by many of the societies of the eighteenth century for answers to questions that the society posed. They gave issue to essays that were published, sometimes separately, and sometimes in series. These efforts, with some stretch of the imagination, may be regarded as a form of sponsored research. In this case, however, the, granting agency chose the topic for investigation rather than the grantee. They thus provided the benefit of having many investigators working on a problem at the same time, without having to fund more than one.

There are large gaps to be filled in the history of the scientific journal. In 1960, estimates of the number of scientific and technical journals ranged anywhere from 20,000 to 60,000 (King, McDonald, and Roderer 1981:319). Some of the reasons for the growth are found in the nineteenth century with the transformation of science in Germany between 1825 and 1900, when departments and institutes in new disciplines were created in German universities and there was intense competition for leadership of these new organizations. The importance of publication as a basis for advancement in academic life also was greatly enhanced. It seemed as if every department chairman or institute director wanted to have his own journal. As the number of disciplines, subdisciplines, and clinical and technical specialties grew, their identities had to be bolstered by new journals. The growth of mission-oriented research in the twentieth century (Adams 1981:224) added another factor, so that all kinds of combinations and permutations of disciplines and clinical and technical specialties were identified, each with its need for a special journal. As each specialized journal is created with an ever-decreasing audience size and a greater scatter of the literature, the distribution system becomes less and less viable. Nevertheless, the scientific journal basically has not changed in format and management from those that appeared in the eighteenth century, although the conditions in which they appear and the technologies that are available to us have changed considerably.

One of the conclusions I reached in my earlier study of the scientific periodical was that it served two primary and important functions: first as a vehicle and then as a depository. This conclusion was not received with any great acclaim because it was, of course, obvious. I felt a little like the Moliére character who, on hearing the definitions of poetry and prose, was surprised to find out he had been speaking prose all his life. The distinction, nevertheless, is an important one. The journal serves first as a vehicle for the dissemination of a collection of individual papers or memoirs that are related in some way or other, although the relationships may not be those that are significant to any individual subscriber. The problem is an old one: classifications are linear while the world is multilinear. As the number of disciplinary and mission-oriented societies grew, the number of periodicals where an individual paper could be published increased, resulting in an ever wider dispersion of the literature.

The other purpose that the scientific journal served, I concluded, was as a haven for individual papers that could be retrieved on demand. The periodical title then serves as a mnemonic to locate a particular paper. The journal also serves other purposes, of course, such as providing a criterion for validating the paper, which is based on the credit and reputation of the journal in which it had been printed. However, if this function could be performed by some other mechanism, any arbitrary system could serve as a locating device. In some sense, this is the way many investigators access the literature today, with the use of indexes and other aids such as *Current Contents*, *Chemical Abstracts*, *Science Citation Index*, and a score of other services. The need for these kinds of aids was recognized very early in the history of the scientific journal. With the growth of experimental science, the individual paper became the significant dissemination unit. It is the unit with which the investigator is primarily concerned.

New technology is making possible the successful introduction of new systems of distribution for research reports and the creation of what can become personally designed journals for investigators. The early journals, like most of our scientific and technical journals of today, were composed of different kinds of ingredients, such as news items, reviews of books, and abstracts of papers, just as most journals are today. There will always be a place for journals that are created for more general audiences, leaving specialized research reports for individual distribution. The issue is not between the use of electronic or paper media for storage or distribution, but of rationalizing and creating economically, socially, and intellectually viable systems that will be able to satisfy all the different options readers desire. These new systems, when they are created, must, of course, not ignore the reward and recognition aspects of the current system, nor sacrifice its methods of quality control and openness, which, as history reveals, were fostered from the very beginnings of the scientific revolution in the seventeenth century.

The study of the history of scientific communication is influenced by many factors, including the prevailing philosophies of the time, the social role of the scientist, the scientist's political environment, and the available technology for communication. There are many disciplines that impinge on these factors, including the philosophy of science, the sociology of science, the history of science, and the rhetoric of science. There is a rich and voluminous literature on all of these subjects and I can claim only a nodding knowledge of any of them. However, all of them must be considered in the study of scientific communication, not only in the early modern period addressed in the papers in this collection, but in the modern period as well. The growing literature on the subjects involved is not encompassed in a single language. I must take credit or blame for many of the translations into English. I only hope they convey if not an elegant, at least a reasonable sense of the original.

The first twelve essays included here were published at different times, in different journals, and for different audiences. The final three have never appeared in print, except for the last one, which was published in an abbreviated form in the *Library Quarterly* [71(1), 2001, pp. 28–43], They are brought together here in the hope that they will serve, in part, as an introduction to this interesting subject.

I owe a large debt of gratitude to the librarians of the many libraries I consulted over the years, both in this country and abroad, but especially to the library of the University of Texas Health Sciences of San Antonio, where I served as director for many years. I am grateful to Scarecrow Press, which has published three other books of mine and made this collection possible, and also to my editor, Nicole Averill, who has graciously seen this book through production.

REFERENCES

Adams S. *Medical Bibliography in an Age of Discontinuity*. Chicago, Ill.: Medical Library Association, 1981. 224 pp.

King, D. W., D. D. McDonald, and N. K. Roderer. *Scientific Journals in the United States: Their Production, Use, and Economics*. Stroudsburg, Pa.: Hutchinson Ross, 1981. 319 pp.

Kronick, David A. *Scientific and Technical Periodicals of the Seventeenth and Eighteenth Centuries: A Guide*. Metuchen, N.J.: Scarecrow Press, 1991. 332 pp.

Chapter 1

Nicolas de Blegny, Medical Journalist

The first medical journal to appear in the vernacular was issued in Paris in January 1679. It deserves interest not only because we can see in its history some of the mechanisms that brought forth such a prolific progeny in the years to come, but also because the life of its progenitor presents such a fascinating glimpse into seventeenth-century French medical life. Like many of the rascals of history, Blegny's character has mellowed somewhat with the passage of time so that he does not appear to be quite the prodigious knave that some of his contemporaries made him out to be. In an age more vulnerable to his talents and frenetic ambitions, he might have been a captain of industry or a mogul of the movies. He deserves a full-length portrait rather than the brief sketch presented here. In his busy and active life in the last quarter of seventeenth-century Paris, he touched every aspect of medical life, from peddling philters to the demimonde, to serving as a physician to the court. He was one of the most litigious of men, and the legal papers that survive, as well as the records that resulted from his continual conflicts with the Paris Faculty of Medicine and other official bodies, provide us with considerable documentation for some of the facts of his life.

The details of his early life are somewhat obscure. In one of the hints about his ancestry, which he provides in the prefaces to some of his books, Blegny claims to be the son of a famous master apothecary in Paris. His birthdate is usually given as 1652, but to several writers it seems more likely that he was born in 1642. His birthplace is also controversial and is given variously as Paris, Lyon, and Chaumont. Chaumont is a geographic

Published in the *Bulletin of the Cleveland Medical Library* 1960: 47–56 subdivision of the department of Haute-Marne, and Blegny was the name of a small village in this district. It seems probable to at least one writer that he may have borrowed this name, as he says, to achieve a dignity that was not his by birth. It is not known at what age Blegny came to Paris. It is generally agreed, however, that he did reside for a time at the Surgical College. "I have a brother," Blegny tells us in his *Remède Anglois*, "who was formerly a beadle at the Saint-Cosme." It is there that he probably learned the rudiments of the medical arts, whether or not he acquired them, as Cabanes says, by "gliding furtively about the rooms in which the lectures were conducted." He seems later to have entered into a partnership with a barber–surgeon as an apothecary, but this could not have been for very long, for a contemporary document speaks of him at this time as being reduced to selling matches, "eau de vivre," and even some sort of remedy against bugs.

A definite turn in his fortune seems to have taken place with his meeting at the Saint-Cosme with one of the king's surgeons, whom he later accompanied into the battlefields of Flanders. It may have been at this time that he was able to meet and ingratiate himself with the duke of Orleans, who later became one of the instruments of his advancement. He did not, however, become surgeon to the duke until 1683, and as he appears to have returned to Paris in 1668 after the peace of Aix-la-Chapelle, the interval was probably taken up with numerous other activities, which established his reputation in the city and his place at court.

One surviving contemporary document now preserved at the Faculty of Pharmacy in Paris speaks of Blegny and his wife as being "driven from their home for an infamous impropriety," unfortunately not cited, on the complaint of a neighbor in 1672. He had married Charlotte Gallois, one of the irregular or unlicensed midwives practicing in Paris at this time, and because Blegny was occupied in some sort of unofficial capacity as a director of their activities, it is possible that they may have met in the course of his official duties. One of the pitfalls of trying to reconstruct a historical character from fragmentary evidence is illustrated by the attempt of one of the writers on Blegny to answer the query: Was his marriage a happy one? It seems quite likely, says this biographer, as in his *Livre Commode des Addresses*, a precursor of the city directory that Blegny published in 1692, he cites the name of his wife as being a very good midwife. Another conclusion we might draw from this favorable comment on his wife's ability is that it may have been prompted more by a desire for increased family income than marital compatibility, but history is silent on this point.

There were two major roads to advancement in the medical world of seventeenth-century Paris. One was through the venerable Faculty of Medicine, which was nominally a part of the university with an authority going back to the time when it was an ecclesiastical body independent of the crown. It was the official medical licensing and regulating body and it guarded its prerogatives jealously. A position on the faculty was gained only through a lengthy and expensive course of attendance at its lectures and ceremonies. The other road, a less arduous one and the only one open to Blegny, who did not even have a provincial university degree—which in any case the Paris faculty refused to recognize—was the court. The monarch and his household had the right to name anyone they wished to the staff of physicians, surgeons, and apothecaries who attended to the needs of the court. The appointments were frequently made on the recommendations of the first physician and surgeon, and Blegny had the good fortune to be in the good graces of two of the leading court physicians of the day, D'Aguin and the Abbé de la Bourdelot.

In 1673 Blegny had published his first book, *L'Art de Guérir les Maladies Vénériennes.* The first edition was quickly exhausted, and it appeared in English in 1676 and in a revised French edition in 1679. His *L'Art de Guérir les Hernies,* published in Paris in 1676, is considered by some historians to be the major seventeenth-century work on hernias. It contains a description of a new kind of elastic truss that the author had invented. He acquired a royal patent for its manufacture and added this enterprise to his many other activities. In the work on hernias, Blegny signs himself "surgeon in ordinary to the Queen," but most of his biographers are in accord in stating that he did not actually enter into this position until 1678. Another of his publications appeared in the same year. When the "secret" of quinine was brought to France by the Englishman Robert Talbot, Louis XIV, anxious to secure this remarkable remedy for his people, paid Talbot a huge sum for it and had the remedy made public through D'Aquin, his first physician. D'Aquin chose Blegny as his medium of communication with the public, and Blegny published *La Découverte de la Remède Anglois* in 1678. The following year, through D'Aquin's and Bourdelot's interventions, he received the royal privilege of publishing a monthly review on medicine, which resulted in his major innovation—the one for which he is most remembered.

Journalism in France at this time was under the direct control of the royal censors and was confined to three major publications: the *Gazette,* the organ for politics and news, and the creation of another French physician, Theophraste Renaudot, whom Blegny is sometimes accused of having taken as his model; the *Mercure Galant,* the medium for the court and society; and *Journal des Sçavans,* which shares with the *Philosophical Transactions* of the Royal Society of London the distinction of being one of the first scientific or learned journals. In accepting the privilege of publishing his journal, Blegny had to enter into an agreement with the Abbé de la Roque, the third editor of *Journal des Sçavans*, that he would confine himself to medicine and would not infringe on the journalistic terrain that the journal had marked out for itself. De la Roque soon had occasion to regret his compliance in this scheme. In the sixth issue of his new journal, Blegny takes the editor of *Journal des Sçavans* to task for reprinting a medical observation from his journal without indicating the source. The immediate success of Blegny's journal may have added to de la Roque's burdens, for he soon joined the Faculty of Medicine in their attempts to have this unorthodox publication suppressed. When they finally succeeded three years later, de la Roque quickly stepped into the breach with a medical journal of his own, obviously hoping to capitalize on the readership Blegny had created. His *Journal de Médecine ou Observations des Plus Fameux Méecins, Chirurgiens et Anatomistes de l'Europe*, however, proved an unsuccessful venture, despite its imposing title. It had, in the words of Levy-Valensi, "six months of life, three years of lethargy, six further months of existence, and that was it."

In addition to the journalistic climate in which Blegny began his medical periodical, another major influence was the academies that flourished in Paris and Europe at the time. The academy, as a formal or informal assemblage of literati or intellectuals devoted to discussions of the new learning, had existed in Europe since the early Renaissance. In the seventeenth century, some of these groups began to devote their attentions more exclusively to the new "natural philosophy" that had resulted from the breakdown of the authority of the old learning. Their activities in England and France had already given rise to unofficial and official organizations of scientific amateurs and professionals. Jean Baptiste Denis, most renowned in the history of science for his early experiments with the transfusion of blood, conducted meetings at Paris between 1672 and 1674. These meetings gave rise to a monthly publication called *Ménoires Concernant les Arts et les Sciences*, which was issued for a time as a supplement to *Journal des Sçavans*. Its interests, however, were much broader than the field of medicine. The Abbé Bourdelot, one of Blegny's most influential patrons, conducted a series of meetings at about the same time and published the results at Paris as *Les Conversations de l'Académie de M.l'Abbé Bourdelot*. Blegny's interests, however, were exclusively medical and he called the meetings that he began to conduct at his dwellings sometime around 1678 the Académie des Nouvelles Descouvertes en Médecine. The medical publication he issued was nominally a result of the proceedings of the meetings of this academy. However, although the academy may have provided him with some of the material for the journal, it served principally as a springboard for his publishing activities. His sources of information were much broader and more diverse, through his contacts with the court and with the various kinds of medical activity in Paris, ranging from the numerous charlatans and empirics who frequented the city, to the premier physician himself

The first issue of *Les Nouvelles Descouvertes sur Toutes les parties de la Médecine* appeared in January in 1679. In his preface addressed to the reader, Blegny says that it is his intention to bring together all those discoveries, experiences, and comments that may be found useful in the art of medicine. In order that he may continue to do so, he urges all physicians, surgeons, and apothecaries, both Galenic and chemical, those residing abroad as well as those in France, to send him their discoveries. They should be sent, he adds, in good time so that they may not lose their usefulness. Each issue, he says, will be sold for five sols, which, he adds, is just sufficient to reimburse him for the expenses entailed in publication. He anticipates also that the issues may be counterfeited abroad as well as in France, and therefore suggests that he may sign each issue himself in order to assure the reader that he has an authentic copy in hand. Furthermore, he offers a reward of twenty louis d'or to anyone who will report to him those booksellers or other individuals who have sold counterfeits.

The issues are cast principally in the form of letters to a physician in the provinces very much in the form that had so recently been distinguished by Pascal. Blegny's journal does not consist, in the main, of separate and individual contributions; it is rather a running report on medical activities in the capital along with news of new books and unusual cases and remedies. Occasionally there does appear a contribution whose authorship is attributable to some other writer, for example the letters of Bourdelot in the first volume, and the exchange of letters by Fagon and Bourdelot on the death of La Rochefoucauld, the celebrated author of the Maxims, in the second volume. The journal has its quota of monsters and reports of incredible phenomena in nature that appear so often in the medical publications of this century. He reports an unusual case of plastic surgery in which a man whose nose had been cut off was advised to use that of his valet, which evidently suited his own physiognomy well. The operation was successfully performed, at least for the recipient of the new nose, until the death of the valet, at which time it was discovered that the nose in question had also died. This story apparently had a fair amount of currency in the seventeenth and eighteenth centuries, for it is also encountered in other sources. From time to time, Blegny announces new remedies that are reported in the city and occasionally provides information about where the reader can procure them. One of these early medical advertisements, for this is what they essentially are, reads with a fluency that could inspire even a contemporary ad writer. In the second volume of the journal, Blegny apprised the reader of a cosmetic "huile de talc"—which is, he says, almost like water. It is unctuous but not greasy like other oils. It is entirely without odor and therefore not at all disagreeable to use. Those who are curious will find it at the establishment of M. Lemery, apothecary to the king, on the Rue Galand near the Place Maubert. It was, moreover, not sold for its real value, but only for little more than it actually cost to manufacture.

In the last issue of the first volume, that of December 1679, Blegny announced that the title of the journal henceforth would be *Le Temple d'Esculape, ou les Despositaires des Nouvelles Descouvertes Qui se Font Journellement Dans Toutes les Parties de la Médicine*. In the preface to this volume, he stated his reason for selecting this title and gave us some insight into both his views of the function of the journal and his ideas of the role and the status of medicine in his day. Nothing served as well to establish medicine firmly, he said, as the tablets that were once deposited in the temples of the pagans in which Aesculapius was worshipped as the god of medicine. These tablets contained not only the names of those who had been cured, but also an exact description of the extraordinary remedies that had been used. It was his hope that, in a similar fashion, this publication of new discoveries in medical science would contribute to the development of more certain principles and more infallible rules for the art of preserving life. Blegny may have had more utilitarian motives than those he stated in his preface (the journal spread his reputation throughout the city and throughout Europe), but in effect he did create a wider audience for medical writing than had hitherto existed. The Faculty of Medicine carried on all its activities in Latin, the language of the scholar. They made no attempt to publish their proceedings, and their records of this period are still preserved in manuscript. There were few more than a hundred physicians in Paris in 1675 who were recognized by the Faculty of Medicine. This number did not include, of course, the numerous other practitioners, the surgeons, the apothecaries, those with provincial degrees in medicine, and others who were practicing medicine without the Faculty's sanction. It was to the latter group that Blegny addressed his journal, first by publishing in the vernacular and thus making it accessible to many of those who had not had the formal classical training of the university, and secondly in admitting to the discussion new ideas and remedies that were not fully in accord with the traditional and sanctioned practice of the Faculty. In doing so, he incurred their bitter enmity and they set about with determination to harass and eliminate him.

The journal appeared in Paris for only one more year—a few months before the Faculty finally succeeded in having Blegny's privilege revoked. The journal was issued in 1681 under still another title, *Nouveautis Journals concernant les sciences et les arts qui fontpartie de la midecine*; Blegny evidently had a distaste for conformity of any kind. The enmity of the Faculty was not based solely on the fact that Blegny had defied their authority in his journal. He had provoked them in many other ways as well. He had blatantly dedicated one of his books to the "physicians of the provincial and foreign faculties practicing at court." It came to the attention of the Faculty that he was conducting dissections at the academy he had established, and that he had unlawfully procured for these purposes the cadaver of a six-year-old child with the aid of his assistant Désnoues. The constabulary was alerted and the cadaver was seized in the name of the Faculty. After a long series of legal procedures, both Blegny and Désnoues were sentenced—Désnoues to be whipped and Blegny to be banished from Paris. Blegny's influence at court was still effective, however, and he managed to have these sentences considerably reduced.

Blegny used the medium of his journal to respond to these unwanted attentions on the part of the Faculty. In a discussion of the "English remedy" in one of the issues, he asks: What does the physician of discretion, that is, a member of the Faculty, do in the presence of a "fièvre tierce." He simply reasons in accordance with the doctrine of Hippocrates, of whose works he recites two or three aphorisms in Greek. He speaks in Latin, makes eloquent speeches, prescribes two or three blood-lettings, several purgations, and the fever degenerates into "double tierce" after the seventh attack. The English physician, on the other hand, cures the illness in three days without speaking either Greek or Latin. Which of the two shall then be called a charlatan? One might respond without outraging reason that it is the physician of discretion. And to this heavy sarcasm he added the statement that the true physicians are those who do know how to cure without vain ostentation. The "bonnet" in itself does not provide its wearer with wisdom. Greek and Latin are more suitable for academic disputations than treating diseases. Hippocrates, no more than any of the authorities, knew everything, and finally, there are doctors of medicine who are truly "médecins," and also "médecins" who have no doctorate in medicine.

The Faculty obviously could not endure such attacks for long without striking back, and they soon presented a petition to the chancellor of the realm to have Blegny's privilege revoked. In it they summed up all the complaints they had against him. "The Faculty of Medicine," they said, "came to throw themselves at the feet of your eminence in order to demand justice against one named Blegny, a former beadle of the Master Surgeons and of the midwives of Paris, who has abused the privileges accorded him, and who has attempted similar enterprises to those of the originator of the Bureau d'Adresses and the inventor of *Gazette de France.* He has promised publicly, in evil books that he has engaged authors to write for him and in which he has signed his own name, to provide medical care for the poor who are ill, to lecture in public on medicine, surgery, and chemistry, and to supply pretended secret remedies with which he abuses the public—all of this without the approval of the Faculty and in defiance of the many regulations of the government. He has used the privilege accorded him in an outrageous manner to attack in a most cruel and vulgar fashion the Faculty of Medicine and the Community of Master Surgeons of Saint-Cosme. His works are full of impostures, calumnies, false facts, wicked railleries, and the most tasteless but also injurious jibes against medicine and surgery. In one of his publications, he even had the temerity to contest the qualifications of the Dean as a physician in ordinary to the King. For these reasons, the Faculty beseeched your eminence to revoke the privilege of the so-called Blegny because he used it in such an atrocious and outrageous fashion."

This moving request of the Faculty, to which the Abbé de la Roque added his testimony that Blegny had not kept his promise of avoiding subjects reserved for the *Journal des Sçavans*, finally resulted in the withdrawal of Blegny's privilege on March 24, 1682. From the first of the year the journal had been appearing again under an altered title, *Journal des Nouvelles Descouvertes Concernant les Sciences et les Arts de la Médecine.* However, the cessation of his publication by no means ended Blegny's activities in Paris, where he continued to carry on his various enterprises for more than another decade. He attempted to revive his journal by following a practice common to Parisians who ran into difficulty with the censors. The journal appeared in Amsterdam under the title, *Mercure Sçavant* and under the auspices of a "Gauthier, médecin de Niort," who may have been a Dutch physician with whom Blegny entered into collaboration, but just as likely may have been a nom de plume that Blegny assumed for the occasion. This publication, however, was of an entirely different character than his earlier journals, and only appeared for two issues.

In this sketch, we shall have to content ourselves with a brief resume of Blegny's activities between the period of 1682 and 1695, when he was finally banished from the city. In 1683 he was appointed surgeon to the Duke of Orleans. Evidently, despite his public disdain of academic qualifications, he still smarted from the criticism of the Faculty, for in 1683 he secured a sort of mail order degree in medicine from the University of Caen, which Levy-Valensi calls "one of those schools which manufactured doctors in several days for exportation." In gratitude, he dedicated his next work to "Messieurs les Doyens et Professeurs Royaux de la Faculté de Médecine dc Caen" and signed it Nicolas de Blegny, "docteur en médecine." This publication was his *La Doctrine des Rapports de Chirurgie,* which Garrison cites in his *History of Medicine* as an important work in the history of medico-legal relations in surgery. In 1685 he purchased the position of apothecary at the court and became embroiled in a series of suits that attempted to force him to confine his activities to one profession or the other. In 1689 he opened a sort of hospital and sanitarium whose virtues he extolled in the September issue of the *Mercure Galant* in 1689.

The Faculty accused him of of conducting all kinds of nefarious activities there, and even of using it as a house of assignation. However, his downfall was brought about by one of his boldest and most imaginative schemes. It seems to have arisen, like many of his other enterprises, from a quest both for status and wealth. The ancient order of Hospitalers of Saint-Esprit had fallen into general inactivity, and in 1672 it had been united by Royal edict with the order of Saint-Lazare. In 1691, Blegny attempted to have the order revived and install himself as the General Commander and administrator of the hospitals of the order. He seemed to have fabricated some sort of credentials for himself and set about selling diplomas to those who wished to be appointed chevaliers to the order. He also compounded remedies made according to secret recipes of the ancient hospitalers for sale to the members of the order. These activities eventually led to his arrest in 1693 and to his deportation in 1695. In the trials that ensued, all the old accusations were brought up again, but his principal influence at court, D'Aquin, had by this time fallen from favor, so his eloquent pleas availed him nothing.

In a printed document of this period, presumably issued by Blegny under the title *Au Roy et à Nos Seigneurs les Commissaires Deputez par Sa Majesté pour l'Exécution de l'Édit du Mois de Mars 1693, Concernant l'Ordre Hospitalier du SaintEsprit*, Blegny summed up the charges and presented his defense. Among other things he had been charged with carrying on heretical religious discussions at his conferences. Blegny argued that he had dealt with nothing but the "secrets of medicine," and that the conferences had been open to the public for more than fifteen years. Furthermore, he had been accused of corrupting the clergy in his house at Popincourt, and of having encouraged them to revolt against the authority of the church. In reality, he said, he had only rented part of his hospital to a certain Abbé des Housseaux who, it is true, could actually not have been very orthodox. But should a landlord be held responsible for the opinions of his tenants? And what a tenant— a gentleman who had furtively stolen his furniture after having caused such an extraordinary scandal in his house. The situation of landlords, he judiciously observed, would be very unhappy indeed if they were held responsible for the conduct of their agents and their tenants. Surely, in circumstances such as these, no one would wish to be a landlord. He had also been accused of constructing an unauthorized hospital and accompanying chapel. As for the hospital, he had donated it to the Ordre Hospitalier du Saint-Esprit for the use of His Majesty's troops and for the poor. The so-called chapel was nothing more than a laboratory that he had never intended for any kind of religious use.

"I am accused of practicing all the medical professions," he said. "I find this more flattering than abusive. It is true, that I have at one time or another, practiced surgery, sold drugs, worked in chemical laboratories, as well as in anatomical theatres. I have practiced obstetrics, reduced fractures, treated hernias and hemorrhoids, women's diseases, and all sorts of indispositions. However, in doing so, I have only been following the distinguished example of Galen, Hippocrates, Guy de Chauliac, Aetius and other celebrated authors who practiced the curative arts in the broadest extent. All of these insinuations which have been directed against me are the result of jealousy and the wish to destroy a man whose influence is feared by his enemies."

He went on to recite the many titles and honors that have been heaped on him, but stopped himself to ask whether it was necessary to recount them all. Wasn't the fact that he was a physician sufficiently ennobling? He compared himself with some of the heroes of history who disdained to seek vengeance against their enemies but rather returned ill-will with good deeds. However, for the sake of the honor of his family, he found it necessary to demand justice and full reparation for all the injuries that had been done him. This eloquent plea seemed to have had no effect on the King's counselors because in 1695 he was exiled to Angers in the company of a vendor of prohibited books. It is likely he traveled in Italy for a while, but in 1699 we find him settled in Avignon in pontifical territory where he practiced medicine until his death in 1722.

There seems no doubt in the face of overwhelming historical evidence that Nicolas de Blegny was indeed somewhat of a rogue. Although his contributions in the field of medicine are minor, his innovations do provide him with some importance as a social force, which contributed to the eventual breakdown of the authority of the Faculty of Medicine of Paris and of traditional medicine. His pioneer venture in the field of medical journalism created a precedent that was not followed successfully until the eighteenth century. His journal was translated into Latin by Theophile Bonnet as *Zodiacus, Medico-Gallicus* in Geneva, and in Hamburg as *Monatlicbe neueröffnete Anmerckungen über alle Theile der ArtzneyKunst* in 1680–1683. It is said that the Amsterdam issues of his journal inspired Bayle to create his *Nouvelles de la Républic des Lettres.* It is interesting to note that Blegny, like Renaudot, was a rebel against the established authorities in medicine, and that they both found in journalism a strong ally to aid them in the pursuit of their goals.

REFERENCES

Cabanes. "Le Premier Journaliste Médical et ses Multiples Avatars," *Lancette Française, Gaz. de bop.* Paris: 1924, vol. 97: 141–45.

Chereau, A. *Essais sur les Origins du Journalisme Méical Fraçais Suivi de sa Bibiliographie*, Paris: Bureau de l'Union Médicale, 1867.

Fauvelle, R. *Les Étudiants en Médecine de Paris Sous le Grand Roi.* Paris: Steinheil, 1899.
Lacassagne, J. "A Propos de Nicolas de Blegny." *Bulletin de la Société Française d'Histoire de Médecins* (1934) 28: 68–69.
Levy-Valensi, J., and P. J. Tellier. "Nicolas de Blegny Journaliste." *Aesculape* (1934) 24: 170–73.
Roshem, J. "Un Journal de Médecine au xviie Siècle." *Paris Médical,* 1912–1913, Suppl., 663–69.
Tellier, P. J. *Un Aventurier Médical au xviie Siècle; Nicolas de Blegny.* Paris: Arnette, 1932.
"Une Ètrange Requéte du Sieur de Blegny, Fondateur du Premier Journal de Médecine, d'Après un Document Inconnu de ses Biographies." *La Chronique Médicale* (1902) 9: 3–12.

Chapter 2

Scientific Journal Publication in the Eighteenth Century

In his *History of Tom Jones*, Henry Fielding made a statement regarding the nature of periodical publication, which is of basic relevance in the history of scientific periodical publication. He was connected for short periods of his life with journal publications of his own—the *True Patriot*, which lasted for only a few months, and the *Covent-Garden Journal*, which existed for eleven months (Herd 1952:45)—so he speaks with some authority when he says: "Such histories as these do, in reality, much resemble a newspaper, which consists of just the same number of words, whether there be any news or not. They may likewise be compared to a stagecoach, which performs constantly the same course, empty as well as full" (Fielding 1964:38).

The analogy between the newspaper and the stagecoach is interesting because the most immediate ancestor and model for the scientific periodical was the newspaper, and the stagecoach provides an excellent characterization of the type of function a periodical plays. Just as the stagecoach plying a scheduled route retains some compunction to remain full of passengers, the periodical, which also travels a more or less regular and prescribed (or rather subscribed) route, creates the necessity of constantly providing a new payload for it to carry. The newspaper comparison is also interesting because some of earliest scientific or learned periodicals did very much resemble newspapers in the brevity of their reports and in the repetitive nature of many of the publications. The repetitive quality is underlined by the fact that so large a proportion of the scientific or learned periodicals of the eighteenth century were derivative publications

---

Published in the *Papers of the Bibliographical Society* 59 (1965): 28–45

in the form of abstract and review journals, and by the fact that they borrowed from each other so freely. This last characteristic is preserved by the newspapers of today, except, of course, that instead of simply borrowing stories, accounts, and articles from other newspapers without leave, they acquire them by subscription to a centralized news service.

Although there have been several studies of the eighteenth-century periodical, there does not appear to be enough data to allow any extensive comparisons between the functions, organizations, readerships, and economics of the earlier prototypes and the scholarly journal as we know it today. This kind of comparison may be of some interest in trying to understand the role that the scientific periodical plays today because it is frequently pointed out that, in many ways, the editorial policies and management procedures have not changed very much. One area in which it is particularly difficult to collect data is relating to the economics of eighteenth-century journal publication and distribution, which are important facets of the problem. This chapter, accumulated during the course of studying the eighteenth-century scientific periodical (Kronick 1976), is therefore presented to outline this area of interest and to offer some tentative conclusions regarding the economics of eighteenth-century scientific periodical publication.

One of the significant differences between the eighteenth century and today in the economics of printing is in the price of paper, as the introduction of the Foudrinier paper machine in the nineteenth century considerably lowered the price and offered a steady supply of paper (Plant 1939:330). This benefit was offset to a considerable degree by increases in costs of composition and printing because of increases in the cost of labor. It would seem apparent that the conjunction of these two factors would make it much less profitable in modern times to print small editions than it was in the eighteenth century. This is not borne out, however, by a comparison of costs of printed materials in these two periods.

One of the most extensive historical studies of book prices is that of Krieg (1953), who tried to surmount the problem of comparing fluctuating currencies, commodity prices, and changes in economic structure by trying to compare prices in terms of staple commodities. We learn from Krieg, for instance, that a *Reis* of paper (about 250 sheets) cost one guilder in 1552, at a time in which one guilder could purchase 1,300 eggs—whereas in 1636 the same quantity of paper could be purchased for the equivalent of 100 eggs and, in 1727, for the equivalent of 125 eggs.[1] I am not aware of any factor in the technology of paper production that would have accounted for this decrease between 1552 and 1636. Even as an amateur in economic history, I must confess that I have some suspicion that the decrease in quantity of eggs required for a similar quantity of paper may just as likely be due to an increase in the value of eggs as a decrease in the value of paper. Krieg uses this same kind of data to prove that an incunabulum cost more as a new book when it was printed in 1482, than it did when it was purchased as a rarity in 1912. He points out that Reuchlin's *Vocabularius*, printed in Basel in 1482, sold for 2.80 guilders rubricated and bound, and for 1.25 guilders without rubrics and unbound. In 1912 it was listed by dealers between 160 and 200 marks, when 2.80 guilders could purchase 2,550 eggs in 1482. In 1912, 200 marks could buy only 2,250 eggs (Kreig 1953).

A study of the price of paper in the early years of the scientific journal does have some significance for our understanding of the role it plays in the communications process, particularly in relation to the cost of composition and presswork. It would seem apparent that the higher the price of paper (factors involving materials) in relation to some of the other costs of production (factors involving labor), the more economical it might be to duplicate and reprint information rather than print larger editions. It would increase the possibility of a publication returning its costs when printed in small editions. Unfortunately, we seem to have very little economic data concerning the early scientific periodical. Joachim Kirchner (1928a, 1928b), who is one of the closest students of the journal publications of this period, has marshaled some data relating to the German periodicals of this era. He points out that although there are a great many comments in the literature extolling the growth of the periodical as an instrument of the Enlightenment, or inveighing against them because of the intolerable burden they place on scholars, there is, however, hardly any information available in the eighteenth-century literature about production costs—details about costs of composition or printing, the payment of honoraria to authors, the size of editions, the methods of distribution, or whether or not journal publication was profitable.

Kirchner was able to dig out some information from the archives of the Breitkopf Company, a venerable publishing firm whose records are preserved back to the middle of the eighteenth century. For example, for a review journal of Russian literature printed in German by Breitkopf in 1772 for a firm in Riga,[2] he found the following data: of a total of 59 thaler 20 groschen for printing the issue in an edition over fifteen hundred, 23 thaler 8 groschen represented the cost of printing (composition and press work) while 23 thaler 33 groschen represented the cost of the paper, and 2 thaler 27 groschen represented the amount billed for the other costs, such as proofreading and licensing. In other words, almost half of the entire cost of producing the issue was due to the cost of paper. As a basis of comparison, Kirchner provides us with some figures for the costs of printing an edition of similar size in Germany in 1913: Out of a total of 465 marks, 330 marks represented the costs of printing, of which 260 marks were for composition only, while the cost of the paper was only 70 marks and other miscellaneous costs were 65 marks. That is, the cost of paper in 1913 represented only about 15 percent of the entire cost of producing the issue, as compared with almost 50 percent in 1772. Kirchner also attempted to work out some comparisons between costs in 1772 and 1913 based on a standard commodity (in this case rye). He concluded that the 1772 journal cost just a little more than half of what it would have cost to produce in 1913, the price of paper on this basis having been reduced 45 percent in the interim while composition and printing costs had increased 180 percent.

Arthur Young started his *Annals of Agriculture* in London in 1784 as a nonprofit venture to advance the cause of scientific agriculture in England. In the beginning, it was rather more of a nonprofit venture than he had

Table 2.1. Costs and Income Statement *Annals of Agriculture 1784*

	Costs				
	Issue Number	*Size of Edition*	*Total Costs*	*Page Count*	*Price Per Copy*
Printing and paper	1	1000	£17.19.6	88	4.3d
	2	1000	9.18.0	48	2.4
	3	500	11.02.6	84	5.3
	4	500	9.07.6	72	4.5
	5	500	12.18.6	96	5.8
Stamp office	Total		£63.16.0		
Carriage, postage			3.15.0		
4 percent commission			4.0.0		
Advertisement*			19.4.0		
Grand total			£90.14.3		

*Young adds that: "This is a small part as the county accounts and part of the London accounts are not come in." It isn't clear whether this represents income or expenditures.

Income			
Issue Number	*Copies Sold*	*Total Income*	*Income Per Copy*
1	500	£22.10.0	10.8d
2	500	14.00.0	6.7
3	400	18.00.0	10.8
4	300	10.06.0	8.2
5	300	13.10.0	10.4
Total		£78.16.3	
Balance beside advertising		11.18.3	
Grand total		£90.14.3	

hoped, and at the end of the sixth issue of the first volume, he presented his readers with an account of the year's publishing activity, that is of the first five issues. In the first issue he had promised a reduction in the cost of the publication as soon as conditions permitted. He had also promised additional illustrations. In order to justify his inability to provide either, he presented an account of his cost and income, which provides us with some indirect evidence of the costs of periodical publication in late eighteenth-century England. (See table 2.1.)

When Young reduced the size of the edition from 1,000 for the first and second issues to 500 for the third, having sold only 500 of the first two issues, he increased his cost per copy by only a small amount. The cost of paper and printing for an edition of 1,000 for an eighty-eight-page issue was 4.3 pence per copy, while the costs of an edition of 500 for an eighty-four-page issue was 5.3 pence per copy. The second issue, which also came out in an edition of 1,000, consisted of only forty-eight pages and cost 2.4 pence per copy for the paper and printing. It also seems that periodical publishers may have been more ethical than periodical publishers today, because the price per copy seems to be predicated on the number of pages—that is, issue number 1 at eighty-eight pages, brought in 10.8 pence per copy, while issue number 2 at forty-eight pages brought in only 6.7 pence per copy. This may not have been the general practice with all periodicals of the period. We know that the price of a single copy of *Philosophical Transactions* varied from sixpence to one shilling in the period between 1669 and 1677 (Rostenberg 1952:12). It is rather surprising that over a hundred years after the introduction of learned journalism, the differentiation in pricing policies between books and periodicals was still not more sharply made.

Young's *Annals of Agriculture* lasted over thirty years, an unusual duration for a periodical of this time, but I do not know whether at any time during that period it proved to be a truly profitable venture. There is very little data on whether many of the scientific periodical publications of the seventeenth and eighteenth century were profitable. In a number of instances in the prefaces to new volumes, the editors take the opportunity to confide to their readers that they are seeing the publication through the press at a great sacrifice of time and money. Although we may tend to discount some of their statements, there is undoubtedly some validity to many of them. The German physician Ernst Gottfried Baldinger,[3] who had a penchant for starting journals, complains in *Medicinisches und Physisches Journal* of 1790 that between April 1 and June 11, he has had to entertain sixty-nine foreign visitors, and that although he had been writing letters every morning from four to seven, he is still some three hundred letters in arrears. He is working for four different printers, he says, and all this effort is largely in the service of scholarship because his journals cost him more than they bring in.[4] "Pierre Bayle's celebrated *Nouvelles de la*

République des Lettres, published in the late seventeenth century, apparently brought in only enough to satisfy the booksellers." In fact, in many cases periodicals were started and supported by booksellers as a means of publicizing their wares, following a tradition that associated the literary journal with the old *Mess-Katalogen* published in association with the German book fairs.

Booksellers, on the other hand, frequently tended to confuse book and periodical as modes of publication. In a discussion printed in an early-eighteenth-century periodical, *l'Esprit des Journaux*, the editors state some of the reasons why periodicals are not as profitable as books. Very few, they note, are issued in subsequent editions, from which a publisher derives his real profit. In addition, each part of the periodical must be separately submitted to the censors, and each unit be separately advertised in the newspapers—while this need be done but once with a book. Furthermore, the parts sell unevenly. If a first issue sells well, a publisher may get eager and overprint for the second, which then may not sell well at all. So it is untrue, the essay concludes, that many booksellers get rich by selling journals.

One indirect piece of evidence we have concerning the profitable or unprofitable nature of periodical publications in the eighteenth century, which requires little interpretation or qualification, is their duration. Profit and loss may not be the only factor involved in the continuation or cessation of a periodical. But in the eighteenth century, when so many of them were privately financed ventures, it was probably a much more significant factor than it is today, when so many of our learned periodicals are subsidized. Between 1665 and 1790, approximately two-thirds of the periodical publications were independent of any official sponsorship. In this same period, almost 60 percent of the periodicals existed for only five years or less and only 10 percent were able to survive twenty-five years or more. Kirchner's data (1928a), covering a shorter time span (from 1665 to 1716) but less restricted in subject (covering as it does the entire population of German journal publications in that period), reveals an even greater brevity of duration. Of the 167 titles listed in his bibliography, one-third lasted less than one year, half ceased publication before the end of the second year, and three-fourths did not last five years. Of the twenty-five of this group, which lasted over ten years, only nine survived the twentieth year. (See table 2.2.)

Even though the majority of the learned periodicals were privately sponsored, there were, of course, indirect subsidies to journals, just as there are today, in the form of volunteer, unpaid, editorial, and managerial assistance, and support given by organizations not directly responsible for the sponsorship of the publications. The two earliest publications of the class with which we are concerned, *Journal des Sçavans* and *Philosophical*

Table 2.2. Duration of Scientific-Technological Periodicals, 1665–1790

Duration, in Years	*Number of Titles*	*Cumulative Totals*
1	277	
2	140	417
3	86	503
4	74	577
5	49	626
6	36	662
7	35	697
8	31	728
9	23	751
10	21	772
11–15	86	858
16–20	42	900
21–25	27	927
26–30	25	952
31–	93	1,045

Transactions, although they had no official sponsorship in their early years, both received indirect support. A historian of *Journal des Sçavans* says: "The evidence is lacking, but it is almost certain that the publication was subsidized by the State. We have not been able to discover what the subscription price was, but it is very probable that it was not enough to cover expenses. Besides it was in Colbert's interest to keep the journal under economic control" (Morgan 1929:96).

Another kind of indirect subsidy is that in which the editors or contributors to journals receive some form of support from the state or an officially organized body. This was certainly true in France where many of the members of the Académie des Sciences received stipends from the government (Bertrand 1865:62). Even in England, where this kind of direct subsidy to scholars was apparently against the tradition of intellectual independence, the state provided official positions in their government for some of the scientists who were involved in the publication of *Philosophical Transactions,* as Robert Hooke was for a while. It was not until 1753 that the Royal Society of London took over official responsibility for the journal. In the beginning, Henry Oldenburg, the first secretary, assumed the financial responsibility and these responsibilities were taken over by a succession of secretaries and editors after his death. Tradition has it that the Duke of Saxony subsidized *Acta Eruditorum,* another one of the famed scholarly periodicals of the end of the seventeenth century. It has been reported that he gave the editors some 200 thaler yearly to support it (Kirchner 1928b:65). There is no indication, however, how long this subsidy lasted, and as *Acta* remained in the family of Otto Mencke, the first editor, for several generations, it may be that it finally achieved a self-supporting and even profitable existence.

There are also other scattered examples of subsidies of scientific journals in the eighteenth century. The proceedings of the various scientific societies, which began to spring up about this time, were, of course, for the most part supported by their respective organizations. On some occasions, journals were able to attract individual benefactors. *Memorie di Matematica e di Fisica* of the Società Italiana della Scienze of Verona and Modena, for example, was issued in 1770 at the expense of the director of the military school at Verona, who not only undertook to pay the cost of the publication during his lifetime, but also established a fund or an endowment to carry on this work after his death.[5] There were at least two military medical publications available for general readership in the eighteenth century published at the government's expense. *Journal de Médicine Militaire*, edited by the medical consultant to the French Army from 1782, was "printed at the expense of the King and distributed without cost to the medical officers of hospitals and regiments."[6] The Austrians followed the French example with *Bibliothek der neuesten medicinisch-chirurgischen Literatur* issued in 1789, which, as indicated in the first issue, was "to be made available to military surgeons at the closest military pharmacy."[7] There are other examples of subsidies for independently sponsored scientific journals. There is even one example that might be called a form of subsidy in reverse—that is, in which subscriptions to a scientific journal were solicited on the basis that the proceeds would be used to support a school for orphans.[8] On the whole, however, it is fairly safe to conclude that the independent periodicals had to depend for their support largely on their subscribers, publishers, or editors.

Additional evidence on the economics of the eighteenth-century scientific periodical publication is provided by the data we have on the size of the various editions. One thing we must keep in mind is that in this period the size of an edition is much less likely to be synonymous with its list of subscribers than it is today. Although there does not seem to be much information about distribution methods for periodicals of this early period, the inference can be made that many journals depended on bookstores and other methods of distribution for a large part of their sales. Verifiable data on size of edition for some of the early periodicals is not always easy to acquire. It has been estimated, for instance, that *Tatler* and *Spectator* on occasion sold as many as 2,000 copies. Samuel Johnson, on the other hand, estimated that, according to the amount of stamp tax paid to the government, the average edition was about 1,700 (Collins 1926:285). We have, however, fairly reliable evidence about *Gentlemen's Magazine*, which, despite its title, is directly relevant to this study because it frequently included material of scientific interest. It has been reported that it had a circulation of 10,000 in 1739 and that it reached 15,000 only a few years later (Plant 1939:57). These were general readership periodicals, whose circulation figures were considered exceptional in their day. A periodical of more limited readership, like *Botanical Magazine*, was considered a huge success when it achieved a readership of 3,000. William Curtis, the founder and editor who began *Botanical Magazine* in 1787, has been quoted as saying that an earlier periodical publication on botany called *Flora Londinensis* that seldom sold more than 300 copies may have brought him fame but *Botanical Magazine* brought him pudding (Hunkin 1946).

The size of the edition of the eighteenth-century periodical, and to be sure, an eighteenth-century book as well, was limited by the capacity of the eighteenth-century press. Publisher Charles Knight estimated that on the seventeenth-century press, two men working an eight-hour day could print 1,000 sheets a day—that is, a thirty-two page issue of a periodical in about 500 copies in octavo (Plant 1939:89). This does not include the time necessary for composition and whatever proofreading, folding, cutting, and other processes may have been necessary. Of course, the workday was considerably longer than eight hours in those days. According to the printing regulations of Frankfurt in 1660, the working day was from four in the morning until nine at night. By 1701 it had been reduced from 17 hours to 16 hours, that is, from five in the morning to nine at night, which may have represented a considerable labor gain (Goldfriedrich 1908:22). These estimates of press capacity for the seventeenth-century press apply with almost equal validity to the eighteenth-century press; it was not until the introduction of the steam-powered press that the speed of printing was greatly increased. By 1814 a machine was capable of turning out 1,800 impressions an hour, and by the middle of the century, this had been increased to 20,000 an hour, a development that made possible the popular magazines and newspapers of that day.

We shall have to know more about the economics of the eighteenth-century periodical press before we can estimate what the break-even point was for the edition of a periodical publication in that era. Kirchner says that most of the periodicals that survived for any length of time ranged in size between 1,000 and 2,000, although society publications that had, so to speak, captive audiences as well as subsidies could exist for long periods with much smaller editions. Some of the evidence, however, is contradictory or difficult to interpret. A study of Swiss periodical publication by Lang (1939), covering the same period as that of Kirchner's study, says that the editors of *Freymüthige Nachrichten*[9] counted on only 200 to 300 subscriptions when they founded the journal, but does not state whether they achieved their expectations or not. The editor of *Burger-Bibliothek* on the other hand anticipated only fifty to seventy. By contrast, he says *Oberrheinischen Mannigfaltigkeiten* had 639 subscribers at the end of the first year. It is rather anomalous that *Oberrheinischen Mannigfaltigkeiten* closed up shop after three years, while *Freymüthige Nachrichten* continued to appear for some twenty years.

One of the difficulties, I believe, in interpreting these figures, is that the number given for subscribers does not include the bookstore sales in many cases, and thus represents a misleading picture of the readership. Some of the data about size of edition seems fairly unambiguous, on the other hand. Oldenburg, secretary of the Royal Society of London, and editor of *Philosophical Transactions*, writes to Robert Boyle on January 16, 1665, that the first issues have not sold more than 300 copies and this has hardly been enough to pay for the costs of the paper and printing (Rostenberg 1952:13). Kirchner has resurrected the following figures from the Breitkopf archives, which represent actual printings:

Vermischte Beitrage zur Philosophie und der schönen Wissenschafaten, 1762–1764, 835–1,580
Bibliotheca physico-medica, 1776–1777, 500
Mineralogische Belustigungen, 1768–1771, 1,000
Gothaische Magazin der Kunste und Wissenschaften, 1776–1777, 2,000
Leipziger Magazin der Naturkunde, Mathematik und Oekonomie, 1781–1788, 1,000

Kirchner has estimated that, taking into consideration contemporary costs, a publisher would have to dispose of an edition of at least 1,000 to make a profit (1928b:49). It is difficult, therefore, to interpret some of the information relative to size of edition that we find in the publications themselves. In the preface to the first edition of *Journal der Physik*, a natural history journal edited by Albrecht Carl Gren, for instance, only eighty subscribers are listed,[10] a1most all of them from Germany. *Magazin für die gerichtliche Arzneikunde und medizinische Polizei*, a public health journal issued in Stendal from 1781, lists a total of ninety-six subscribers in their first issue, again almost entirely from Germany.[11] On the other hand, Tode, in his *Sundhedstidende*, a popular health journal with a fairly limited readership since it was published in Danish, lists over 400 subscribers in the first volume.[12]

A similar publication in Germany was *Der Arzt, eine medicinisehe Wochenschrift* edited by Johann August Unzer. In the preface to the second edition published in 1769 of the volumes that had appeared between 1759 and 1764, a total of 1,700 subscribers are listed, and what is even more interesting, they are listed along with their occupation and place of residence, so we are provided not only with a kind of readership survey of the publication, but also an indication of its distribution. One difficulty is that the term "subscriber" here should probably not be interpreted the way we would today in relation to a periodical, but rather in relation to what we would call a subscription book, so that the number of subscribers does not necessarily coincide with the size of the edition. Among the subscribers to *Der Arzt* are listed some seventy book dealers in various cities, most of them subscribing to multiple copies, of which the highest number of an individual dealer is thirty-five copies, and the total dealer subscription is 400. Physicians and surgeons represent 11 percent of the subscribers, and clergymen 14 percent. The rest are divided among officials and other laymen.[13]

Lorenz Crell, one of the earliest of the chemical journalists, provides us with similar information for the first volume of his *Chemische Annalen*, in which he lists about three hundred names,[14] confined almost entirely again to German-speaking areas. He tells us nothing, however, about its distribution, which is one of the most significant aspects of the economics of the eighteenth-century periodical. This is also true of the scientific periodical of today, for without proper and adequate distribution, the scientific periodical cannot carry on its mission of disseminating information. We must not, however, make the mistake of equating distribution of literature with the dissemination of ideas. Distribution merely contributes one of the necessary means. One of the examples that is frequently cited of a failure in scientific communication is Mendel's classic paper on heredity that is said to have lain fallow for almost a half a century before its ideas were cultivated. *The Union List of Serials* shows twenty-one American libraries had volume four of the proceedings of their collections in which Mendel published his paper. The ten libraries that responded to a query about when they had acquired it all indicated they had it in their possession long before it was "rediscovered" in 1900 (Dorsey 1944). Publication in German probably was a much greater barrier to the transmission of information to a wide audience in Crell's day than it was in Mendel's day. A great deal of the periodical literature of the eighteenth century was translated, abstracted, and reviewed in journals in languages other than the original. *Journal des Sçavans*, for example, was published in a Latin edition at Leipzig and in an Italian edition at Rome. It was also issued in several pirated editions, particularly in a rather badly printed duodecimo edition in Amsterdam. In the preface to the March 1684 edition of *Nouvelles de la République des Lettres*, Pierre Bayle says: "If one did not print the *Journal des Sçavans* in Holland . . . it would remain unknown forever to the booksellers of the north and of Germany."[15]

Although we do not seem to know a great deal about the distribution of the eighteenth-century periodical, the evidence that we do have indicates that it may have left a great deal to be desired. In the preface to the

January 1680 edition of his *Temple d'Esculape*, Nicolas Blegny, physician, editor, and publisher, addresses his readers as follows:

> Those as soon as they are printed and who do not have convenient communications with Paris for this purpose, can assure themselves of doing so by sending an *ecu* to the author (truly very little) at the beginning of each year, because he then will not fail, following this, to send each month by messengers who will be designated to him, the newly imprinted issue, with cost to no one. (Roshem 1912:663)

Some of the ambiguity in these instructions may be due to what they lose in translation; it is difficult to determine, for instance, whether the purveyors of the individual issues were to be engaged by the publisher or by the subscriber, or would just be someone who happened to be traveling to Paris from the provinces, but it seems evident that there was no well-established system for distributing the journal. It is probably also true that there was no great sense of urgency in obtaining current issues as there is today, and that subscribers were quite content in some cases to wait for the end of the year when an entire volume might be made available at the booksellers.

John David Michaelis, who edited *Göttingsche Anzeigen von gelehrten Sachen*, addressed the preface in 1754 almost entirely to the problem of distribution. First, he notes the complaints of readers who say that they have to pay too much to receive the journal at distant places. He suggests that they might reduce their costs by purchasing only the biannual (two volumes were published each year) or even the annual issue. The journal cost only one-half louis d'or at the subscription place. The price otherwise was three Reichstaler and four groschen. Surely, he goes on, most of the large German provinces have one traveler or more who might stop at Göttingen at least once a year and who would be delighted to pick up copies for his friends. In order to make things easier for their readers, he says, they have begun to send copies to the Leipzig Fair, and these could be bought eventually from any traveling bookseller. On the other hand, those who wished to receive the journal as it was issued, two or three times a week, had to apply to the post office in their own area. There is no point, the editor assured his readers, in writing to him, since he has nothing to do with sending out the journal, and can do nothing without a post office contract. In the same way, those who did not receive a particular issue need not write to him, but should direct their complaints to the post office. He can, he says, provide individual issues, but does not like to encourage the practice because he does not like to break up sets of issues. Nevertheless, individual issues would be made available at the end of the year. However, as a scholar, he says, he prefers not to be annoyed with this sort of thing.

It is impossible from these and similar bits of evidence published in the journals themselves to reconstruct any clear picture of the methods of distribution of the seventeenth- and eighteenth-century learned periodical. Part of the answer is to be sought in the history of the national and international postal services in particular, and in the history of transportation in general. All these developments have brought about improvements through which the speed of transmission of scientific information as embodied in the scientific journal has increased to a phenomenal degree over that known in the eighteenth century. When we talk of the speed of transmission, we are of course considering the periodical largely an artifact and not an embodiment of ideas. The kind of transmission we have been discussing deals with transmission only along one of the dimensions we have mentioned: the dimension of space. When we deal with the transmission of scientific information in time, we are no longer primarily concerned with artifacts, and discussions of things like publishing formats become more or less irrelevant.

In studying the economics of any system of disseminating and storing scientific information, all the factors we have mentioned in relation to the eighteenth-century periodical must be considered. Among them are the cost of production, the basis of support, and the means of distribution. The effect of the relationship between the costs, labor, and materials in printing may have produced in the eighteenth century the kind of vehicle for the dissemination of information that is inconsistent with our requirements today. Another and even more important question we can ask of this historical analysis is how well does it serve as a vehicle for the transmission of information and how well does it serve the purposes of the storage of information, a role that the periodical plays for the remainder of its life after it has had its brief fling with its editors and readers?

It seems very likely that the scholarly journal in the eighteenth century was as much an object of subsidy as it is today. The growth of advertising revenue may have had some impact on the nature of support of some publications and the growth in the total audience may have created potential for others; but on the whole, the specialization and fragmentation of knowledge have probably kept the size of the audience relatively stable and advertising subsidies have not been significant for most publications of this kind. Before these questions are answered, however, additional historical and comparative studies will be required.

NOTES

1. This is confirmed by Plant (1939:205), who says that prices were higher for paper in the eighteenth century than they were in the seventeenth.

2. *Russische Bibliothek zur Kenntnis des gegenwartigen Zustandes der Literatur in Russland.*

3. Baldinger was successively a professor of medicine at Jena, Göttingen, and Cassel and eventually at Marburg. He was at one time or another involved with six different periodical publications (Martin 1924).

4. *Medicinisch-chirurgische Zeitung* 4 (1790): 200–204.

5. *Repertorium der medicinische Literatur* 1790: 394–95.

6. *Journal de Médecine Militaire* 1782: 23.

7. *Bibliothek der neuesten medicinisch-chirurgischen Literatur* 1789: 160.

8. *Journal des Sciences et Beaux Arts.* See *Observations sur l'Histoire Naturelle* 11 (1778): 80.

9. A literary review journal published between 1742 and 1763 (Lang 1939:40).

10. *Journal der Physik*, vol. 1, 1790.

11. *Magazin für die gerichtliche Arzneikunde und medizinische Polizei*, vol. 1, 1781: 795.

12. *Sundhedstidende*, 771.

13. *Der Arzt, eine medicinisehe Wochenschrift*, second edition, vol. 1, 1769.

14. *Chemische Annalen*, vol. 1, 1784.

15. *Nouvelles de la République des Lettres*, 1684, preface to the March issue.

REFERENCES

Bertrand, Joseph L. *L'Académie des Sciences et les Académiciens de 1666 à 1793.* Paris: 1865. Reprint edition Amsterdam 1969.

Betz, Louis P. *Pierre Bayle und die Nouvelles de la République des Lettres.* Zurich: Albert Muller, 1896.

Collins, A. S. "Growth of the Reading Public during the Eighteenth Century." *Review of English Studies* 2 (1926): 284–94, 428–38.

Dorsey, M. J. "Appearance of Mendel's Paper in American Libraries." *Science* 99 (1944): 199–200.

Fielding, Henry. *The History of Tom Jones.* New York: Random House, 1964.

Goldfriedrich, J. *Geschichte des deutschen Buchandels, 1648–1740.* Leipzig: 1908.

Herd, Harold. *The March of Journalism: The Story of the British Press from 1622 to the Present Day.* London: Allen & Unwin, 1952.

Hunkin, J. W. "William Curtis: Founder of the *Botanical Magazine.*" *Endeavor* 5 (1946): 13–7.

Kirchner, Joachim. *Die Grundlagen des deutshcen Zeitschriftenwessens.* 2 vols. Leipzig: Hiersemann: 1928a–1931.

Kirchner, Joachim. "Zur Entstehung und Redaktiongeschichte der *Acta Eruditorum.*" *Archiv für Buchgewerbe und Gebrauchsgraphik* 65 (1928b) Sonderheft 4: 78–88.

Krieg, Walter. *Materialien zu einer Entwichlungsgeschichte der Bücher-Preise und des Autoren-Honorars vom 15 bis zum 20 Jahrhundert.* Vienna: Stubenauch, 1953.

Kronick, David A. *A History of Scientific and Technical Periodicals: The Origins and Development of the Scientific and Technical Press, 1665–1790.* 2d ed. Metuchen, N.J: Scarecrow Press, 1976.

Lang, Carl Ludwig. *Die Zeitschriften des Deutschen Schweiz zum Ausgang des 18ten Jahrhundert (1694–1798).* Leipzig: Harassowitz, 1939.
Martin, Alfred. "Ernest Gottfried Baldinger (1738–1804): A Portrait from the Period of German Nationalism." *Medical Life* 31 (1924): 97–103.
Morgan, Betty. *Histoire du Journal des Sçavans Depuis 1665 Jusqu'en 1706.* Paris: Presse Universitaire de France, 1929.
Plant, Marjorie. *The English Book Trade: An Economic History of the Making and Sale of Books.* London: Allen & Unwin, 1939.
Roshem, J. "Un Journal de Médecine au xviie Siècle." *Paris Médicale* 13 (1912–1913) Suppl: 663–69.
Rostenberg, Leona. "John Martyn, 'Printer to the Royal Society'." *Papers of the Bibliographical Society of America* 46 (1952): 1–32.

Chapter 3

Studies of the Early Scientific Journal

The Basic Source Lists

The history of the scientific journal that so dominates the channels of scientific communications today dates from the appearance in the seventeenth century of two extraordinary and innovative publications. Each in its own way brought public recognition to the social and intellectual phenomenon that has been called the scientific revolution. They also heralded the evolution of a new method of disseminating scientific information that developed into a unique and characteristic type of scholarship. These two periodicals still survive today after some three hundred years, despite considerable modification of their format and their mission. They are, of course, *Philosophical Transactions* of the Royal Society of London and *Journal des Sçavans*, both of which began in 1665. From these embryonic beginnings, the periodical started its meteoric rise to the dominant position it holds today in the communication of scientific ideas. It went through a period of amorphous development in which the periodical format was not clearly distinguished from the other communications formats that were available at the time. As is true with the development of other communications media, earlier forms tend to coexist with the more technologically advanced forms. The scientific journal as we know it today, however, bears little resemblance to some of its archetypal ancestors. In fact, in some ways we may have to seek quite a different lineage for our scientific primary journal.

In spite of some superficial specific resemblances in format between today's scientific journal and the scientific journals of the seventeenth century, a close study may reveal some significant differences. These differences are

Published in *Texas Reports on Biology and Medicine* 32 (1974): 61–74

not merely a result of the evolutionary process but represent significant differences in styles and modes of scholarship in such matters, for instance, as attitudes toward authorship or, perhaps better, toward authority, toward originality and ownership of ideas, and toward practices of attribution. These early forms of scientific journalism are interesting objects of study, not only because they are invaluable sources for the history of the development of ideas and for the history and sociology of science, but also as a basis for studying how changes in technology can affect the techniques of scholarship. It is not clear whether new communications techniques contribute to the development of new styles of scholarship, or whether indeed these new styles call forth new technologies. What were the effects, for instance, of the mechanization of printing on the way in which scholars collected and synthesized ideas? Or, going back further in history and reversing the question, what changes in the need to store and manipulate information called forth the development of the cuneiform alphabet? It may be demonstrated that the introduction of the scientific periodical led eventually to such a significant technological change, and therefore to a significant change in the style of scholarship.

The first step in such a study is to collect as many of the various species of communications formats as possible that meet the criteria for this new media. In the 1950s, I undertook an analysis of this body of literature (Kronick 1962), which the illustrious medical historian Henry Fielding Garrison identified as "The Medical and Scientific Periodicals of the Seventeenth and Eighteenth Centuries" (Garrison 1934). Because of the problems of separating the applied from the pure sciences, particularly in this early period, the inventory from sources other than Garrison included those periodical publications that were concerned with the growing application of science to technology. The inventory ended at 1790 rather than 1800, which was Garrison's terminal date, for two reasons. First, the journals, even in this early period, were demonstrating the familiar phenomenon of exponential growth and the addition of the decade 1790–1800 expanded the inventory beyond that which was feasible for analysis. The second reason is that the year 1790 does seem to mark a watershed in European history. Several of the other periodical bibliographies consulted in compiling the inventory concurred because they too found an effective breaking point at the year 1790 (Hatin 1866; Kirchner 1928).

The question can be raised whether the periodicals of science and technology from 1665 to 1790 constitute in any way a well-defined body of literature as an object of study. In the sense that it can be considered in isolation from the rest of the scientific and technical literature of its time, the answer must be no. The periodical provided an important means of access to the general literature because in the beginning it largely played a parasitical role. The introduction of the periodical did nothing to impede the growth in the number of published books. On the contrary, if we regard the periodical as a vehicle for one stage in the process of the progressive synthesis of ideas, the growth in the journal literature should contribute to the growth in the number of books. The periodical had just begun to play this kind of a role before the end of the period. It can be more easily demonstrated rather that the periodical was stimulated by the growth in the number of books. In fact, in the period in question, it is extremely difficult in many cases to distinguish the periodical from the book, and the evolution of the scientific periodical may in some ways be described as a process of differentiation from the book. This process has taken place so completely today that a huge artificial barrier has been built up between the two media, which in many ways inhibits scholarship. Starting with 1665, each medium over the centuries has developed a distinct apparatus of bibliographic access. Since the differentiations are not nearly as clear in the seventeenth and eighteenth centuries, the first step in defining this body of literature is to attempt to separate out from the literature of the period those literary artifacts that fall in the evolutionary level of development of the periodical.

The beginning of this process was based on an initial list compiled from a series of sources (Bolton 1897; Garrison 1934; Kirchner 1928; and Scudder 1879). These yielded about a thousand titles that the compilers of these lists had identified as periodicals. It was possible to locate only about half of them in the United States, and because of their wide geographic dispersion it was possible to examine less than three hundred. There is ample evidence that the compilers of these lists were not able to examine many of the titles either, and many of them appear to be admitted primarily on the evidence of the words in the title, which are frequently deceptive or misleading. Bolton, whose list covered scientific and technical periodicals up to the end of the nineteenth century, was able to provide locations for only 3,160 of the 8,600 titles he cites. Garrison was able to locate many of the titles in his list in the United States, but for many of his titles he provides no other source than citations in Callisen (1830), an early-nineteenth-century medical bibliographer who provided the useful service of citing reviews of many of the periodical issues in other critical and review journals of the period. Although Kirchner made a great effort to locate many of his titles in German research libraries, for many of the titles he had to cite only Beutler or other periodical bibliographies. As a result, many titles, particularly with introductory terms such as *Beiträge, Magazin, Journal,* or *Recueil,* find their way into these lists despite the fact that they do not meet any of the criteria for a periodical publication.

In Garrison's case, it appears that this may have been carried out with clear intent to include nonperiodical titles. As a medical bibliographer his primary interest was in medical periodicals, but he also clearly recognized the importance not only of the general scientific journal, and journals in other disciplines, but also the general literary periodical as sources of scientific information. This was particularly true in the seventeenth and eighteenth centuries because the audiences to a large extent were also undifferentiated and the scientist in the modern sense had hardly yet appeared on the scene. Garrison's principal motivation in compiling the list seems, however, to be to identify a body of literature to which adequate access was needed for historical purposes—any titles, such as periodicals, that need indexing for one reason or another He thus includes collections of miscellaneous works by single authors, single volume collections of case histories, and academic writings published without a view toward continuity and resembling periodicals only in that they are comprised of distinguishable bibliographic units to which some sort of content access was required. If more conferences and symposia had been published in the seventeenth and eighteenth centuries, they would have been included as well. There are other entries, however, that fail to meet even these criteria.

I called attention to a number of these discrepancies in Garrison's list in 1958 (Kronick 1958), but many also occur in the other periodical source lists. These become apparent only when one is able to take the particular volume in hand, so a great many of them escaped my notice until I was recently given an opportunity to examine some of these items in some of the great libraries in France and Great Britain. The survey was by no means exhaustive because many of the titles in the basic list still continue to be elusive. Out of the first eight hundred titles checked, for instance, over two hundred could not be found in any of four major sources listed in the references; nor could they be found in the *Union List of Serials*, which covers most of the research libraries in the United States and Canada, the *British Union Catalogue of Periodicals*, which does the same for the British Isles, or the *Catalogue Collectiv des Periodiques*, to which most of the major libraries in France contributed. Some of the missing items turned up in the card and sheaf catalogs of the British Museum in London and the Bibliothéque Nationale in Paris because, for one reason or another, they had not been recorded in the periodical inventories. Others are provided with locations in German libraries by Kirchner. Some of the titles may represent problems of bibliographic entry, but there still remains a fairly large residue of titles for which the only trace seems to be a citation in one of the sources cited by Garrison or Kirchner, or in some cases a citation in a periodical of the period itself. Some of these bibliographic problems and difficulties may be demonstrated by citing a few of the titles that were seen.

The *Bibliothek Auslandischer Chemisten*, for instance, published in four volumes in Nürnberg from 1781 to 1784, differs very little from a large number of serial anthologies or collections found as title entries in most of the standard periodical bibliographies. It does not, however, seem to be listed in any of them beside Callisen (vol. 23: no. 1346). It was discovered in the British Museum sheaf catalogs only because its editor, Johann Hermann Pfingsten, was a prolific editor of periodicals and it appeared as an entry under his name. It was published in one volume each year, and there is no indication that Pfingsten did not plan to publish it indefinitely. For the most part, it contains lengthy contributions on chemistry and mineralogy from a number of sources, which are sometimes indicated and sometimes not; it is difficult to determine whether some items are published for the first time. In several instances, the editor indicates that the article is a translation from another language even if the author of the original is not always named. One can frequently assume that the editor is also the author of an unsigned contribution, but this is by no means always the case. Quite a different item is *Adversaria Medico Practica* that Kirchner (no. 1394) cites as a medical periodical. It was published in Latin at Leipzig in three volumes of four issues each between 1769 and 1772 and consists of a large number of case histories, of which only a part are attributed to any particular author.

The elusive nature of some of these early periodical publications is demonstrated by the frequency with which titles were uncovered that apparently had not come to the attention of periodical bibliographers. Such a title is the curious *Curieuse Natur Staats und Sitten Praesente*, of which only one issue (January 1709) exists in the British Museum. The title page bears the notation *Zweite Jahrgang*, so it obviously must have appeared for at least one year. It consists primarily of hygienic and medical advice on a number of topics and apparently was written by a single author. It is obviously directed to a specific audience because the preface is headed "Madame." It can be considered as falling in the category of publication that the Germans call *moralischen Wochenschriften*, or "moral weeklies," a widespread literary format in this period for which *Tatler* and *Spectator* are the recognized prototypes. The single issue is accompanied by a supplement that deals with a singular remedy, "the so called *Rammoneur*, or the *Weyland Obristen Moncken* powder."

The foregoing title may not be considered any great contribution to the history of science, except perhaps in providing some insight into the social history of the time. It is hard to imagine, however, how the following title escaped the attention of the bibliographies. The famed Académie des Sciences in Paris published many books under its imprint in the seventeenth century that represented the work of its illustrious members. It was widely known, however, that they did not try to issue any periodical publication until after the turn of the century when their *Histoire et Mémoires* began to appear. In the seventeenth century, the society apparently had been content to let whatever contemporary notice was made of work appear as brief notes in such publications as *Journal des Sçavans*. In January 1692, however, they began to publish these brief notes in monthly issues of sixteen pages, each under the imprint of the Royal Printers as *Mémoires de Mathematiques et de Physique tirez des Registres de l'Académie Royale des Science*. The collection at the British Museum continues through December 1693 and contains brief notices of the work reported at the academy's sessions. What is even more extraordinary is that they were reprinted almost simultaneously in London under the title *Physical and Mathematical Memoirs, Extracted from the Registers of the Royal Academy of Sciences at Paris, to be Continued Monthly*. Only one issue under this title, that of January 1692, exists in the British Museum so it may have represented more of a pious hope than an accomplishment.

A significant feature of the seventeenth- and eighteenth-century scientific periodical is that to a large extent it was treated in the same way as a book. This is demonstrated by the frequency with which journal issues are reviewed in the same context as books, and by the frequency in which they were issued in new and revised editions, often with editorial emendations or other textual changes. This difficulty is compounded by the fact that books, beginning about the same time that the periodical appeared, began to be issued in parts. Although Wiles' study (1957) is limited geographically to the British Isles, many of its generalizations about the publication practices in the book industry can be applied to most of the countries of Western Europe. The practice of issuing books in parts was stimulated by one of the same factors that stimulated the growth of the periodical: the growth of the reading audience in the seventeenth century. "Publishers discovered," says Wiles, "that hundreds even thousands of people not previously interested in books would buy them if they could get them in inexpensive parts, piecemeal" (Wiles 1957:1).

One of the earliest examples of such a book in parts, cited by Wiles, is the well-known Moxon's *Mechanick Exercises or the Doctrine of Handy Works*. It was published in 1677 in the form of monthly parts, each with a separate title page and a serial number. The rest of the title, which reads "began January 1, 1677 and Intended to be Monthly Continued," clearly states this intent. Wiles describes it as follows: "It is not a periodical, and it is not a series of separate pamphlets, for it has continuous pagination and its signatures run consecutively through the fortieth number of Volume I, starting again in the second volume" (Wiles 1957:71). Yet it can to some extent be considered as one of our first technical journals, representing the beginnings of the Baconian design of applying science to the "handy works" of man. Nevertheless, it is not a periodical on the basis that it lacks what Kirchner calls "collectivity" or "heterogeneity and diversity of offerings" and that he regards as significant characteristics of the periodical format. This is true in regard to the authorship of the articles that are from Moxon's own pen and to the material covered in each issue, which is usually confined to a single subject.

Titles of this and similar types occur frequently on the basic lists of seventeenth- and eighteenth-century periodicals, particularly when their titles contain a word suggesting a periodical format. Sometimes they are indistinguishable from bona fide periodicals unless, for example, the editor indicates that he plans to publish only a specified number of issues, and thus violates the periodical criterion of expectation of continuity. An example is *Botanische Unterhaltungen*, which is cited as a periodical by Bolton (no. 1033), Garrison (no. 557), Kirchner (no. 3026), and Scudder (no. 3113). It came out in Munich in twenty-two monthly issues between June 1784 and April 1786 and consists of a series of dialogues and conversations (a popular literary style of the time) on botanical subjects. There is no indication that the editor, or here more likely the author, one Georg Anton Weizenbeck, intended to continue it. On the other hand, the issues do not have that unity, or integrity of design, that would mark it as a closed work. *Modern Husbandman*, an early agricultural magazine published by William Ellis in monthly issues between 1743 and 1744, has many more of the attributes of the periodical but is cited only by Buttress (1950). Buttress, however, provides an example of the other kind of trap the unwary periodical bibliographer can fall into, that of being misled as to the subjects' scope by a title. Among those he cites as agricultural titles is the *Country Gentleman*, published in Dublin from 1726. The British Museum copy is in their "Burney Collection," a collection of early newspapers and news sheets that are among the progenitors and companions of these early science journals in which we are interested. It is a typical single sheet newspaper of the period and contains primarily political news and literary essays. Buttress also lists *Corncutter's Journal* (London, 1733–1734) and *The Country Journal or Craftsman* (London, 1727–1750). These are also in the Burney Collection and are similar in content to *Country Gentlemen*. These titles are, of course, occasionally not without interest in the history of science and technology. *The Country Journal*, for instance, in addition to political news and essays, contains some medical advertisements that must have outraged the regular practitioners even of those more lenient times. One example among many is the advertisement in the January 5, 1733–1734, issue of a remedy "for preventing and dissolving the Stone and Gravel in the Kidneys . . . and to be had at Mr. Noye's a Painter."

There are numerous other examples of titles cited in periodical lists that are essentially books in parts, or "continuations," as modern bibliographers describe them. *Spectacle de la Nature* (Garrison no. 550) is an important eighteenth-century work that influenced and reflected the growing interest in natural science in this period. Hahn characterizes it in the following terms, which catch a little of its flavor. He says it "turned the lessons of nature to moralistic ends" (Hahn 1971:88). It is clearly an example of a work that, although issued serially, was designed as an organized whole to cover a specific subject area. The preface to the first volume (Paris, 1733) begins with an encomium on the benefits of nature study to form the minds of the young, and goes on to outline the work systematically, beginning "with animals of the smallest bulk," that is, insects. It appeared in a series of dialogues published in nine volumes between 1733 and 1749 and drew its material from the best scientific literature of the time. Its popularity is attested to by the large number of editions that appeared in France and the number of translations into English (at least six between 1733 and 1770), as well as in editions in German, Italian, and Dutch.

Dictionnaire Raisonée Universel d'Histoire Naturelle edited by Bomare, published in Paris in six volumes in 1764, is another example of an encyclopedic work by a single author that is cited by Garrison (no. 532). It was translated into German, or rather used as the basis for a new work by Friedrich Heinrich Martini, who issued it in eleven volumes from 1774 to 1793 under the title *Allgemeine Geschichte der Natur* (Garrison no. 532a). Martini is careful to explain in the preface to the first volume that he had been collecting material for such a work for years when he was asked to translate Bomare's work, and his is essentially a new work.

The last two titles are examples of works issued in parts that appear on periodical lists, although each varies from the other in some respect or another. *Deliciae Medicae et Chirurgicae*, a work in German despite its title, is another kind of serial. It was published in Leipzig in ten volumes from 1703 to 1705, and is cited by both Garrison (no. 1) and Kirchner (no. 95) as a medical periodical. An examination reveals it to be a kind of medical encyclopedia or rather compendium, for its seeming lack of organization warrants some skepticism about calling it an encyclopedia. It seems to be a continuous narrative discussion of various aspects of medical and surgical practice.

Another example of a book issued in parts is *Physikalisch-Chemisches Magazin für Aerzte*, which sounds very much like a bona fide periodical title and is indeed cited that way by Bolton (no. 3462), Garrison (no. 720), Kirchner (no. 1618), and Scudder (no. 2364). The two volumes that make up this work were both issued in Berlin in 1780 and consist primarily of essays on applied chemistry by J. A. Weber, who does indeed identify himself in the first volume as the author. As an illustration of the ambiguous and amorphous quality of some of these titles, Weber's second volume includes some contributions of other authors, although Weber does not indicate whether they are new or reprinted papers. Joseph Morenheim's *Beobachtungen verschiedener chirurgischer Vorfälle* (Garrison no. 330), published in two volumes from 1780 to 1783, is a collection of Morenheim's own papers. *Chymische Nebenstunde*, also published in two volumes between 1762–1768 in St. Petersburg and cited by Kirchner (no. 1454) as a natural history journal, is a collection of the papers of Johan George Model. Kirchner (no. 603) also includes as a natural history periodical *Beyträge zum Wachstum der Natur und Grössenlehre*, of which only two issues appeared in Dresden and in Leipzig in 1747. It is obvious, however, from the author's (Gottfreid Heinrich Gummeth) preface that he intends this as a publication of his collected papers. "I am determined," he says, "to place the observations I have made on various matters in physics and mathematics on view." However, he continues, since cosmology and some parts of medicine are closely allied to natural history, one should not be surprised to find observations on these subjects as well. He intends to issue the work in parts, he adds, without being bound to any fixed time schedule. As soon as enough parts have been issued to comprise a volume, he promises, he will issue an index, and ends with the pleasant greeting, "*Lebe wohl!*"—or "farewell!"

A number of titles are identified as periodicals but represent other varieties of compilations. Kirchner (no. 3186) cites on his list of periodicals a title under the name *Deutsches Theatrum Chemicurn*, issued in three volumes in Nürnberg between 1728 and 1732. The issues consist of a collection of separate alchemical tracts each published with its own title page and brought together under a collective title. Despite the fact that each tract has its own imprint, the pages are numbered consecutively through the entire volume, indicating that this is either a reprint or that it was designed from the beginning for collective publication. On the other hand, Kirchner (no. 82a) says that *Der Aus dem Reiche der Wissenschaften*, which appeared in twelve parts in six volumes in Augsburg from 1750 to 1759, is a book in parts, although it is cited as a periodical by Bolton (no. 745) and Scudder (no. 2894). Kirchner was apparently unable to examine it because he cites a bibliography as his source for his judgment. The title is in many ways indistinguishable from many of the typical periodicals of the times. It contains brief articles, frequently only a page or less and seldom over ten, on a wide range of subjects including science and technology. It is derived from many sources including the English *Spectator*. On the other hand, *Monatlich-herausgegebenen Insekten-Belustigungen*, issued in Nürnberg in four volumes from 1746 to 1761, is cited by Bolton (no. 3100) as a periodical while it is obviously a book in parts. It consists of a series of colored plates of insects, issued in eight-page sections and published in a classified order.

Another breakdown in the distinction between periodicals and books in the seventeenth and eighteenth centuries occurs in the fact that journals sometimes included fairly lengthy treatises published in parts in successive issues. Such a treatise is the long dissertation on potash that appeared in Johann Hermann Pfingsten's *Journal für Forst, Bergwerks Salz* (Hanover, 1786–1790, 3 vol.) over the course of three successive issues (Bolton no. 2476; Kirchner no. 2747). This kind of publication practice could also occur with a title that in superficial appearance looks like the proceedings of a scientific society. *Abhandlungen of the Ober-Sächsische Bergs-Akademie* (Dresden, 1746, 3 vols.) is a collection of separate papers ostensibly by Carl Friedrich Zimmermann, the editor, and constitutes an approach toward a textbook on mining (Kirchner no. 602). There are a number of other occasions in which journals presented surveys in the form of a series of articles on the state of the art that could just as easily have appeared in the form of a monograph. An interesting example is *Journal de Marine* (Bolton no. 2371; Scudder no. 1090), issued under the auspices of the Royal Marine Academy at Brest from 1778 to 1780 and designed to deal with all aspects of seamanship, as its full title indicates. Each issue deals with "the Actual State of the Science of Navigation" in a series of articles that easily could have comprised a monograph on that subject.

Because of the brief life of a great many of the serial publications of this period, duration is not always a good criterion by which to distinguish a periodical from a book in parts. The intention to continue indefinitely is, however, a distinguishing characteristic if this can be determined. This may be the case, for instance, with *Builder's Magazine; or Monthly Companion for Architects, Carpenters, Masons, etc.* published in London in 1774, although only a single number seems to have appeared. On examination it does not look very much like a bona fide periodical, however. The first three hundred pages are devoted to an alphabetic description of terms used in building, followed by an abstract of current building laws, and a series of architectural plans and sketches. Other periodicals of the period frankly did not make a fetish of maintaining a rigid schedule of publication. *Hamburgisches Magazin* (Bolton no. 2025; Garrison no. 657; Kirchner no. 683; Scudder no. 2755), devoted largely to natural history and related disciplines, of which twenty-six volumes appeared from 1747 to 1767, makes this clear in the preface to its first volume. The editor had considered trying to meet a monthly schedule, he says, but found this entirely too restrictive in character. He promises to provide at least one issue every two months, with four issues to make up a volume.

The extent to which books in parts and other nonserial, nonperiodical titles are represented on the basic list compiled for this study cannot be determined until every one of the titles has been located and examined. Some titles may even represent errors in transcription such as *Monumente Antichi Mediei* [*sic*], which Garrison (no. 693) cites as a scientific journal issued in Rome from 1784 to 1789. A title at the British Museum with the same dates reads *Monumente Antichi Inediti* and consists of a series of descriptions of Roman monuments, apparently issued on a monthly basis. There may also be some ghosts in the lists as well, such as *Recueil, Académie de Montauban* that Garrison (no. 444) shows as a "Scientific Transaction of the 18th Century" issued from 1742 to 1750. The British Museum is cited as a source. Both the British Museum and the Bibliothéque Nationale list the volume as being published in 1868, covering the year 1867 as the earliest volume in the series. There is, however, a volume published by the academy in the collections of the Library University of Paris published in 1750 under the title *Mélanges de Poesie, de Literature et d'Histoire*, which may relate somehow to this title.

Bolton (no. 2437) cites *Journal des Observations Mineralogiques* as being issued at Nancy in 1782. It is not a periodical at all but the prize-winning essay of Espirit Pierre de Sevry awarded by the academy of Nancy in that year. *Naturalist's Journal*—cited both by Bolton (no. 3252), an issue in 1767, and Scudder (no. 3889), who cites an issue in 1775—is indeed a periodical, but not at all in the context we have been considering here. It is a blank workbook ruled for daily entries in which a naturalist could record his meteorological and other observations of the day. It was issued by Daines Barrington, a prolific eighteenth-century writer on travels and natural history. There are also titles that are misleading in other ways. *Taschenbuch für das Verdauungsgeschaft*, published in 1785, sounds like an exciting discovery in the early history of the physiology of digestion. Garrison (1934:336) correctly identifies it as one of his category of "facetious and fantastic" items, a humor magazine containing verse and riddles appropriately published in Spashausen (Leipzig).

It is difficult to know to what extent to include in an inventory of scientific and technical periodicals such titles as *Miscellanea Curiosa or Entertainment for the Ingenious of Both Sexes* (York, 1734–1735, 2 vols.) cited by Bolton (no. 8563). It appeared quarterly and included enigmas, paradoxes, and mathematical questions. There are numerous other similar publications, predominantly in English, although there are also many titles in German whose intent seems primarily to entertain rather than to instruct or convey scientific information. One must be cautious, however, about reaching these kinds of conclusions about such titles as *Memoirs for the Ingenious* (London 1693), to which Dobell refers as "a curious monthly magazine in imitation of the *Philosophical Transactions*" (1932:61). Dobell also calls attention to Leeuwenhoeck's famous letter on the capillary circulation of the blood, which was published for the first time in this journal (May 1693). The editor, Jean Cornard De La Crosse, commented on that occasion that he had four other such observations, "which I shall publish in due time," but he apparently was not able to do so. These titles all serve to indicate that the lines between the professional or vocationally dedicated scientist and the amateur and dilettante did not begin to be drawn until much later in the eighteenth century.

This analysis has emphasized the periodical or nonperiodical nature of some of the titles included in the basic list in order to illustrate a few of the bibliographic problems involved in the study of the early scientific and technical journal. While this may be an important consideration as it relates to the study of the evolution of the scientific periodical as a means for transmission and dissemination of information, it tends to create an artificial barrier in relation to the study of the development of the literature and the process of synthesis that is characteristic of science. Perhaps Garrison was inspired to include a large number of nonperiodical titles because one of his primary motivations may have been to provide access to the authors and subjects of a body of literature that is characterized by the inclusion of a number of separate works within a single bibliographic unit. Despite the fact that they represent to some extent a different kind of medium than the periodical as we know it today, they cannot be ignored in the history of scientific journalism. First, because they represent in some ways transitional forms; but second and most important, they represent a class of publication to which the same kind of bibliographic access is required as that which is accorded the periodical. This is one of the dilemmas into which modern bibliographic practice has fallen. In fact, modern adherence to a policy of making a fast and rigid distinction between periodicals and nonperiodicals has severely penalized a whole body of literature that, for some arbitrary or chance reason, is incorporated into collected volumes of papers rather than in what is regarded as bona fide journals—if indeed such a creature exists. The result has been that journal articles and articles that appear in collections such as symposia and proceedings of societies for the most part have been indexed separately and reduced the efficiency of the information retrieval process.

REFERENCES

Beutler, J. B. C., and J. C. F. Gutsmuth. *Allgeineines Sachregister über wichtigsten deutschen Zeit- und Wochenschriften*. Leipzig, 1790.

Bolton, Henry C. *A Catalogue of Scientific and Technical Periodicals 1665–1895*. Vol. 440. 2d ed. Washington, D.C.: Smithsonian Collections, 1897.

British Union Catalogue of Periodicals: A Record of the Periodicals of the World from the Seventeenth Century to the Present Day in British Libraries. 4 vols. London: Butterworth, 1955–1958.

Buttress, Frederick A. *Agricultural Periodicals of the British Isles, 1681–1900 and Their Location*. Cambridge: University of Cambridge, School of Agriculture, 1950.

Callisen, Adolph Carl Peter. *Medicinisches Schriftsteller-lexicon der jetzt lebenden Aerzte, Wundaerzte, Geburtshelfer, Apotheker und Naturforscher aller gebildeten Völker*. 33 vols. Copenhagen, 1830–1845. Reprint, Nieuwkoop: B. de Graaf, 1964.

Catalogue Collectif des Périodiques du Debut du XVIIe Siècle à 1939 Dans les Bibliothèques de Paris et Dans les Bibliothèques Universitaires Des Départements. 4 vols. Paris: Bibliothèque Nationale, 1967–1977.

Dobell, C. *Antony van Leeuwenhoek and His Little Animals.* New York: Harcourt, Brace & Co., 1932.

Garrison, Fielding H. *The Medical and Scientific Periodicals of the 17th and 18th Centuries. Bulletin of the Institute of the History of Medicine* 2 (1934): 285–343.

Hahn, Roger. *The Anatomy of a Scientific Institution, the Paris Academy of Sciences, 1666–1803.* Berkeley: University of California Press, 1971.

Hatin, Eugène. *Bibliographie Historique et Critique de la Presse Périodique Française.* Paris: Éditions Anthropos, [1866] 1965.

Kirchner, Joachim. *Die Grundlagen der deutschen Zeitschriftenwesens.* 2 vols. Leipzig: Hiersemann, 1928–1931.

Kronick, David A. "The Fielding H. Garrison List of Medical and Scientific Periodicals of the 17th and 18th Centuries, Addenda et Corrigenda." *Bulletin of the History of Medicine,* 32 (1958): 456–74.

Kronick, David A. *A History of Scientific and Technical Periodicals.* New York: Scarecrow Press, 1962.

Scudder, Samuel H. *Catalogue of Scientific Serials of All Countries, Including the Transactions of Learned Societies in the Natural, Physical, and Mathematical Sciences: 1665–1876.* Cambridge, Mass.: Library of Harvard University, 1879.

Union List of Serials in Libraries of the United States and Canada. 3d ed. 5 vols. New York: Wilson, 1965.

Wiles, R. M. *Serial Publication in England before 1750.* Cambridge: Cambridge University Press, 1957.

Chapter 4

Toward a Typology of the Seventeenth- and Eighteenth-Century Scientific and Technical Periodical

ABSTRACT

Starting slowly in the middle of the seventeenth century, the number of scientific and technical periodicals had reached a total of over a thousand titles by 1790. To deal with them as a body of literature and to provide an effective framework upon which to construct generalizations concerning them, it is desirable and necessary to develop a classification of the formats in which they tended to appear. Three classifications of scientific periodicals are examined: one that originated at the turn of the nineteenth century, another that appeared in 1839, and Fielding H. Garrison's classification of 1934.

Various formats in the period before 1790, including those dealing with society publications, academic literature, and the essays resulting from the awarding of prizes by scientific societies, are examined in the light of the classification in which they appeared, and some of the issues they raise are discussed. It is fairly clear, for instance, that by the end of the eighteenth century the periodical has still not been segregated in bibliographic access and treatment, either as a vehicle for or as a repository of scientific information. Miscellaneous writings of individual or multiple authors are in many cases regarded in the same light, and are sometimes indistinguishable from titles that can be regarded as bona fide periodicals.

The basic differentiation to be made is that between an original and a derivative literature, without begging the question of originality in the conceptual sense. Borrowing, translating, extracting, and reprinting were so prevalent that distinctions are difficult to make in this early period. A

Published in the *Serials Librarian* 1 (1971): 385–99; 2 (1977) 81–166

classification of formats (types of component unit publications) is offered, derived from an analysis of some of the titles on the lists, and some of the other facets under which the scientific and technical periodical of this era should be examined are listed.

The scientific and technical periodicals of the seventeenth and eighteenth centuries appeared in a wide variety of formats, which to some extent defy classification. Evolving slowly toward the significant role they play in modern scientific communication, they represent in a sense the archetypes upon which their modern counterparts are based. They can also be regarded as a kind of experiment in the art of embodying and conveying information of various kinds, to various audiences, for various reasons—not the least of which may have been to keep the printing presses of the period busy. Although they were designed to fulfill newly emerging needs, they were also based on contemporary and preexisting formats, primarily the newspaper. Inevitably they were all influenced by the printed book, which had been in existence for over two hundred years when the first learned periodicals began to appear in 1665. By 1790, well over a thousand titles had appeared that are classed in various bibliographies as scientific or technical periodicals.

There are some interesting and serious questions about what differentiates a periodical from any other kind of publication. The distinction is made today with a certain amount of arbitrary rigidity that separates this class of publication from the rest of the literary corpus, and influences to a great extent the nature and quality of bibliographic control of information. These distinctions were not made nearly as sharply in the seventeenth and eighteenth centuries. The periodical was integrated with much greater uniformity into the bibliographic apparatus than it is today.

Considerable attention has been paid to the problems of the formal definition of the periodical,[1] but not much attention has been paid to the effect the introduction of the periodical had on the way information is disseminated and organized. The factors of periodicity and regularity of issue in parts, maintenance of a continuous title, and multiplicity of authorship are among those cited as most distinctive. These are factors that relate principally to the periodical as a *vehicle* for the transmission of information. In this respect, it most resembles its ancestor—the newspaper. As a vehicle it serves as a medium to carry information in various forms to a more or less self-designated or selected audience. This is, however, only the initial role the periodical plays. After performing its function as a primary vehicle, it spends the rest of its life as a *repository*, which includes that literature in other kinds of formats.

As is true today, the periodical literature of science and technology in the seventeenth and eighteenth centuries did not constitute a unified medium. It varied in its relationship to information dissemination roles and functions and in levels and types of audience it addressed. One of the first steps, therefore, in analyzing the periodical in this period is to develop a working taxonomy that would help us to define some of these functions. Several authors and bibliographers have developed classifications of this body of literature either to organize it in some fashion or to study its development historically. Before trying to develop a typology of our own it may be helpful to look at some of them.

ERSCH'S CLASSIFICATION

An example of a more or less contemporaneous classification is that of the indefatigable German bibliographer Johann Samuel Ersch (1766–1828). Although he cannot truly be called the "father of modern bibliography" (Bertheau 1914), he was responsible for an extraordinary bibliographic progeny. His monumental *Handbuch der deutschen Literatur seit der Mitte der 18 Jahrhunderts* was published in Leipzig in eight volumes from 1812 to 1814 (Ersch 1812). It was based on his earlier *Allgemeiner Repertorium der Literatur*, published in Jena in 1793 (Ersch 1793). In this work he adapted a scheme of classification that had been introduced by Schütz and Hufeland (Schneider 1934:207). It is this classification that Ersch uses in the *Handbuch* and in the sections devoted to particular subjects such as the *Literatur der Medizin*, on which this analysis is based.[2]

It would require great diligence and perseverance to attempt to reproduce Ersch's complete classification. Schneider calls it "a constant source of amazement because of the number and incomparable fineness of the subdivisions" (Schneider 1934:207). There are at least eleven levels of analysis in some of the classes, representing both form and subject classes. Periodicals do not appear as a separate class but are cited at various points in the classification as *Vermischte Schriften*, which may be translated as "miscellaneous" or "mixed writings." The term "anthology" might also fit here, although it is usually reserved today for collections of literary pieces drawn from a number of sources, a definition that Ersch might indeed find acceptable. It is obvious, at any rate, that he does not make any clear or sharp distinction between collected writings of any kind, whether they are issued in periodical formats or form part of a single volume or multiple volume collections. He does, however, sometimes distinguish between such works by individual or by multiple authors.

Ersch first divides the literature of medicine into two parts—one dealing with the subject in general and another dealing with special aspects of the subject, under which some of the subject and form divisions are repeated (see table 4.1). His next major breakdown is into two classes: (1)

history and bibliography, and (2) substantive (*abhandelnde*) works. We could call these two classes "secondary" and "primary," respectively, if it were possible to define items in terms of the works that Ersch cites under each rubric. History and bibliography are secondary in the sense that they are usually derived from other recorded sources. However, Ersch also cites in the second class, or among the substantive works, titles that are abstract and review journals, anthologies, and collections from other published sources. We can only guess that Ersch may make the distinction on

Table 4.1. Ersch's General Classification

- I. General (covering all or more than one part of medicine)
  - A. History and bibliography
    - 1. History
      - a. Without regard to period or country
      - b. Individual periods or countries
        - (1) Ancient times and peoples
        - (2) Recent times and peoples
    - 2. Bibliography
      - a. General systematic and alphabetic works
        - (1) Systematic
        - (2) Alphabetic
      - b. Miscellaneous works
        - (1) Older and scarce works
        - (2) Recent bibliography and critical journals (partly with original essays and extracts)
- II. Substantive works
  - A. For physicians
    - 1. Formal *Encyclopädie* and older introductory works
      - a. *Real-Encyclopädie* and dictionaries
      - b. Miscellaneous works
      - c. Medical geography and anthropology
  - B. For nonphysicians (popular medicine)
      - a. Introductory works and theory
      - b. Popular works
        - (1) General
          - (a) Systematic and lexographic
          - (b) Miscellaneous works
        - (2) For works on aging, sex, and hygiene, see dietetics and other parts of the practical literature
- III. Separate parts of medicine
  - A. Medicine in itself
    - 1. Human body in its normal aspects (anatomy, physiology, etc.)
    - 2. Human body in diseased states (pathology, surgery, therapy, etc.)
  - B. Medicine in association with other sciences
    - 1. Philosophy
    - 2. Jurisprudence, etc.

the basis of the degree that one needs to refer to the original source. A review or abstract of a periodical article may serve as a substantive source in the degree that it substitutes for the original article. This is hardly in accord with the modern use of the term "primary," which would exclude any abstract or review journal from that category. The term "substantive" is usually reserved, moreover, for those publications that do not occupy themselves with news or other administrative aspects concerning a particular discipline, but which convey "substantive" information concerning the discipline itself. There are very few periodicals of a nonsubstantive nature in this sense in the seventeenth and eighteenth centuries. Although Ersch does not specifically identify any title as a periodical, they do tend for the most part to fall in specific classes.

HISTORY AND BIBLIOGRAPHY

There is only one title entered under "history" that can be called a periodical: Philip Ludwig Witwer's *Archiv für die Geschichte der Arzneykunde.* It appeared in only one issue in Nuremberg in 1790, although the intent was to continue to publish historical articles on the physicians of antiquity, as well as biographies of contemporary physicians. It was prevented from doing so by the death of its editor (author). It could easily be classed with the substantive periodicals except in the sense that all historical works may be regarded as "secondary."

The bibliography section is divided into two parts: systematic and alphabetic bibliographies. The systematic section is again divided into two parts, although not under separate rubrics. The difference is apparently that the first part is retrospective, containing such monumental systematic bibliographies as those of Albrecht von Haller, while the other contains well-known contemporary current bibliographies. Although they differ to some degree from the review and abstract (extract) journals that Ersch cites later under "miscellaneous works," they fulfill essentially the same kind of role and function. *Die Entdeckungen der neuesten Zeit in der Arzneygelahrtheit* was issued in four volumes in Nordlingen from 1778 to 1788. It can also be considered retrospective to the extent that the first volume covered the literature published from 1770 to 1771. Although it contains abstracts, extracts, or reviews of the medical literature both monographic and serial, like the abstract-review journals that Ersch cites later, it differs in that the entries are arranged in a classed order.

The distinction between abstracts—or "extracts" as they are sometimes referred to in the literature of that period—and reviews is a difficult one to make. When the contents of a journal article are described along with some critical comments, we can call the description a "critical abstract."

When the same process is applied to a book we call the product a "review." Without comparing the original text of the article or book with the abstract or review, we cannot tell to what extent the account is given in the language of the original author or in the words of the abstractor or reviewer. The fact that the term "extract" is used so frequently in the earlier period implies that in many cases the text may have been an extract transcription of the original text, or what can be called an "apograph" as defined by the lexicographers.

In the preface to the second volume of *Entdeckungen*, which covers the literature from 1772 to 1773, the editor, Augustin Phillip Gesner, describes the process like this: "The impossibility of getting one's hand on all the new publications, of reading all of them and abstracting them [*durchzulesen und auszuziehen*] makes it necessary to use other journals for this purpose" (Gesner 1782:vol. 2). Periodicals are cited throughout all four volumes and reviewed as separate items, frequently as separate issues of the same journal. Of the 350 items in the last volume, at least 77 can be identified as reviews of complete journal issues of such titles as *Philosophical Transactions, Roux'Journal de Médecine*, and the proceedings of Dutch, French, German, and Swedish societies.

Another title in this section is *Repertorium der medicinischen Literatur*, which was published in five volumes in Zurich between 1790 and 1795 by the physician Paul Usteri. It comes closer to Ersch's classification as a current systematic bibliography or index. It reviews the literature for the previous year and provides brief annotations for each book or journal issue cited, including the individual journal articles. An author index is included in each volume, citing the authors of individual journal articles as well as authors of monographs, so that it can be regarded as a kind of forerunner of the periodical index. The main body of the work is in a classified arrangement under twenty rubrics of which several are form rather than subject classes. These resemble to some extent form divisions used by later bibliographers:

1. Biographical notices
2. Academic reports
3. Critical journals
4. Journals and periodic publications
5. Collections of academic and other publications
6. Collections of medical observations
7. Medical disputations

Both Gesner's and Usteri's publications can to some extent be considered transitional forms between the abstract-review journal and the true periodical index, although a few collective indexes to general periodicals had made their appearance earlier (Kronick 1976:268ff). The Danish physician, poet, journalist, and social reformer Johann Clemens Tode reviewed the first volume of Gesner's *Entdeckungen* in his own journal. He recognized the differences between that type of work and an abstract-review journal like his own in saying that Gesner's publication "does not make the journal indispensable to the reader" (Tode 1779, vol. 5:364).

ABSTRACT-REVIEW JOURNALS

The second part of Ersch's section on bibliography carries the heading "miscellaneous works," although it does not seem to have much meaning in this context. The section is subdivided further into two parts; the first headed "older and scarce books." There do not seem to be any titles that are periodical in nature in this group. The second heading is very comprehensive and broad: "newer bibliography, critical journals [partly with original essays] and excerpts therefrom." This section contains a great many of the German medical abstract-review journals we know from this period. It is hard to see, however, how some of these titles differ from those titles that Ersch later classifies under substantive journals.

This class also includes such titles as the *Commentarii de Rebus in Scientia Naturali et Medicina Gestis*. It was published in Leipzig in three volumes between 1752 and 1798.[3] The nineteenth-century historical article on medical periodicals (*Medicinische Argos* 1839, vol. 1:61) calls it a "critical-reference periodical" and says it is more reference than critical. It adds, however, that it is "for its period the . . . best bibliographic aid we have," despite its difficult Latin style. Sudhoff confirms this view by calling it the best repository of medicine for that period (Sudhoff 1903). Callisen provides it with an unusually lengthy annotation in his bibliography: "In France, where complete sets are seldom to be found, the work costs 300 francs. Every volume consists of four parts, which come out irregularly. It contains domestic as well as foreign literature and for the last half of the previous century it is well nigh complete. The reviews are anonymous. . . . Besides reviews it contains biographies and personal notices" (Callisen 1836, vol. 25:no. 1517a). It also contains long reviews of scientific books, dissertations, and journal issues and articles, as well as transactions of scientific societies.

The first of a series of three medical review journals published in Göttingen included here is *Medicinische Bibliothek*, edited by Rudolph Augustin Vogel from 1751 to 1773. The first two volumes and nineteen issues actually appeared in Erfurt and Leipzig, and it was not until 1753 that it was moved to Göttingen under a new title, *Neue medicinische Bibliothek* and eight additional volumes were issued. It also contains long book reviews as well as reviews of academic disputations and programs. The first issue, however, also includes an original case history on the treatment of heart disease and medical news from various cities cited from such sources as the August 1750 issue of *Journal des Sçavans*. The first volume, which consists of ten issues, concludes with an index in three parts: (1) a general author index; (2) an author index to the case histories, eleven in all; and (3) a subject index. In the preface to the second volume, also in ten issues, Vogel announces a small change in policy. Dissertations "which do not contain anything new" will be indicated only by title. This policy will also be followed for "brief manuals and small tracts, which are not created for eternity." The editor also includes a list of what he calls *Ephemerides, Monats-und Wochenschriften*, which is a rather mixed bag of transactions of scientific societies, general scholarly review journals, and German translations of foreign medical periodicals. It can hardly be regarded as a complete list, even of titles that must have been known to Vogel, because it contains only eighteen items. If it is selective, the basis on which the selection was made is unclear. What is more interesting, however, is the editor's description of his working methods: "I will provide a complete extract [*Auszug*] from the newest medical books. In the process I will take pains to render the writer's meaning completely and correctly. I will, therefore, often use his own words, in order that I might less often be accused of error. Nevertheless, I will not always report on the entire book in the order of and including all the sections and chapters" (Vogel 1752, vol. 2:preface).

The second in the sequence of Göttingen medical review journals is Johann Andreas Murray's *Medicinisch-praktische Bibliothek*, published between 1774 and 1780 in three volumes of four issues each, thus in two issues a year. It included literary and personal notices as well as announcements of subjects offered for competition for the prizes that were being increasingly offered by the learned, scientific, agricultural, and other societies of the period. The annotation in the *Medicinische Argos* (1839, vol. 1:61) calls attention to the increased emphasis on practical medical problems that is reflected in the title. Murray himself, in the preface to the first volume, discusses the redefinition of scope as follows: "The medical journals which have appeared in Germany up to now, despite their many good qualities, have this deficiency that they attempt to cover all of medicine, so that in many cases the contributions can be only briefly noted" (Murray 1774, vol. 1:preface). His primary purpose, he says, is to bring to German physicians literature that might otherwise be inaccessible. The first issue, for example, includes reviews of a volume of *Philosophical Transactions* and twenty-three books, of which only three are in German.

This policy is continued by the third in this sequence of Göttingen medical review journals, *Medicinische Bibliothek* of Johann Friedrich Blumenbach, published in three volumes between 1783 and 1795. The journal, says the editor in the preface to the first volume, will be especially devoted to foreign works that are little known in Germany. The emphasis will be on "practical medicine, but also on new discoveries in anatomy and physiology so that the practical physician will be familiar with them" (Blumenbach 1783).

It is interesting to note that these three titles represent essentially the same publication effort in terms of scope, objectives, place of publication, and even of auspices. However all three editors were professors at Göttingen University and each title was regarded in a sense as the property of their editors, as indicated by changes in title and volume numbering. In fact, all of them are to a large extent written by the editors themselves, so that they can in this respect be regarded as authors. This is not true of another contemporary review journal also published in Göttingen, the famed *Göttinger gelehrte Anzeigen*, published from 1753 to 1834, and which can be said to have been more firmly under university auspices. The celebrated Swiss poet-physiologist Albrecht von Haller was its editor for several years. Callisen says that until Haller's departure from Göttingen in 1774, it was the most important repository of physical and medical literature in Europe (Callisen 1836, vol. 23:no. 1092a). Despite the fact that Haller wrote an inordinate number of reviews for the journal (Rohlfe 1880), there were a great many other contributors.

The other titles included in this section of Ersch's bibliography, although they vary somewhat in format, frequency of issue, and place of publication, follow generally the pattern of the three Göttingen journals described above. An exception is a title by Johann Clemens Tode that is also included here. In addition to his many other activities, Tode published a whole series of medical periodicals in Copenhagen in the last quarter of the eighteenth century. The journals differ one from another with respect to the audience to which they are addressed and their editorial policy, except in the sense indicated in the comment in *Medicinische Argos* that they were all slanted toward Tode's own particular point of view (*Medicinische Argos* 1893:1:61). He did not strictly have a review journal in mind even with his *Medicinisch-chirurgische Bibliothek*, published in nine volumes between 1775 and 1781 (Kronick 1976:196–98). In the preface to the first volume, Tode says that despite the "large swarms" of learned journals and *Bibliotheken* in Germany, medical journals are not as plentiful as might be desired. He has no aspirations to achieve the eminence of a Murray or a Richter, but he will attempt to show what the Danish are doing in medicine. The first issue is divided into:

1. Book reviews and notices, with emphasis on Danish imprints.
2. Reviews and notices of academic publications, particularly dissertations.

3. Medical news and information. For instance, in this issue there is a list of the physicians who graduated from the University of Copenhagen beginning with the earliest available records in the fifteenth century to 1699. Only thirty-five names are listed for this period along with their thesis subjects. The second issue continues with physicians graduated from 1700 to 1771 and lists seventy-two.
4. Data on the incidence of disease among the poor in Copenhagen (an indication of Tode's interest in social reform).
5. Medical case reports (two in the first issue).

Ersch, therefore, with equal validity could have included this title among the substantive journals.

Critical review journals could also, of course, differ with respect to quality. Tode is vehement in his criticism of Johann Christian Traugott Schlegel's *Medicinische Literatur für praktische Aerzte*, published in Leipzig from 1781 to 1794. Tode is known for the vigor of his judgments, but his opinion in this is confirmed by Callisen, who says rather blandly that Schlegel's journal lacks candor (Callisen 1863, vol. 23:no.483). Another title that Ersch includes in this section varies in another way because it can be regarded as published for a special audience. The title, which differs in other respects as well, is the *Bibliothek der neuesten medicinisch-chirurgischen Literatur für die K. K. Feldchirurgen*. It was published in Vienna from 1789 to 1793. Johann Adam Schmidt, who was one of its editors, was also the secretary of the Viennese Surgical Society. The ambitious prospectus that appeared in the *Taschenbuch für die deutsche Wundarzte* (Taschenbuch 1789:162–68) identifies it as an activity sponsored by the society that, in addition to publishing its proceedings and other publications, had decided, the prospectus continues, to publish an abstract journal. It includes, in addition to the usual book notices and reviews, military-surgical news, obituaries, and notices of prize essay announcements. In the preface to the first volume is a list of twelve journals that the editors promise to review systematically.

MISCELLANEOUS WORKS

Although there are a few serial works listed in the earlier portion of the section on substantive publications in Ersch's classification (see table 4.1), there seem to be none that could justifiably be called periodicals. Exceptions are some of those classed under miscellaneous works and under that same heading under publications for nonphysicians.

Miscellaneous works or collections, then, can be considered as a generic class under which periodicals constitute one species. To some extent, the distinction is also specious. It makes no difference to an individual who is performing a retrospective search of the literature for a particular contribution whether it is found in an original periodical, an anthology, a symposium, or any other form of collection of "miscellaneous writings." There is, however, a significant difference in the way it is disseminated—that is, the way it is made public or published, because the medium selected may vary considerably in the way it performs as a vehicle and in the way retrospective access to it is provided.

Collections can be brought together and organized in many different ways:

1. They can be derived from different kinds of contributions, such as dissertations and other academic writings, case histories (as in medicine), observations and experiments, travel accounts, or prize essays. These are all prevalent kinds of publications in the seventeenth and eighteenth centuries.
2. They can represent differences in publication status—that is, they may be published in the collection for the first time (a characteristic of the primary journal), or they may be derived from other published sources.
3. They can also vary as to the source from which they are derived (periodical literature, monographic literature, pamphlets, letters, etc.).
4. They can differ in their relationship to the original contribution in language (as in translations), or in part to whole relationships—that is, whether they are abstracted or abridged.
5. There can also be differences in scope, chronological coverage, subject, or geographic area covered. Another difference relates to whether they are closed collections—that is, whether the intent is to complete the collection in a specified time or in a specified number of issues, or whether the intent is to continue them indefinitely.

Miscellaneous works of all these types can be found in the scientific and technical literature of the seventeenth and eighteenth centuries, and many of them appear under this rubric in Ersch's classification.

The primary division that he makes under the heading of miscellaneous works is between (1) works of multiple authorship and (2) works of individual authors. The characteristic of multiple authorship, which is considered an indispensable characteristic of the modern periodical, is by no means found invariably in what can be considered bona fide seventeenth- and eighteenth-century periodicals. In some cases the reason why the periodical is a product of an individual, rather than multiple authors, is that the editor or publisher was not able to attract other contributors—although this was clearly the intent. In other cases, however, it was quite clearly the objective not to admit more than a single author into the collection. Retrospective collections of either individual or multiple authors, although they are sometimes in this early period issued in series and at fixed intervals, can hardly be considered periodicals under any terms of reference. Nevertheless, titles of this kind are cited from time to time in periodical bibliographies. An example cited by Garrison is *Artis Medicae Princeps*, issued in Lausanne and Basel in eleven volumes from 1769 to 1774 (Garrison 1934b:319). It is a collection of the medical writings of classical authors, such as Hippocrates and Galen. It is only when we come to the section on academic publications that we find titles that can be considered periodicals in nature.

ACADEMIC LITERATURE

Academic dissertation literature and other publications associated with academic protocol and ceremony played a pervasive role in the seventeenth and eighteenth centuries out of proportion to the role they play today. They were a part of the formalistic, ritualistic aspect of university life, but they deserve closer attention than we can afford here in order to better understand how they functioned in seventeenth- and eighteenth-century scholarship. Ersch includes many serial titles relating to academic literature in this section, many of which can be found on periodical lists. The domestic (German) titles are subdivided according to whether (1) they deal with foreign as well as German dissertations, (2) they come from more than one German university, or (3) they come from individual universities. The first two titles Ersch cites are not on any periodical lists, and some of the others could not be located for analysis. Nevertheless, it is possible in several cases to determine the scope and function of the publication from their titles.

An example is *Gründliche Auszüge aus der neuesten medicinischen und chirurgischen Disputationibus, welche auf den Hohenschulen gehalten worden*. It was issued in Stuttgart in four volumes between 1749 and 1756. Kirchner lists it as a periodical but identifies only three issues between 1749 and 1752 (Kirchner 1928:no. 787). It was not listed by any of the other periodical bibliographies consulted. It was apparently a current abstract journal for medical and surgical dissertations presented at German universities. Ernest Gottfried Baldinger's *Auszüge aus den neuesten und besten Dissertationen der mehrsten Akademien die zur Naturgeschichte, Arzneywissenschaft, Chemie und Physic gehören* appeared in six issues from 1769 to 1772 in Berlin and Stralsund, and is cited by at least three of the periodical bibliographies. It was also current but more selective in quality and broader in subject scope. C. Jacob Mellin's *Auszüge aus den besten medicinischen Probeschriften der 16ten and 17ten Jahrhunderts*, although it also claims to strive for qualitative selectivity, is narrower in scope and retrospective in nature. It appeared at Altenburg and Offenbach at wide intervals between 1771 and 1786, and among the periodical bibliographies is cited only by Garrison (1934b:no. 113, 2263).

Some of the series cited here by Ersch were issued with a view toward continuity and some were not. Christian Gottfried Gruner, who was very productive in this period as a medical author and journalist, issued in this class the *Kritische Nachrichten von kleinen medicinischen Schriften in-und auslandischer Akademien*, in Leipzig in three volumes between 1783 and 1788. Neither Garrison nor Kirchner who list it were apparently able to locate it (Garrison 1934b:no. 13, 291; Kirchner 1928:no. 2859). The Bibliothèque Nationale in Paris holds all three of these volumes. The first covers the literature from 1780 to 1781 and was published in 1783. The second volume continues the coverage for the same period, and the third, published in 1788, in addition picks up those dissertations that appeared from 1780 to 1783. The first volume contains abstracts of 204 dissertations of German origin and 36 from foreign sources, but does not include France and Italy. In the preface to the second volume, Gruner comments on the difficulty of obtaining copies of the dissertations for analysis. He complains that he has not been able to get any from Göttingen and Strasburg, and that he is still awaiting some from Copenhagen through a friend from whom he has had no word.

The number of titles in this section of Ersch's classification do not by any means measure the impact or at least the extent of publication of medical dissertations in this period, because he cites many other titles under special subject rubrics in the classification. Dissertations also pervaded the rest of the scientific and technical serial literature including primary journals, abstract and review journals, and as other serial collections. These collections can all probably be regarded as derivative publications because most academic writing tended to be published first as separate works.

PUBLICATIONS OF SOCIETIES

The third category under "Miscellaneous Works of Multiple Authorship" contains most of the other genres of periodical publications that Ersch includes in this bibliography. He divides the category basically into German and foreign publications. The distinction is made on the basis of original language of publication because the "foreign" titles are all cited in German translations. Under German publications, he includes two categories: (1) "Society Publications," and (2) "Journals and Collections," which is subdivided further. The society publications and their derivatives cover a wide range of formats, and it is difficult, particularly in the earlier periods, to recognize any uniformity of goal or purpose. There are variations in such factors as the nature of the relationship of the publication in question with the particular group it is supposed to represent, and the edition status of a particular volume and its relationship to the original publication from which it is derived. Good examples of some of these variants are the publications associated with the first society Ersch cites, the earliest German scientific society and one that predates even the famous scientific societies of London and Paris. It was founded in 1652 in Schweinfurt as the Collegium Naturae Curiosorum. In 1687 it became the Academia Caesarea-Leopoldina Naturae Curiosorum, a name that has overtones of both the Renaissance humanistic society and imperialistic grandeur. Its last name change came in 1922 when it became the Deutsche Akademie der Naturforscher.

Ersch lists nine titles published before 1790 as being related to this society. However, they are by no means all the titles in this period that have been so associated. Garrison (1934b) lists five additional titles. The first title chronologically is the *Miscellanea Curiosa Medica-Physica*, published in Latin from 1670. It could just as easily be considered a primary journal because, like *Journal des Sçavans* and *Philosophical Transactions*, it provided an open medium for contributions from all over Europe (Kronick 1976:196–98). Buess (1953), who analyzed the first ten decades of publication, says the first volume was the work largely of a single Breslau physician, Jacob Sachs von Lowenhaim, who contributed, he says, sixty of the first sixty-three commentaries. Of the 278 authors in these volumes, 180 were not associated with this society, and they contributed the overwhelming number of papers (Buess 1953:5). The publication served, says Barnes (1934:215), "as a repository for short original papers and communications which were sent to the editors who prepared and supervised their publication." Other publications in a direct line of descent from this ancient publication have continued with brief interruptions down to our very day. Without a more detailed analysis it is difficult to determine what role this particular title played in seventeenth-century science. It was largely medical in nature because most of the society's members, at least initially, were physicians. Published in Latin, it was accessible only to the more learned members of the reading public, but on that account it was also accessible to an international audience.

Ersch cites three other titles in this period as associated with this society, *Acta physico-medica*, published in eighteen volumes in Nuremberg from 1727 to 1791, *Auserlesene medicinisch-chirurgische Abhandlungen*, published in twenty volumes in Nuremberg between 1755 and 1771, the *Aüszuge*

Medicinisch-chirurgischer Beobachtungen, published in two parts in Nuremberg in 1789 and 1790. The last two titles are translations into German of selected portions of *Miscellanea* and *Acta*. The *Aüszuge* covered the literature that appeared between 1727 and 1754, and *Union List* (1965) is probably correct in identifying it solely with *Acta*. The first title, however, is not correctly described because *Union List* identifies it with *Acta's* successor, *Nova Acta*, while the preface to the first volume clearly indicates that it was drawn from some of the society's earliest publications. In fact, one of the "articles" included is a one-paragraph observation addressed to Sachs von Lowenhaim in Breslau by the famous seventeenth-century Danish physician Boerhaave. Sachs von Lowenhaim added a three-and-a-half page annotation to the communication, a practice he followed with many of the other contributions to *Miscellanea*, which may account for Buess attributing most of the writing in these early volumes to him. Another title that Ersch cites here is not a periodical, but Andreas Buchner's history of the society published in Halle in 1756.

The relationship of the other titles in this period that have been associated with Academia Caesarea-Leopoldina, largely by Garrison (1934b), is rather more tenuous than the ones cited above. First, there is *Historia morborum*, which was published in Breslau over the course of the first decade of the eighteenth century. Neuberger (1937) cites issues under this title from 1699 to 1702 and from 1706 to 1710 and associates them with a *Verein Breslauer Aerzte*, which may indeed be the same as the Academia. A copy under this title in the British Museum has the imprint "Breslau, 1706," and covers the years 1699 to 1701. It consists largely of reports of morbidity, as the title suggests, occurring in certain areas covering the years in question. The British Museum Catalog also cites editions in 1701 and 1711. Interestingly enough, the copy described above is bound with two short medical tracts, both with imprints in 1706 but without attribution of authorship. These publications, however, are quite outside the sphere of scientific communications as understood in this study of the scientific periodical. They belong more to the kind of enumerative and statistical reports of official or quasi-official bodies such as those represented by the English *Bills of Mortality*.

A different kind of title that is also associated with the Academia is the *Sammlung von Natur-und Medizin*, also published in Breslau from 1717 to 1725. It does not appear to have been officially sponsored by this society, although its editors, such as Johann Kanold, were also for the most part members. The contents of the early volumes consist primarily of morbidity data, reports of scientific news, and meteorological information. The study of what can be called medical meteorology or geography goes back at least to the Hippocratic era. In the seventeenth and eighteenth centuries, however, it seems to have assumed epidemic proportions; in fact, it provided standard fare for many of the scientific periodicals of all kinds. It is not hard to formulate a hypothesis that this represented an attempt to replace astrology as an environmental cause of disease, as astrology was being more and more discredited as a valid theory, at least in scientific circles. It is not fair to make a judgment on *Sammlung* or its successor, *Miscellanea medica-mathematica*, published in Erfurt from 1731 to 1734, on the basis of the examination of only a few volumes. They differ from the titles that Ersch associates with the Academia (e.g., they are in the vernacular, rather than in Latin). The annotation in *Medicinische Argos* (Hacker 1839:82) describes the contents more fully as consisting of: (1) weather data; (2) the relationship between the weather and disease; (3) the state of agriculture; (4) observations on physics, chemistry, medicine, and so forth; (5) reports of new discoveries in science; and (6) literary news (i.e., book notices).

There are also several publications that are derived from these series associated with the Academia that are cited by Garrison (1934b:117) and others. One title that Garrison cites as being published in Breslau between 1736 and 1834 is *Schatzkammer der Natur und Kunst* (Kirchner 1928:no. 3182, cites only a 1736 edition). *Schatzkammer* was identified as another edition of *Sammlung* (British Union Catalogue 1955–58) as well as a translation of *Miscellanea* (Garrison 1934b:117). Another title is obviously a selection or abridgment from these sources on the basis of narrower subject interests, *Vermischte ökonomische Sammlungen*, published in Leipzig in two volumes in 1750. There is in addition one other title, *Abhandlungen der Hallischen Naturforschenden Gesellschaft*, of which only one volume was published in Dessau and Leipzig in 1783. Garrison (1934b: XIII, 483), Kirchner (1928:no. 2478), and Scudder (1879:no. 2493) all cite it as an independent society, but *Union List* (1965) identifies it also with the Academia.

The titles cited by Ersch under "Society Publications" by no means exhaust the complete range of different kinds of publications issued by or associated with scientific or technical societies. It is difficult to generalize about them at this point, as Ersch deals primarily with medical societies in Germany or society publications in German. Such generalizations will be possible only when we have also analyzed the more general societies in this period, as well as the societies in other special areas such as agriculture, mining, commerce, and even beekeeping. *Akademie der Wissenschaften* in Berlin, for instance, published at least ten different serial titles before 1790 in Latin, French, and German. Ersch cites only one here, *Physikalische und medicinische Abhandlungen*, which was translated from the Latin and French proceedings by Johann Ludwig Konrad Mumler and published in four volumes in Gotha between 1781 and 1786. Garrison (1934b:XIII, 487h) cites another German translation under the title *Anatomische, chymische und botanische Abhandlungen*, translated from the

French by W. B. A. Steinwehr, covering the years 1692–1737 of *Akademie* and published in nine volumes in Breslau between 1749 and 1762. Garrison cites Callisen (1836, vol. 25:no. 150) as his source, but Callisen is obviously incorrect because *Akademie* did not receive its charter until 1700, and did not begin to publish its transactions until 1711 (Ornstein 1928: 193). Ersch cites it correctly as a translation of the memoirs of the Parisian Academy of Sciences (Ornstein 1928: 332). Mumler, in the preface to the first volume of the compilation, indicates definitely that he is following Steinwehr in publishing his translations from the Latin and French proceedings of *Akademie*. Ersch could also have cited *Sammlung der Deutschen Abhandlungen*, a collection of the contributions made in German to *Akademie* from 1789 to 1803. These were printed in French in *Akademie Mémoires*, and are reprinted in their original German form in *Sammlung*.

Publications of three other "societies" cited by Ersch merit attention. The first is the *Abhandlungen und Beobachtungen aus der Arzneygelahrtheit von einer Gesellschaft von Aerzte* in Hamburg, of which only one volume appeared in Hamburg in 1776 (Ersch 1812:291). The editor, P. D. Giseke, in his preface attributes the papers to a group of physicians who had been meeting since 1772 to exchange observations and discuss cases. It is interesting that the papers are presented more or less anonymously because they are signed only with a single initial. The editor argues that this is done so that the reader will not be tempted to make any prejudgments on the quality of the contributions. This may seem like a strange position to take, in view of some of the questions relating to authority, reliability, and the assignment of responsibility that can be raised. The position may also, however, have some justice, if we argue that ideas should be evaluated on their own merit. These are, of course, issues that are central to the whole process of scientific communication, and on which we find much different and contradictory testimony in this early period. The process of publication under the auspices of a "society" in itself lends authority to a work because it implies the application of organized judgment of a recognized group, which is, therefore, less subject to error than the judgment of an individual.

Ersch also cites *Medicinischer Briefwechsel von einer Gesellschaft Aerzte*. It was edited by A. G. Weber in two issues that appeared in 1785 and 1786. It can be considered the product of a "society of physicians," perhaps only in the sense that any contributor to a learned journal is a member of the "republic of letters." It contains the standard mixture that is found in many of the medical periodicals of the time: case histories, observations, news, and reviews. It is a "correspondence" only in the sense that the contributions are largely in the form of letters. The review in *Neue literarische Nachrichten* (1786) says, moreover, that it is as useful for the layman as for the physician.

The last society publication Ersch cites in this section that falls into the period with which we are concerned is *Abhandlungen* of the Roemisch-Kaiserlich-Königlich, Josephinische Medicinisch-Chirurgische Akademie. Richter, in the first volume of his surgical review journal (*Chirurgische Bibliothek* 1788), hailed it as the first German surgical society. The society published two volumes in Vienna in German under the title *Abhandlungen* in 1787 and 1801, and one in Latin in 1788 under the title *Acta.* The Latin volume was translated into Italian in 1790 and into French in 1792, which makes one wonder whether Latin was not already beginning to lose its grip as the lingua franca of scholarship. Surgeons, of course, were not expected to be as literate as physicians in the eighteenth century were. It was customary to publish surgical works in the vernacular from the time in the sixteenth century when Ambroise Paré wrote his important surgical treatises in French. *Abhandlungen* is, however, a bona fide society publication, at least in the sense that most of the contributions were by members of the society, whether they represented actual presentations at the meetings or not.

COLLECTIONS FROM GENERAL PERIODICALS

The second part of Ersch's section on miscellaneous writings is called simply "Journals and Collections." It is divided again into two parts: (1) "Extracts from Miscellaneous Journals of Medical Articles," and (2) "Journals and Collections for Medicine and Partly for Natural History." The first consists of only three titles, all of which are collections of articles related to medicine gathered from a series of general periodicals. It is interesting to note that all of these titles were published before 1790. This may be interpreted to indicate that after that date, the periodical press had become specialized enough that physicians no longer needed to use the general scholarly journals as a vehicle for their observations and communications. This same phenomenon is reflected in Colonial America at a later date, when we still find many articles of largely medical interest published in general periodicals.

The first title under this heading in Ersch's list is *Auszüge Amneywissenschaftliche Abhandlungen*, edited by August Gottlieb Weber, who was also the editor of *Medicinischer Briefwechsel*, cited earlier. *Auszüge* appeared in two volumes in Halle in 1788 and 1789 and brought together the articles of a medical nature that had appeared in *Wochentlichen Hallischen Anzeigen* from 1729 when it was founded until 1784. *Hallischen Anzeigen* is representative of a host of regional scholarly and literary journals that appeared all over Germany, and particularly in the university towns. Sudhoff (1903:457) cites it as one of the "half-learned-half political newspapers which dealt with medical material and not always of a popular nature."

The second title cited in this section brings together the medical articles published in a much better known scholarly journal covering a period of fifteen years. It is *Auswahl der medicinischen Aufsatze und Beobachtungen aus dem Nürnbergische gelehrten Unterhaltungen,* published in Nuremberg in two volumes in 1787 and 1788. It covers *Commercium Litterarum ad Rei Medica et Scientia Naturalis* for the entire period it was published in Nuremberg, from 1731 to 1745. *Commercium* appeared in Latin in weekly issues and contained original contributions on medicine and natural science as well as reviews of books and periodicals. The fact that medical articles were reprinted a half-century after they were written is a clear indication that currency was not a very important aspect of eighteenth-century medical journalism. We can almost surmise by some of the evidence presented by these publications that throughout much of the medical writing of this period, there is a strong residual feeling that the greater the antiquity of an observation, the greater its validity.

The third title in this list is drawn out of a series of periodicals published from 1755 to 1791 and called, successively, *Hannoverische Gelehrte Anzeigen, Hannoverische Beiträge,* and *Hannoverische Magazin.* The subtitle to the last title in the series provides something of the flavor of the three publications: "in which brief transactions, individual thoughts, reports, proposals and experiences related to the improvement of nutrition, rural and urban commerce, trade, manufacture and art, physics, culture, and the fine arts are collected and presented." It appeared biweekly, at least in the beginning, and each issue contained an original contribution as well as news and other items. For a period in 1759, the journal included mortality reports from abroad as well as at home. We learn, for instance, that of the 17,576 deaths in London in the previous year, 5,971 were under the age of two and 1,795 between two and five, or a total of over 40 percent, which gives us some insight into the tremendous rate of infant mortality in those years. The eighth issue for January 26, 1759, contains an article on a species of glandular tumor by the physician Rudolph Augustin Vogel, whose own journal, *Medicinische Bibliothek,* did not begin to appear until 1773. Although nearly as sharp a division did not exist between the medical literature written for the laity and that written for the physician, some of the articles included seem to be clearly on a professional level.

The article by Vogel, along with many others, appeared in a collection drawn from the series *Sammlung medicinischer und chirurgischer Original-Abhandlungen aus dem Hannoverischen Magazine von 1750 bis 1760,* published in Hannover in three volumes between 1785 and 1787. Kirchner (1928:no. 2748), says the extracts cover 1750 to 1780, but the British Museum copy reads "1750 bis 1786." This kind of anthologizing is also found in subject disciplines other than medicine throughout the eighteenth century.

JOURNALS AND COLLECTIONS

Ersch heads his next section "Journals and Collections for Medicine and Partly for Natural History," because to some extent "journals" and "collections" are not easy to separate in this period because the practice of borrowing articles from other periodicals, particularly those published in other languages, was widespread. He divides it again into two parts, the first consisting of twenty-five titles, of which only five were published in 1790 or earlier. Even these five do not by any means form a homogeneous class. Looking at them as representative of the German medical journals of this period, one is tempted to say that there were no periodicals in this period that published original contributions in the form we know them today. Original contributions were more likely to be presented through other channels, such as the proceedings of scientific or medical societies, collections by individual or multiple authors, or as monographs. The first title that Ersch includes in this category, for instance, is Johann August Philip Gesner's *Sammlung von Beobachtungen aus der Arzneygelahrtheit und Naturkunde*, published irregularly in five volumes in Nordlingen between 1769 and 1776. In some respects, it differs significantly from his *Entdeckungen der neuesten Zeiten*, cited earlier. In the preface to the first volume of *Sammlung* Gesner invites original contributions, offering the authors the option of signing their papers or not as they prefer. The first four volumes are essentially a commentary by Gesner on the current literature, to which he appends an account of some of his own cases. It resembles to some degree a modern scientific literature review journal in that he reviews and summarizes the literature on specific topics, citing those authors who have made contributions to the subject. The fifth volume, however, is made up almost entirely of the account of an epidemic in Nordlingen, signed with the initials J.A.P.G., which do not exactly conceal the author's identity.

The next two journal titles were both edited by Ernest Gottfried Baldinger, a prolific medical journalist and author of this period. They are both practically personal vehicles for the author. The first is *Magazin für Aerzte*, which contains in the first and second issues historical articles on the classical medical authors and translations from the Latin. Each issue also contains one original article. The second title is not much different in emphasis but is less formal in its presentation. In the preface to the first issue of *Medicinischen Journal*, published in Göttingen in 1784 and continued with various changes of title to 1802, Baldinger says that his goal is to provide a better knowledge of the ancient physicians as well as reports of modern developments in medicine. Kirchner (1942:162) distinguishes between the two journals by indicating that the first was directed toward the practitioner and student whereas the second was intended more for the scholarly physician. The distinction is hard to discern, except that *Medicinischen Journal* contains comments by Baldinger on the classical authors, extracts from various publications, as well as news and book notices. One might be tempted to call Baldinger's journal a "commonplace book," if that is an appropriate term for what seems to be a miscellaneous collection of items related to medicine and set down almost at random.

A similar title is *Medicinische Wochenblatt für Aerzte*, published in Frankfurt with various changes of title between 1780 and 1791. It also consists of reviews and commentaries on the literature, although it occasionally prints what may be regarded as an original contribution. *Aesculap: Eine medicinische-chirurgische Zeitschrift*, of which only one volume appeared in Leipzig in 1790, purports in its subtitle to be "by a society of German medical practitioners." The announcement of the new journal, which appeared in the *Medicinisch-chirurgische Zeitung* (1790) however, does not mention any society. The contents of one issue, as reviewed in the *Allgemeine deutsche Bibliothek* (1791), consist of two articles by the editor of *Medicinischer Briefwechsel*, cited earlier, one by another editor, Markus Philip Ruhland, and one review.

In a separate part of this section, Ersch cites two titles that are distinctly different from the others. The first, *Geist und Kritik der medicinischen und chirurgischen Zeitschriften Deutschlands*, is cited by Ersch as being published between 1789 and 1806. However, both Kirchner (1942:244) and Engelmann (1965) agree that it was not published until 1798, which would put it outside the scope of the present study. It resembles, at least in its title, periodicals that had begun to appear in the general literature as well as in other disciplines, which implied in their title, such as *Esprit* and *Journal der Journale*, that they were conveying the essence of the literature. Kirchner regards *Geist und Kritik* as representing a new departure in medical journalism. It would be interesting to determine whether the journal does in fact differ from the bibliographic periodicals cited earlier by Ersch. It is hard in any case to justify Kirchner's view, which recognizes it as a significant innovation in medical periodical journalism.

"FOREIGN" PUBLICATIONS

Ersch divides the next group, translations of works originally published in languages other than German, into two classes: (1) miscellaneous works, covering more than a single country; and (2) society publications and journals of a single country. They fall generally into three groups: (1) selections from multiple languages or multiple sources; (2) selections from a single language or a single source; and (3) cover-to-cover translations. These categories can be subdivided further in many cases according to whether the title is a society or nonsociety publication.

Ersch cites three titles in the first group, the earliest *Journal der ausländischen medicinischen Literatur*, published in Berlin from 1802 to 1814 under various titles. It does not differ to any great extent, however, from a number of other German titles published before 1790, some of which Ersch cites in other sections of his bibliography. An example is *Sammlung auserlesener Abhandlungen zum Gebrauch praktischer Aerzte*, edited by Christian Ehrhard Kopp in twenty-four volumes in Leipzig from 1774 to 1807 (Ersch 1812:no. 2622). It derives its contents from translations of both books and journals published in languages other than German. The first issue, for instance, contains translations from three French obstetrical treatises, as well as from a French translation of a Dutch prize essay on a related subject. It came out in quarterly issues of about two hundred pages each. The second issue is rather more varied than the first, containing selections from *Philosophical Transactions*, the publications of the Académie des Sciences of Paris, selections from the proceedings of a Dutch Society, and translations of articles from the Parisian *Journal de Médecine* as well as from monographs on a variety of medical subjects. The reviewer of the fifth volume published in 1799 (*Frankfurter medizinischer Wochenschrift* 1780) speaks of its success in bringing little-known works of foreign authors to the attention of German physicians. Its currency is demonstrated by the fact that this volume includes a translation of an English monograph published in the same year, an example we would find difficult to duplicate today.

The relationship of any of these translations to the original texts cannot be determined without collating and comparing them. There are numerous examples in which the translators and editors took advantage of the opportunity to comment upon, amend, or even to correct the text. This was, of course, a common practice of the manuscript period that preceded the introduction of book printing. The copyists apparently exercised these options frequently in transcribing the text from one copy to another. The printing press reduced these hazards by reproducing an author's ideas or words in multiple identical copies. It may also have influenced the author's attitudes toward his proprietary rights in his printed utterances in a way that does not seem to have been thoroughly investigated.

The influence of copyright upon the practices of reprinting, translating, and anthologizing in this period, particularly from works produced outside the country, was probably negligible. In a sense we might say there were fewer restrictions in the traffic of ideas in the "republic of letters" of those days than we have today. These issues are, of course, all closely bound to the economics of printing and distribution processes, which require a different line of investigation than has been attempted here. The translations in the titles that Ersch cites here are perhaps more likely to have stood in lieu of the originals than similar titles that he cited earlier. If we are to believe the comments made by some of the editors in the prefaces to these volumes, about the competence of most German physicians in foreign languages, some of these titles provided their only access to the foreign medical literature.

Ersch cites eighteen titles in the section for foreign journals and society publications published in 1790 or earlier. They are arranged according to country of origin as follows: Italy, one; France, five; Holland, two; England, six; Denmark, one; Sweden, one; Poland, one; and Russia, one. Ersch also includes titles with similar formats and objectives under other subject rubrics, demonstrating what a widespread practice it was, at least in Germany, to translate the foreign literature. An examination of the titles translated from the French may serve to demonstrate a few of the variations of format that occur in this genre. The first is *Anatomische, chymische und botanische Abhandlungen*, drawn from the serial publications of Académie des Sciences in Paris. It was published by W. B. A. Steinwehr in Breslau in nine volumes between 1749 and 1760. Steinwehr also edited a collection of contributions in physics drawn from the same repository. It appeared in thirteen volumes between 1746 and 1758. The French society has a long and complex publication history through the seventeenth and eighteenth centuries, particularly in terms of the related variant editions and extracts and translations drawn from its publications. It is generally acknowledged that the Académie did not attempt any periodical publication until it began to publish its *Histoire et Mémoires* in 1700. There is, however, one antecedent periodical publication that has generally been overlooked. The Académie did make an effort to apprise the public of its activities on a monthly basis by publishing *Mémoires de Mathematique et de Physique Tirez de l'Académie Royale des Sciences* in 1692 and 1693 (Kronick 1974:66). Steinwehr apparently draws from this publication, as his extracts begin with 1692.

Christian Gottfried Gruner began a similar attempt to translate on a continuing basis the proceedings of the Société Royale de Médecine in Paris. His series appeared in Halle under the title *Sammlung der gemeinnitzige praktischen Aufsatze und Beobachtungen aus den Schriften den Königlichen medicinischen Gesellschaft zu Paris mit Anmerckungen*. Only one volume appeared, however, in 1784, which covered the society's proceedings of 1776. Translation in this form, of course, did not preclude the appearance of translations of extracts from any of these society publications in other German extract, review, or "primary" journals.

The French medical periodicals of this period also attracted the efforts of the German medical journalists. *Sammlung auserlesene Wahrnehmungen aus der Arzneywissenschaft, der Wundarzney-und Apothekerkunst aus der Französischen*, despite the breadth of scope indicated in its title, is largely a translation of the Parisian *Journal de Médecine*. It was published in Frankfurt and Leipzig in nine volumes between 1757 and 1765, covering the issues of the French journal from 1754 to 1761. It was continued for an additional ten volumes from 1766 to 1777 under the title *Neue Sammlung* and drew its material primarily from *Journal de Médecine*, but also included occasional contributions from such French publications as *Mercure de France* and *Journal Economique*. *Auszüge aus dem besten franzosischen periodischen, medicinischen, chirugischen, pharmaceutischen Schriften* attempted to continue the series in Leipzig under new auspices, but continued only for five volumes from 1780 to 1784. In addition to *Journal de Medecine* and the other sources, it added the Parisian *Gazette Salutaire*, which began to appear in 1761, as well as the proceedings of the French medical societies.

The last title in this group that appeared in 1790 or earlier is *Französische medicinische Literatur*, which, according to its subtitle (*oder Anzeigen und Auszüge aus der neuesten Französischen Werken über Physik, Medizin und Oeconomie*) attempted to cover a number of other disciplines in addition to medicine. Only two issues appeared in 1790 in Heidelberg, edited by Heinrich Tabor. It was apparently not greatly missed after its disappearance, as Usteri is very critical of it in his review in his *Repertorium* (1790). Despite its title, he says, it contains only a few borrowed and sterile extracts from works on physics and natural history.

These titles by no means represent the range of resources the German physician had to draw upon to learn about the French medical literature. As we have seen, the general medical review and abstract journals drew heavily from the French literature, and even the general scholarly or regional journals did not hesitate to print an occasional extract or review in this field. What this review serves to emphasize is the large extent to which the German medical literature of this period was derivative. While the number of German medical periodicals looms large in any statistical bibliography (Kronick 1976:129–32), it would be a serious mistake to equate this with medical activity or original investigation or observation in medicine taking place in Germany.

INDIVIDUAL AUTHORSHIP

What follows now in Ersch's bibliography are the collected or miscellaneous works of individual German authors, first those publishing in Latin, and then those publishing in German. This section deserves our attention, even though works of single authors seem to be so much outside the purview of a study of periodical journalism. First, Ersch includes a few titles in this section that are found in standard periodical bibliographies. Second, the distinction between individual and multiple authorship in serial publications is frequently blunted in this period. One title that appears in several periodical bibliographies, for example, is Christian Gottfried Gruner's *Almanach für Aerzte und Nichtaerzte*, published in Jena in fifteen volumes between 1782 and 1796. It appeared annually and included many of the familiar appurtenances of the typical almanac of this period: important dates, phases of the moon, and the like. However, it also included reviews of the literature and polemics on various medical subjects as well as subjects announced by various societies for prizes and other news. It is hard to say what kind of an impact it may have had either on physicians or nonphysicians, but it was popular enough to provoke a satire that appeared in 1787 under the title *Almanacum perpetuam für Aerzte und Nichtaerzte.*

Almanacs are a genre of publication unto themselves, with a history and a tradition all their own. They appeared in vast numbers beginning almost with the very origins of printing. With the introduction of the scientific-technical periodical, some of the almanacs began to assume some of the characteristics of these new publications. A number of these titles that can be defined as almanacs either in name or in nature and that are of interest to the historian of science and technology can be found in this period. The British Museum classes a large variety of publications under the heading *Ephemerides*; in fact, the term seems to be regarded there almost as a synonym for "annual." Although an obsolete meaning of the word *ephemeris* is "an almanac or calendar," the term is usually reserved for astronomical diaries or almanacs. In the seventeenth and eighteenth centuries, these publications were also likely to include original contributions, meteorological and other data, and news. Many of them were of a popular character, addressed to artisans and other special groups. Although they may not contain much significant new information, they do provide important indications of the social and intellectual impact of science and technology in this period, and should not be ignored in any study of scientific and technical journalism.

POPULAR PERIODICALS

Gruner's *Almanach* could also be considered a member of the last section of Ersch's bibliography—the section under general medicine in which he lists works "For Non-Physicians and Popular Medicine." *Der Arzt, ein Wochenschrift*, published in Hamburg between 1759 and 1764 in twelve volumes by Johann August Unzer, is perhaps the best-known German title in this field. Like the *Almanach*, it was written entirely by Unzer, although he did solicit contributions from time to time (Runge 1937:984). Although it seems to be directed primarily to a nonprofessional audience, it did number physicians among its subscribers. The preface to the second edition, published in 1769, lists 1,700 subscribers, of whom 150 can be identified as physicians. Some 400 of the subscriptions are designated for booksellers, and, of course, some of these may also have found their way into the hands of physicians, but the overwhelming number of subscribers is obviously lay people.

Gazette de Santé, or *gemeinnütziges medicinisches Magazin für Leser allen Ständen,* published between 1783 and 1786 in Zurich some twenty years later, took its title from the French journal of the same name, although they are quite different in character. The French *Gazette* was a kind of medical news journal issued every ten days in Paris between 1773 and 1829. Johann Heinrich Rahn's publication is more substantial, issued only six times a year and containing original essays on hygiene, human anatomy, and physiology, as well as extracts from such journals as Tode's *Unterhaltende Arzt,* which Ersch also cites here. Johann Clemens Tode had already published a whole series of popular health journals in Danish when he began to publish his *Unterhaltende Arzt über Gesundheitspflege, Schönheit, Medicinal-Wesen, Religion und Sitten* in Copenhagen from 1785 to 1789. The publication carried on his practice of printing anecdotes and essays relating to medicine and hygiene, rather broadly defined as indicated in his title, but obviously intended for a popular audience.

This last group thus introduces another element that needs to be considered in developing a typology: the audience to which the publication is directed. In the title just cited this distinction is fairly easy to make, although even in the case of Unzer's journal, as we have seen, a substantial number of physicians saw fit to subscribe to his "popular" journal. In many of the scientific and technical periodicals of this time, the distinction is difficult if not impossible to make. Where there is a well-defined professional or occupational group, such as in medicine, it is easier to identify a specific body of literature as uniquely associated with them. But even here, the tradition that medicine was a part of humanistic education and belonged in the province of any well-educated man persisted until medicine began to be established more firmly as a scientific discipline. In physics, chemistry, astronomy, or in general science, the boundaries between the specialist and the layman were necessarily less distinct. There were not very many professionals or professional groups associated with these disciplines, except in the case of some of the continental scientific societies where the academicians who made up their membership were in the employ of the state. Specialization, however, can also to some extent be defined in terms of intensity of interest, and the development of the ability on the basis of education and training to cope with some of the advanced concepts in a discipline. In this sense, specialized groups of interested and knowledgeable amateurs, or "virtuosi," began to develop early in the history of the scientific periodical.

Groups of "miscellaneous works" appear under many other subject rubrics in Ersch's medical bibliography as well as in the other subject bibliographies that make up his *Handbuch der deutschen Literatur* (Ersch 1812–1814). While this analysis has by no means exhausted all the species of periodical publications that appeared in this period, it has provided us with an entrée to most of the major categories we need to consider. To refine and augment them, we should look at least to two additional classifications.

CONTRIBUTIVE AND DERIVATIVE LITERATURE

For another view of the periodical literature, we can turn now to a classification that appeared in a historical review of German medical periodicals published in 1839 in *Medicinische Argos*, which was cited earlier. Despite the fact that it covers only a relatively small number of titles (for the period 1790 and earlier it includes less than fifty titles), it develops two concepts that are important in any classification of periodical titles. The first is the distinction between what the review calls *gebende* and *nehmende* journals, literally "those that give" and "those that take." Another way to translate these terms in this context might be to call them "original" and "derivative," or "primary" and "secondary." On the face of it this seems like an easy distinction to make, but on reflection some interesting questions can be raised. In modern usage, "secondary" journals are those derived from other journals and published in the form of indexes, abstracts, or reviews. For the most, part they require reference to the primary or original journal for full information. How, then, would one classify a collection of articles from a number of periodicals that are printed in full in the original language or in translation, and that require no reference to the primary source? The review in *Medicinische Argos* (1839) actually makes this distinction because it divides the journals into three categories: (1) journals with original articles, (2) journals that collect articles from other journals, and (3) critical-review journals.

There are, of course, other connotations of the terms original-derivative and primary-secondary. A critical review or a literature synthesis can be just as original or creative in some ways as the report of original experimental results. It is also true that every piece of scientific reporting is in a sense derivative because it is always based on previous work. In fact, it probably *should* be based on previous work to be good science. Carrying the idea to its extreme, one could say that the use of language of any kind to embody a concept, no matter how new it was in itself, must be a derivative act, as language itself involves a condition of prior consensus as to the meaning of words. One might, I suppose, easily get into a lengthy metaphysical discussion of the problem, although I must admit that the hazards of doing so are fairly negligible in this analysis. The simplest solution is to use the terms in a descriptive rather than an evaluative sense, and to classify any new combination of words published for the first time and not directly derived from another published work as original or primary. The question of originality and redundancy in seventeenth- and eighteenth-century scientific journalism is an interesting and important one to raise, as indeed it would be in relation to our current scientific literary output, but not one we will be able to solve with the measures available to us here.

If we ignore the society transactions, there were very few journals in the seventeenth and eighteenth centuries that were made up entirely or to a large degree of original contributions in any sense. Most of them, as the review in *Medicinische Argos* implies in labeling the group "journals which include original articles," were admixtures of original and derivative materials. The eighteen titles published in 1790 and earlier and listed here in this group represent a rather curious mélange. The brevity of the list seems to indicate that there could not have been a wide diffusion of German medical periodicals in this period, as apparently so few of them that were published before 1790 are recognized in this review. Nevertheless, only half of these titles are cited in Ersch's bibliography. For some of them, such as *Sammlung von Natur und Medizin* and *Miscellanea physico-medica, Oder eingenehme, curiose und nutzliche Nachrichten*, which were connected even if tangentially with *Academiae Naturae Curiosorum*, as indicated earlier, there may be an explanation. It is possible that Ersch may have considered them too general in scope to include in a medical bibliography.

The next two titles in the *Medicinische Argos* list represent bibliographical problems. The title *Medicorum Silesiacotum Satyrae*, said to have been published in eight parts from 1738 to 1742, is associated by Garrison (1934b: XIII, 503) with *Schlessische Gesellschaft für vaterlandische Kultur*, which apparently did not begin to publish until after 1800. Callisen (1836:vol. 25, no. 658b), however, associates it with other publications related to the aforementioned *Academia*, so they may all involve societies in Breslau somehow associated historically. *Acta medica Berolinensium*, published in Berlin between 1717 and 1731, resembles *Sammlung von Natur und Medizin* in format, which was published about the same time. Callisen (1836, vol. 25:no. 657) associates it with the following two titles on the list, although there is no discernible connection except that they were all published in Berlin. Ersch does not cite either of them, but it is obvious from their titles that they are both also general in nature.

They were both edited by Friedrich Heinrich Wilhelm Martini. The first was the *Berlinisches Magazin, Oder gesammelte Schriften und Nachrichten für Liebhaber der Arzneiwissenschaft, Naturgeschichte und der angenehmen Wissenschaften*, which appeared from 1765 to 1767. The second was *Berlinische Sammlungen zur Beforderung der Arzneiwissenschaft, der Naturgeschichte, der Haushaltungkunst, Cameralwissenschaft und der dahin einschlagenden Literatur*, which was published from 1768 to 1780. It is difficult to determine why they are included here, as the annotation in *Medicinische Argos* itself indicates that "original papers are seldom found; there are mostly extracts from foreign journals and short critical lists of new books" (Kirchner 1928–1931; Sudhoff, 1903). The editor does not help us very much in determining his objectives because, in the preface to the first volume of the *Berlinische Sammlung*, which is unsigned, he says that prefaces are more a habit than a necessity and that he will let his title speak for itself. The first issue is certainly not very medical in nature, because it includes, in addition to miscellaneous news and book notices, an unsigned essay on constructing wooden buildings. It also includes a memoir on the sexuality of plants that was taken from the proceedings of a Danish society, and an observation on a species of flying rodent in Surinam, also unsigned and unattributed.

Ersch also does not include the following title: *Commercium Literarium ad Rei Medica et Scientia Naturalis Incrementum*, a general scientific review journal published in Nuremberg from 1731 to 1745, although he did include *Auswahl der medicinischen Aufsätze und Beobachtungen* derived from it, as indicated earlier.

The next three titles on the *Medicinische Argos* list are all by Ernest Gottfried Baldinger. Two of these, *Magazin für Aerzte* and *Medicinisches Journal*, are cited by Ersch. The third title is *Artzeneien, und physikalische und medicinische Monatschrift*. Sudhoff compares it to Unzer's *Arzt*, although it is much more limited in scope as its subtitle indicates: *zum Unterricht allen denen welche der Schaden der Quacksalbers nicht kennen*. Ersch does not include this title at all, although he does include another related work of Baldinger: *Historia Mercuri et Mercurialium Medica*, published in Göttingen from 1783 to 1785. Baldinger makes it clear in his preface to the first volume that his sole objective for the publication is to alert his readers to the perils of mercury.

Ersch does cite most of the other titles in this section, albeit under other subject rubrics than general medicine. All in all, the *Medicinische Argos* list is a very meager one that does not make a very strong case for German medical journalism before 1790. However, in listing and discussing the secondary literature, the review introduces another conceptual classification, according to the role or combination of roles a periodical elects to play. They fall into three classes:

1. Those that follow a single path: They contain only abstracts or reviews, or they are purely compilations from other sources.
2. Those that follow two paths: They are primary compilations, but also include original articles.
3. Those that try to follow three goals: They are primarily critical reviews, but also include compilations and original articles; or they are primarily compilations, but also include critical reviews and original articles.

There are other possible combinations and permutations, of course, but for the purpose of the review, they are divided into two classes: (1) compilations, collections, or anthologies of original contributions in their entirety or in substantial extracts; and (2) what the review calls "critical reference or review journals."

The first class is divided into those collections that derive their materials primarily from German sources and those from foreign sources. It is interesting to note that there are no titles published before 1790 indicated in the first part. To some extent it helps to confirm our suspicion that there was not a great deal of significant literature in German medical journals of that period. On the other hand, this provides more justification for the titles in the second part, those compilations that drew their material primarily from foreign sources. The entire list of "critical reference or review journals" consists of fourteen titles published in 1790 or earlier, and covers essentially the same titles that Ersch had included among the current periodical bibliographies, as a class under secondary or "nonsubstantive" works.

GARRISON'S CLASSIFICATION

The two bibliographies we have examined have been productive in suggesting periodical publication formats, but they are limited in scope. First, they are confined largely to medical publications, and then only those published in German or in Germany. Fielding H. Garrison, on the other hand, compiled one of the most comprehensive lists of scientific and medical serial publications published before 1800. It covers publications in all languages, in most of the subjects relating to science, except for certain areas of technology (agriculture, mining, and the like). The list encompasses 1,003 items, 906 in the original language and 97 in translation. Counting continuations, abridgments, and some separate publications of scientific societies that are also included, the total is 1,199. His classification is pragmatic rather than strictly logical, mixing chronology, format, and subject. The primary emphasis, again, is on medical periodicals, which Garrison arranges in a separate section divided by decades. Audience is used as a criterion at least in one section: "Personal hygiene and popular medicine." One of his principal objectives seems to be to provide assistance to the subject-oriented historical researcher, as demonstrated by his use of so many subject classes.

Putting aside, however, any consideration of chronology, subject, or audience, Garrison seems to identify five specific formats: (1) periodicals; (2) transactions; (3) repositories; (4) serial collections; and (5) directories. He divides the list into the following categories:

1. Medical periodicals and transactions of the seventeenth century
2. Scientific periodicals and transactions of the seventeenth century
3. Medical periodicals and transactions of the eighteenth century
4. Personal hygiene and popular medicine
5. Pharmacy, pharmacology (including materia medica)
6. Vaccination
7. Veterinary medicine
8. Animal magnetism
9. Anthropology and psychology
10. Brunonian doctrine
11. Repositories of medical excerpts, reprints, reviews, and translations
12. Serial collections of dissertations, cases, and observations
13. Scientific transactions of the eighteenth century
14. Biology
15. Botany
16. Zoology
17. Entomology
18. Scientific and belletristic periodicals containing medical matter
19. Medical directories
20. Miscellaneous (facetious and fantastic items)

The term "format" has been used here with specific reference to the kind of bibliographic unit that is the primary component of the publication. These "formats" can be defined from Garrison's list, but only by inference and then with difficulty. We could say, then, that "periodicals" are those publications that contain primarily original contributions by authors acting independently, and "transactions" contain original contributions made in relation to some formally constituted organization or society. "Repositories," which according to Garrison contain primarily "excerpts, reprints, reviews, and translations," are definitely derivative publications. The translations that Garrison includes here are not those of complete volumes or series, but of items derived from different sources, and thus akin to the repositories of excerpts, reprints, and the like. Translations of complete volumes of periodicals or transactions or even of parts of series translated selectively are included with the original titles. The term "serial collection" seems to have much in common with Ersch's miscellaneous writings, as it includes collections by individual as well as multiple authors, frequently, but not always, published in more than one volume.

It is not wise to generalize too much on any of these formats without a more thorough analysis of each class than can be attempted here. First, they cover a span of over a hundred years that witnessed much evolution and change in each class. They also exhibit at any single period in time a great deal of variation both in terms of the type of bibliographic component unit (format) and other classification factors such as subject and audience. The issues are complicated by the fact that as the review in *Medicinische Argos* indicated, very few of these publications follow a single path. With some modification, however, we can consider Garrison's classification as a framework around which to build our discussion. One of the problems is that Garrison avoids making some of the necessary distinctions by grouping periodicals with transactions, and by including excerpts, reprints, reviews, and translations in a single category under repositories without attempting to define them.

Although Garrison is not able to follow his classification with any consistency, the literature must basically be divided into two categories, an original and a derivative (or primary and secondary) literature. Of course, large numbers of the periodicals of the seventeenth and eighteenth centuries are of a mixed character, much as many are today. The primary difference seems to be that the periodical was much less a vehicle for original contributions than it is today, a role that it developed in an evolutionary process. Another factor that is reflected in Garrison's list, and that distinguishes the early periodical from the scientific periodical of today, is the degree to which the general periodical literature was used as a vehicle for scientific communication. This accounts for the presence on the list of such titles as *Journal des Sçavans* as well *Acta Eruditorum* and the various *Giornali de'Letterati* in Italy that were modeled after it. These were all largely secondary in nature, devoted primarily to reviews of the monographic literature, academic publications, and the growing output of the scientific societies, which still served as the primary avenues for original contributions.

These are all seventeenth-century titles, but the difficulty in classifying a title as primary or secondary persists also through the eighteenth-century literature. Garrison, for instance, classifies with the primary periodicals such titles as Tode's *Medicinisch-Chirurgische Bibliothek*, Arnemann's *Biblioth-chirurgische Zeitung*, all of which Ersch had listed with the nonsubstantive bibliographic publications. Although these titles included some original contributions from time to time, the same may be said for many of the titles that Garrison lists under repositories.

Garrison also paid his respects to the phenomenon of the scattering of subject literature, which Bradford (1950:110) formulated later. Specialization occurred fairly early in the history of the scientific periodical, particularly in the field of medicine, as indicated earlier, but the process of developing specialized publications in other disciplines took place slowly. The general scholarly and literary periodical, therefore, served as an important vehicle for the communication of scientific and technical information. The amount of scientific and technical reporting in these journals also provides an important measure of the development of interest in these subjects among the general population. They also serve as indicators of the level and kind of knowledge to which the literate public was exposed, as Garrison (1934b) recognized in his studies of the medical content of *Spectator* and *Tatler*. Garrison lists some of the important titles of this nature under the heading: "Scientific and belletristic periodicals containing medical matter." These general periodicals were especially concerned with matter relating to health and medicine, but they also found room for contributions and information on other scientific and technical matters. Among the titles listed by Garrison are *Deutsche Magazin*, 1787–1802, *Edinburgh Review*, 1755, and the long-lived and well read *Gentleman's Magazine*, but every provincial German university town had its own literary periodical of a similar nature.

TRANSACTIONS AND PRIZE ESSAYS

"Transactions" as a format are also represented by a large variety of publications. The society transactions or publications associated with societies in the seventeenth century show little resemblance to those that appeared in the eighteenth century, when the patterns for the presentation of the proceedings of scientific societies began to appear. There does not indeed seem to have been any great impetus toward the current reporting of their activities by scientific societies in the seventeenth century. *Journal des Sçavans* had only a very tenuous connection with the Académie des Sciences, and concerned itself with a much broader range of subjects than the Académie. The Académie, in fact, seems to have made no real effort at systematic and periodic reporting until the end of the century (Kronick

1974:66). Even the prestigious *Philosophical Transactions* was not formally taken under the wing of the Royal Society until after the middle of the eighteenth century.

There is a great deal more diversity among the society publications listed by Garrison than is implied in the term "transactions." He seems to list all those publications he has been able to find that were issued in series by a scientific society. The published contributions represent a number of different kinds of relationships with the society: papers presented orally or submitted in writing to the society by members and nonmembers, or simply a collection of contributions published under the auspices of a society. The contributions may be solicited, as in the case of essays submitted in competition for a prize on an announced subject.

The latter category is a genre of literature that proliferated in great numbers in the eighteenth century. Following again the example of the Parisian Académie, the provincial societies and societies all throughout Europe offered prizes for papers on a variety of subjects covering the whole range of contemporary science and technology. In some cases these essays constituted the only source for publications that appeared under society auspices. In some ways the papers seem to be more in a direct line of development leading to the modern scientific paper than the contributions that were published in periodicals. They were also frequently published individually as a commitment associated with the presentation of the award. They also provided considerable substance for the repositories, collections, and other serial publications of the period. They seem to have few modern counterparts today, aside perhaps from reports on contract research, except for a few societies that still sponsor competition for prizes. A significant difference in this case, however, is that the subject in the modern competition is usually of the author's own choosing. The questions proposed for prizes by scientific societies in the eighteenth century form a veritable inventory of the scientific, social, and economic preoccupation of the period and deserve much closer study than they have received.

REPOSITORIES AND COLLECTIONS

The primary characteristic of the publications that Garrison lists under repositories is that they are for the most part derivative; that is, they take their material in one form or another from previously published work. This characteristic is also displayed to some extent by periodicals and "collections." In fact we could easily sink into a semantic morass by trying to make any sharp distinction between some of the titles Garrison lists under repositories and those he lists under collections. The former term has a more static connotation than the latter, in that the items in a repository are less likely to have been brought together under any strong guiding principle. Essentially, however, both terms refer to a number of disparate units that are brought together into a single bibliographic unit. Garrison seems to make the distinction on the basis of whether the items are secondary (repository) or primary (collection) materials, although there are numerous violations of this principle in both sections.

Garrison does not make any effort, under repositories, to distinguish between excerpts, reprints, reviews, and translations. Some of the titles listed indeed include representation from each of these classes. This is a characteristic of any classification, because classifications are linear and the world is not. Garrison includes in this class some of the titles that Ersch lists as nonsubstantive bibliographic works, such as the titles associated with Vogel, Murray, and Blumenbach. The term "reprint" obviously does not refer to its current use to describe another issue or edition of a work in its original form, but to items taken or extracted from another work either in partial or total form, either in altered or unmodified form.

The manner in which Garrison handles dissertations and other academic publications offers an example of the kinds of distinctions he seems to make. In one sense we can regard the literature generated by or derived from the formal aspects of university life as much a separate class of publication as the society transaction. In the seventeenth and eighteenth centuries, this form of publication played a decidedly larger role than it plays today. These were not solely publications of student exercises but also a means of announcing what new ideas were being generated under university auspices. In fact, one author suggests that they provided a primary means of presenting publicly new work that had not yet achieved enough importance to be published in a book (Dulieu, 1958). This is not exactly in keeping with the prevalent idea that universities were generally outside of the mainstream of scientific development in this period. However, the appearance of newness does not necessarily have to equate with innovation or progressive thought. Garrison includes under serial collections some of the publications derived from universities and brought together because they were the product of either (1) a particular institution, (2) a particular individual, or (3) on a particular subject. Presumably, they represent reprintings in full of works published as separate entities, or in some cases perhaps even published for the first time. Under repositories, on the other hand, he lists those titles that represent reviews, extracts, or bibliographic notices of academic publications.

Serial collections, in Garrison's list, are more or less a potpourri of publications, ranging from the collected work of individual or multiple authors, current or retrospective, to selected translations in full or abridged form from a variety of sources. The full heading "Serial collections of dissertations, cases and observations" would to some extent make an acceptable definition of the modern primary scientific periodical if one included the expectation of indefinite continuity of issue, which is one of its hallmarks. Very few of these titles represent conventional periodicals, however, to any extent. Expectation of indefinite continuation of issue is the characteristic that seems to be most significantly lacking from this class of publication in Garrison's list. Most of them appear to be of a predetermined duration and might more appropriately be called "continuations" if they were issued in more than one volume and at intervals in time. Among the representative titles of this class is Johann Theodor *Pyl's Aufsätze und Beobachtungen aus der gerichtlichen Arzneikunde*, issued in eight volumes in Berlin from 1783 to 1793. Examination reveals it to be a collection of postmortem and forensic reports, mostly by Pyl, but a few from other physicians as well. *Beiträge zur gerichtlichen Amneigelahrtheit* by Wilhelm Hermann Sebastien Bucholz is a review of his own postmortem and forensic reports going back ten years. They were published in four volumes in Vienna from 1782 to 1792.

A different kind of collection, if indeed it can be included in the same category with the previous titles, is *Raccolta d'Opuscoli Medicopractici.* It was edited by Giovanni Ludovico Targioni and appeared in Florence in seven volumes from 1775 to 1783. It is a collection of miscellaneous papers written by Italian authors on medical subjects and apparently published here for the first time.

The various formats that are implied or represented in Garrison's list do not by any means exhaust all those that appeared in this period. Although he does include some titles that may be classified with what have variously been called almanacs, annuals, or ephemerides, he does not identify them as a class. Almanacs, as indicated earlier, are represented by widely diverse types of publications that have an ancestry predating the journal by almost two hundred years. In the seventeenth and eighteenth centuries, they included a large number of titles that, although they may be of marginal importance in the development of scientific ideas, are of some significance for their social implications, and as indicators of the development of popular interest in science and technology. They are united primarily by the fact that they were issued annually and provided information—astronomical, meteorological, or otherwise—pertinent to the year in which they were published, but they could also be used to convey scientific and technical information about health, veterinary medicine, mining, gardening, and a host of other subjects. An example cited by Garrison is *Almanach oder Taschenbuch für Scheidekunstler und Apotheker*, issued by F. A. Göttling, C. F.

Bucholz, and others over a period of almost fifty years from 1780 to 1829 in Weimar. In addition to calendar information and book notices, it included selections from the chemical literature and pharmaceutical formulae.

SUMMARY AND CONCLUSIONS

To demonstrate the full range of periodical publication in the seventeenth and eighteenth centuries, one must go beyond the categories suggested here to consider some of the bibliographic problems related to reprinting, reissues, new editions, and translations of specific titles. These processes as related to periodicals took place with much greater frequency in this period than they do today. Widely read and esteemed titles such as *Journal des Sçavans, Philosophical Transactions,* and others were reprinted many times in both genuine and counterfeit editions, many times with emendations and changes. A new edition of a periodical in the eighteenth century was regarded not simply as a means of making a title more broadly available but also as a means of updating and providing critical commentary on work done in the past. The fact that progress in science was not as rapid or as time bound as it is today may also have made reprinting a more feasible enterprise.

This review of the formats of serial publication of this early period, if it has not accomplished anything else, has perhaps indicated some of the difficulties in arriving at a meaningful classification of seventeenth- and eighteenth-century scientific and technical periodicals. A classification of some kind, however, is always a helpful step in dealing with any large number of events or artifacts. The number of periodicals by 1790 had reached such proportions that such a classification by format is a necessary means of dealing with them as a body of literature. The following formats then seem to emerge from the review as being more or less distinctive and may provide one useful framework upon which to continue our discussion:

1. Periodicals (substantive journals)
2. Transactions (proceedings of societies)
   (a) Prize essays
3. Extract (abstract) journals
4. Review (bibliographic) journals
5. Collections
6. Academic (dissertation) literature
7. Almanacs (annuals)

The principal distinction, of course, remains on the basis of whether the publications are original or derivative, primary or secondary. Classifying a publication as primary may seem to beg the question of what is "original" in terms of concept or idea, a question we cannot attempt to resolve here. The issue is complicated by the fact that what is not "original" may still be original in the sense of being "new" to the recipient. Most of the periodicals we encounter in this period are admixtures of both derivative and original elements. These concepts are intimately bound up with the roles the periodical plays as both a vehicle and a repository for ideas.

In any case, concentration on *format* exposes only a single facet of these variegated publications, one related only to one of its formal aspects. The *subject* approach intrudes itself inevitably in each of the classifications we have examined. Even in its early stages, the scientific periodical began to show evidence of subject differentiation that must be taken into account, particularly as the process of specialization accelerated in the eighteenth century. Although the *audience* of the scientific and technical periodical represents a much smaller range in this period than it does today, the process of differentiation began to appear early. Another criterion, although it relates to that of audience, is more social and psychological in character. It is represented by such a question as: What was the editor's or publisher's *motivation* in issuing the journal: entertainment, education, self-aggrandizement, local pride, or solely the need to disseminate new scientific information? The degree of *content access* that a periodical enjoyed may be considered another basis of classification. This factor relates both to the physical access that contemporaries had to the journal, how widely it was distributed, and the intellectual access the reader had to its contents through indexes and other devices. Physical access to these early periodicals is also, of course, important if we are going to make them available to scholars. Content or intellectual access, the ability to determine in which of these periodicals items by particular authors or on particular subjects can be located, is of equal if not of greater importance to them. This kind of access may be either current (provided by indigenous or other means in close chronological proximity to the time it was published) or retrospective (provided at a later date). It can also concern individual periodical titles or multiple titles, either on a cumulative (incorporating other, earlier indexes) or noncumulative basis. Another factor that represents a bona fide criterion for classification is the *geographic*, a factor that relates to the regional distribution of different kinds of titles and to the development of periodical journalism in any area. Examples of most of the types of periodicals on almost every level can be found in most of the countries in Western Europe in this period. All of these criteria represent avenues of exploration, both to organize these materials for purposes of historical research and to study the development of scientific journalism as a mode of organizing and disseminating scientific ideas. There are probably very few titles on any of the lists we have examined that are precisely alike with respect to all of the criteria that have been brought up for discussion. However, they do tend to fall into the major categories that have been listed and that form a useful skeleton on which to organize observations and construct generalizations regarding some of these early periodicals. In the process, we may arrive at a clearer understanding of the role these periodicals play today and in which ways the apparatus of scientific communication can be modified. We may be able to cope better with the gigantic increases that have characterized this medium in the past three hundred years and that threaten to overwhelm us.

NOTES

1. See for instance Osborn (1973) and Kronick (1976).
2. The revised edition by F. A. B. Puchelt (Leipzig:1822) brought the literature up to 1829 and was used in this analysis.
3. Kirchner says it appeared until 1808, but Sudhoff says it lasted until 1805; Erch indicates it was issued in only thirty-three volumes and lasted until 1796.

REFERENCES

Barnes, Sherman B. "The Beginnings of Learned Journalism." Ph.D. diss. Cornell University, 1934.

Bertheau, Franz. *Kleine Chronologie zur Geschichte des Zeitungswesens in Hamburg von 1616 bis 1931*. Hamburg: Lutke and Wulfe, 1914.

Blumenbach, Johann Friedrich, ed. *Medicinsche Bibliothek*. 3 vols. Göttingen: Dietrich, 1783–1795.

Bolton, Henry C. *A Catalogue of Scientific and Technical Periodicals (1665–1895)*. 2d ed. Washington, D.C.: Smithsonian Institution, 1897.

Bradford, S. C. *Documentation*. Washington, D.C.: Public Affairs Press, 1950.

British Union Catalogue of Periodicals of the World. 5 vols. London: Butterworth, 1955–1958.

Buess, Heinrich. "Der Beitrag der Schweitzer Aertze zu den *Miscellanea Curiosa* des *Deutsch Akademie der Naturforscher*." *Sudhoff's Archiv* 37 (1953): 1–22.

Callisen, Adolph Carl Peter. *Medicinisches Shriftsteller-lexicon der jetzt lebenden Aerzte, Wundaezte, Geburtshelfer, Apotheker und Naturforscher aller gebildeten Völker*. 33 vols. Copenhagen, 1830–1845. Reprint, Nieukoop: B. de Graff, 1964.

Dulieu, L. "Le Premier Ancestre de *Montpellier Médicale*." *Montpellier Médicale* 54 (1958): 14–27.

Engelmann, Wilhelm. *Bibliotheca Medico-chirugica et Anatomico-physiologia*. Hildesheim: Georg Olms, [1848] 1965.

Ersch, Johann S. *Allgemeiner Repertorium der Literatur für 1785–1790*. 3 vols. Jena: 1793–1809.

Ersch, Johann S. *Handbuch der deutschen Literatur seit der Mitte des 18 Jahrhunderts*. 8 vols. Amsterdam: 1812–1814.

Garrison, Fielding H. "Medicine in the *Tatler*, *Spectator* and *Guardian*." *Bulletin of the Institute of the History of Medicine* 2 (1934a): 477–503.

Garrison, Fielding H. "The Medical and Scientific Periodicals of the 17th and 18th Centuries." *Bulletin of the Institute of the History of Medicine* 2 (1934b): 285–343.

Gesner, Augustin Phillip, ed. *Entdeckungen der neuesten Literatur in der Arzneygelehrheit*. 3 vols. Nordlingen: Becker, 1778–1788.

Hacker, D. D. and Hohl, eds. *Medicinischer Argos*. Leipzig: Weygand, 1839.

Hoffmann, Friedrich, et al., eds. *Taschenbuch für die deutschen Wundärzte*. 5 vols. Altenburg: Richter, 1783–1790.

Im Hof, Ulrich. *Das gesellige Gesellschaften im Zeitalter der Aufklärung*. Munich: G. H. Beck, 1982.

Kirchner, Joachim. *Die Grundlagen der deutschen Zeitshriftenwesens*. 2 vols. Leipzig: Hiersemann, 1928–1931.

Kirchner, Joachim. *Das deutsche Zeitschriftenwesens*. Leipzig: Harrosowitz, 1942.

Kronick, David A. "Studies of the Early Scientific Journals: The Basic Source Lists." *Texas Reports on Biology and Medicine* 32 (1974): 61–74.

Kronick, David A. *A History of the Scientific and Technical Periodicals: The Origins and Development of the Scientific and Technical Press 1665–1790*. 2d ed. Metuchen N.J.: Scarecrow Press, 1976.

Kronick, David A. *Scientific and Technical Periodicals of the Seventeenth and Eighteenth Centuries: A Guide*. Metuchen N.J.: Scarecrow Press.

Murray, Johann Andreas. *Medcinisch-practische Bibliothek*. 3 vols. Göttingen: Dieterich, 1774–1780.

Neuberger, Max. "Die Entwicklung des wisssenschaftliche Vereinwesen und seinige Bedeutung für den medizinischen Forschug." *Wiener medizinische Wochenschrift* 1 (1937): 666–74.

Ornstein, Marta. *The Role of Scientific Societies in the Seventeenth Century*. Chicago: University of Chicago Press, 1928.

Osborn, Andrew D. *Serial Publications*. Chicago: American Library Association, 1973.

Rohlfe, Heinrich. "Einer literarische Legende." *Deutsches Archiv für Geschichte der Medizin* 1 (1880): 270–72.

Runge, E. "Aus dem Anfangen die Deutsche medicinischen Zeitschriftenwesen." *Medicinische Welt* 1 (1937): 950–52.

Schneider, Georg. *Theory and History of Bibliography*. Translated by Ralph Shaw. New York: Columbia University Press, 1934.

Scudder, Samuel H. *Catalogue of Scientific Serials of All Countries, Including the Transactions of Learned Societies in the Natural, Physical, and Mathematical Sciences: 1663–1876*. Cambridge, Mass.: Library of Harvard University, 1879.

Sudhoff, Karl Friedrich . "Das medizinische Zeitschftenwesen in Deutschland bis zur Mitte des 19 Jarhunderts." *Münchener medizinsicher Wochenschrift* 50 (1903): 455–53.
Tode, Johann Clemens, ed. *Medicinisch-chirurgishe Bibliothek*. 10 vols. Copenhagen: Rothe, 1775–1787.
Union List of Serials in the Libraries of the United States and Canada. 5 vols. New York: Wilson, 1965.
Vogel, Rudolf A. *Medicinische Bibliothek*. 2 vols. Erfurt: Schröder, 1751–1753.

Chapter 5

Indexing of Early Scientific Periodicals in the *Index-Catalogue*

ABSTRACT

We derive our most intimate view of the prevailing beliefs and practices in medicine and science in the seventeenth and eighteenth centuries from the pages of the periodicals that were beginning to be issued everywhere in Europe. Yet today there are still no adequate guides to their contents. One exception, at least for medicine, is the monumental bibliography, the *Index-Catalogue* of the National Library of Medicine. The extent to which it covers this early literature, however, has not been thoroughly explored.

This study is based on an analysis of the periodical holdings of the National Library of Medicine on which its *Index-Catalogue* is based. It displays the relationship of these holdings to the total body of relevant medical literature. Sample volumes from this collection are also examined to determine the extent of coverage and some of the indexing policies.

The periodicals that were beginning to be issued everywhere in Europe, starting with the last quarter of the seventeenth century, provide us with one of our most intimate views of the development of contemporary science and technology. The eminent bibliographers of that period, such as Haller, Plocquet, Reuss, and Callisen, included many citations to this literature, but any thorough analysis of the extent to which they attempted or succeeded in covering it does not seem to exist. The evidence seems to indicate that the periodical had still not been accepted as a unique or different method of communicating scientific information, nor were there many efforts to systematically or comprehensively cover it.

Published in the *Serials Librarian* 4 (1979): 49–59

Haller's productivity and energy are, of course, legendary, even if on occasion subject to some exaggeration. The story of how the 1,200 book reviews that he wrote for *Göttingische Gelehrte Anzeigen* were inflated into 12,000 has been told several times (Rohlfe 1880). He is still credited with "13,000 letters with 1,600 different correspondents writing in French, German, English, Italian or Latin, all of which he used with native ease" (Reed 1915). He is also credited and with having written "13,000 scientific papers" (Armytage 1965:33). To accomplish this extraordinary output, he would have had to write a letter and a paper on average almost every day of his productive life—if he were given, that is, to resting on the seventh day. His four voluminous subject bibliographies, which occupied him during the last part of his life, are considered monuments of scholarship. His nine-volume treatise *Elementa Physiologiae* (Lausanne, 1757–1782) is essentially also a bibliography of the subject. The extent to which these bibliographies cover the contemporary journal literature, however, does not seem to be determined.

Ploucquet, in his medical bibliography (1793), cites a great many of the early journals, including even the general literary periodicals of the period starting with the celebrated *Journal des Sçavans* and *Acta erucutorum.* He also covers many of the regional and less celebrated journals, such as *Frankische Sammlungen* (1756), *Sammlungen von Natur und Medecin* (1717), *Allgemeine Deutsche Bibliothek* (1768), *Allgemeine Literatur-Zeitung* (1785), and even titles that may seem very remote from medical literature such as *Gentleman's Magazine* (1731).

Ploucquet, however, has been subject to a great deal of criticism for his style and accuracy. His entries are highly abbreviated and frequently cryptic, requiring the use of other secondary sources for verification. There is also evidence that he was highly selective and not always accurate. The contemporary French physician Frangois Pierre Chaumeton (1775–1810), however, may have been unnecessarily severe when he said: "Why is it necessary for the number of errors to exceed the number of pages in this compilation, which is, moreover, so unique in this genre? . . . I will content myself by calling attention to a few which are so serious that one finds it difficult to forgive them even in one to whom medical science is a complete stranger" (Chaumeton 1812–1813:154–58).

The importance of Reuss' compilation (Reuss 1801) has been fully recognized as one of the best available guides to the publications of the scientific societies of the seventeenth and eighteenth centuries. He fortunately also provides both a classified subject arrangement and an author index, so both these approaches are possible. His work also, however, suffers from the lack of standardization and incompleteness of citation. In some cases it is not easy to know whether he is citing an original source or one of the many abridged and translated versions of some of these proceedings, which were published in this period. He also does not cover the large quantity of publications that appeared under auspices other than the scientific societies.

Callisen's bibliography (1830), comprehensive as it is for the period between 1780 and 1830, is of very little use for the earlier period. This is unfortunate, as Brodman observes: "It might then complement Ploucquet, which is a subject arrangement without an author index, while Callisen is an author arrangement without a subject index" (Brodman 1954:86). Callisen did recognize the importance of a subject index, which he promised in the preface to his first volume, and raised hopes for once again in the preface to the supplement. In fact, he indicated that it was then currently in progress, having been undertaken by a physician in Berlin (Callisen 1840, vol. 28:9). Nothing, however, seems to have come from this venture either.

In light of the deficiencies of these earlier bibliographies, the importance of the *Index-Catalogue of the Library of the Surgeon-General's Office,* U.S. Army (Washington, D.C., 1880–1961) looms even larger as a means of access to the medical periodical literature of the seventeenth and eighteenth centuries. It has been, with justice, regarded as a monument of medical bibliography. Despite the fact that it is limited to the holdings of a single library, the U.S. National Library of Medicine (NLM), this library contains one of the most comprehensive collections of medical literature in the world. John Shaw Billings, the distinguished American physician, administrator, and librarian, was responsible not only for bringing it into existence in 1880, but also for providing much of the guidance and personal effort in its early years. He also had the organizational skills and industry to undertake what was a formidable task of indexing, retrospectively and currently, all the monographic and periodical literature in the library's collection. Claudius Mayer, who edited the *Index-Catalogue* for many years at NLM, in 1953 described the *Index-Catalogue*, which had printed fifty-seven volumes up to that time, as containing almost three and a half million references (Mayer 1948:6). It represents a major and a primary source for the medical historian who wishes to gain access to the printed record of medicine from its earliest days through the first third of the twentieth century. It is important, therefore, for us to know as much as we can about the extent to which it covers the literature and its policies and styles of indexing.

NLM'S LIBRARY COLLECTION

We should first consider what periodicals were in the collection and available for indexing when the *Index-Catalogue* was compiled. In 1875, Billings estimated that 1,147 medical journals had been published since Blegny's

Nouvelles Déscouvertes sur Toutes les Parties de la Medécine in 1679, and that NLM had acquired 714 of them. In terms of the numbers of volumes, his opinion was that these titles represented over three-fourths of all the volumes that had been printed, as 218 titles had never gotten beyond the first volume (Billings 1965:83).

The library's archives contain the old, handwritten accession books that go up through the first decade of the twentieth century and record the accession number and date these volumes were added. A great deal of retrospective purchasing of files of early periodicals must have taken place before the first volume of the first series of the *Index-Catalogue* was published in 1880, as most of the accession numbers assigned to them fall before this date.

Garrison identified some 755 periodical titles published in the seventeenth and eighteenth centuries as related to science and medicine (Garrison 1934). This list has subsequently been expanded to well over a thousand, but all of them were, of course, not relevant either substantively or qualitatively to the mission of a medical library. There are indeed only 124 titles included in the first list of periodicals indexed in the *Index-Catalogue* that had their origins in this period. Later additions approximately doubled the number of titles in the library, but it is still far short of Garrison's total. The library, nevertheless, is extraordinarily rich in the holdings of these early periodicals.

This is particularly true of the proceedings of some of the early medical and scientific societies. While the library has fairly complete holdings of such titles as *Histoire et Mémoires* of the Académie des Sciences of Paris, its holdings are not very good for some of the provincial societies that also had extensive publication programs such as those at Besançon, Caen, Dijon, Marseille, and Toulouse. This is also true for the publications of other societies such as those in Berlin, Copenhagen, Dublin, Göttingen, Prague, St. Petersburg, Seville, and Uppsala.

The library also acquired some of the early general learned and scholarly journals such as *Berlinische Sammlungen* (1768), *Hamburgischer Magazin* (1747), and *Sammlung von der Natur und Medizin*, cited earlier. However, as we shall see, they were not all indexed or indexed completely. These general journals, nevertheless, are important in this period because in the absence of more specialized media, scientific articles appeared also in the literary and even in news journals. Some of Lavoisier's communications, for instance, appeared in such journals as *Avant-Coureur* and *Mercure* (Smeaton 1956).

Although one of the early library acquisitions was *Hamburgischer Magazin*, the library did not acquire similar regional journals, such as *Leipziger Magazin zur Naturgeschichte* (1781). Neither of two journals that bear similar titles, *Naturforscher, eine physikalische Wochenschrift* (1747) or *Natur-*

forscher (1774), are in the library's collection, nor is what appears to be the earliest journal devoted to psychological medicine, *Gnothi Sauton* (1783). The library did acquire *Annales de Chimie*, founded under the aegis of Lavoisier and Fourcroy and other French chemists, but, as we shall see, did not index it. The chemical journals of Lorenz Friedrich von Crell (1744–1816), which preceded the *Annales*, are not all represented in the library collection either.

Some of these gaps are compensated for by the fact that the library acquired some of these materials in translations, abridgements, and other formats. About 1874, it acquired that celebrated anthology derived from early scientific journals and proceedings, entitled *Collection Académique.* The first volume of *Collection Académique* includes a translation into French of the well-known *Saggi* of Accademia del Cimento, and extracts from the first twenty years of *Journal des Sçavans*, and some of these extracts are checked in the library copy for indexing. Some of the decisions regarding selection for indexing may have been based on whether the original volumes were in the library. Volume two of *Collection Académique,* for instance, which covers *Philosophical Transactions*, is even more selectively checked for indexing because the library already had *Philosophical Transactions* for this period in an abridged edition, which had been indexed. The *Collection Académique* covers *Acta Eruditorum* at least for a part of the period in which it was published. The proceedings of some of the scientific societies like those of Berlin and Turin are also included.

Some of the proceedings of societies are also held by the library in either partial or complete translations. Some of the titles, both of proceedings and of independent journals, are held by the library in multiple editions. Indexing is complicated by the fact that these editions, quite unlike our journal editions that are reproduced by photolithography, may vary from edition to edition. *Medical Essays and Observations* of the Society in Edinburgh, for instance, went through at least five editions in Britain and one in the United States, as well as translations into French and German.

INDEXING POLICIES AND PROCEDURES

There is some information available about the indexing policies and procedures in the early *Index-Catalogue.* Billings stated some of them himself in his writings (Billings 1965:81ff). There had been no periodical indexing at the library before 1873 when Billings began this arduous task. He apparently began indexing on New Year's Day, 1874 (communication from Wyndham D. Miles). Some of the methods he chose have been described by Brodman (1954:115–21). He personally checked with a soft pencil all the articles in the journals he wished to have indexed. These markings still remain in the volumes in the library and can be seen today. In addition, the volumes that had been processed were stamped with a small oval stamp that indicated "Index SGO," although this is not always a sign that any articles in that volume had actually been selected for indexing.

In an unpublished memorandum of 1929 in the library's archives, cited by Brodman (1954:75), Garrison implies that more may have been rejected for indexing than selected. He found some "curious omissions here and there," including "fairly important clinical cases." In general, however, his judgment was that most of the omissions were "quite judicious and well considered"—for instance, "trite public addresses" or "verbose articles with no tendency." He concluded that: "For the medical journals of the third, fourth or fifth rate type Billings showed little consideration." One wonders nevertheless whether Garrison here was speaking, and whether Billings was acting, more as a clinician rather than as a historian. Judgments of verbosity or triteness may be subjective and moreover of little interest to the historian who is trying to reconstruct a contemporary trend or consensus.

Billings had other reasons for being selective in relation to the earlier journals. In his 1876 article on "Medical Libraries in the United States," he stated explicitly: "It would not probably be desirable to extend an index of these journals farther back than 1800, as the works of Ploucquet and Reuss fairly cover all medical periodical literature of any importance prior to that date. A few of the journals would be very difficult, if not impossible to obtain, but these will be for the most part of little practical importance" (1965:83). In an address before the 1881 International Medical Congress, he distinguished between a "dead medical literature" and a "live one," adding that: "The dead is not all ancient, the live not all modern." He also had other things to say, questioning the need to dig "among the dusty folios" (Billings 1965:123). How, then, can one reconcile these observations with the amount and depth of retrospective indexing that one finds in the *Index-Catalogue*? One can only surmise that Billings' scholarly instincts and interests prevailed over his needs as a clinician and administrator, or that he made his judgments regarding importance and relevance in relation to their contemporary setting.

Medical titles such as *Journal de Medécine*, published in Paris from 1754 to 1794, were, as we shall see, indexed almost cover to cover. The title page of the NLM copy of *Giornale de Medicina*, published in Venice from 1763, on the other hand, carries the bold and definite judgment "nothing to index." *Giornale de Medicina* began as an eight-page medical newspaper, and it is true that it is largely derivative; in fact, its first title was *Gazzetta Medica D'oltramonte*. From time to time, however, *Giornale de Medicina* carries an original article that may merit indexing from a historical if not clinical point of view. Some of the early titles in the NLM collection carry no markings of any kind, either in soft pencil on the articles or in title page annotations, and may have been rejected for indexing without notice.

To test these hypotheses, I selected a number of volumes of early periodicals in NLM for closer examination in relation to the *Index-Catalogue.* They were selected on the basis of availability in my personal library or in local resources and to represent a range of formats. The NLM copies were examined to determine which articles were checked for indexing. These articles were then searched in the various series of the *Index-Catalogue* to determine how they had been indexed. A number of articles from each of these sources that seemed to be in scope and that had not been checked were also searched.

Almost all of the articles searched which were checked in the NLM copies, were found under a single subject entry in the *Index-Catalogue.* On the other hand, none of those articles that were not checked, but were searched because they seemed to be in scope and relevant, were found. This indicated that the check marks in the NLM copies were a fairly reliable indicator of whether the article was indexed or not.

A SAMPLE OF SOME TITLES SEARCHED

To explore some of the implications of these findings, let us look at a few specific titles in the *Index-Catalogue* list that I examined.

Hamburgischer Magazin

The accession number on the title page of NLM's copy of *Hamburgischer Magazin* indicates that it was acquired well before Billings began his indexing activity. It is one of many examples of regional German learned journals intended primarily for scholars and amateurs interested in all aspects of natural history, as revealed by its full title. None of the volumes of the NLM copy bore any marks to indicate that any of the articles had been selected for indexing. Those articles that seemed that they could be considered in the scope of the *Index-Catalogue* were searched for but not found. This title is clearly, then, one that was not indexed. There were numerous other examples of original articles of clinical interest in the volumes of *Hamburgischer Magazin* I examined. None of these were either checked in the NLM copies or found in the *Index-Catalogue.* In many of these instances, Billings' reliance on Ploucquet to fill the gap also seemed to be misplaced.

Der Arzt

The editor of the popular medical journal *Der Arzt*, Johann August Unzer (1727–1799), who established himself in practice in Hamburg in 1750, was a frequent contributor to the medical literature. So was Abraham Gotthelf Kastner (1718–1775), a professor of mathematics and astronomy at Göttingen, who contributed articles on optics as well as other subjects (Kastner 1751). Among the contributions by Unzer to the eighth volume is an article on a species of intestinal worm. This article is not found in Ploucquet either, who does include many entries from this journal, as well as over a hundred citations under the specific subject heading for this article, including an article from *Gentleman's Magazine*.

Journal de Médicine

The case is quite different with *Journal de Médicine, Chirurgie, Pharmacie, etc.* This monthly appeared in Paris from 1754 to 1794 and was regarded as one of the most influential medical journals of its day. It achieved several editions in French and translations into Dutch, English, German, and Italian. Other journals drew upon it widely for extracts. A large number of the articles are checked in the NLM copy. Some of the omissions can easily be justified on the grounds of brevity, or on the grounds that the material does not lend itself to indexing—meteorological data and morbidity reports, for instance. Other articles may be regarded as not sufficiently clinical in interest to merit indexing.

Brevity is not always a reason for omission, however, as three and even two page articles are frequently indexed. Longer ones of clinical interest, on the other hand, are sometimes omitted. Such an article is one comparing natural smallpox with the artificial or inoculated form, published in 1762 (Debaux 1762). Billings' criteria in this case are even more difficult to discern because two other articles by the same author on the same subject are included, one from an earlier and one from a later volume of the same journal (Debaux 1761). A number of other likely candidates for inclusion from the 1760 volumes were also not checked for indexing.

Sammlung von Natur und Medicin

Sammlung von Natur und Medicin is a curious general scientific monthly published by physicians who were members of the Academia Caesarea-Leopoldina Naturae Curiosorum, or the Deutsche Akademie der Naturforscher zu Halle, as it was called later, although the title is not formally associated with the society. It began publication in 1718 and concerned itself with agriculture, technology, and natural science in general, although it had a strong medical orientation (Kronick 1976:83–84). Many of the articles or reports in the NLM copy are marked for indexing. A large part of each issue consists of a collection of meteorological observations made in Breslau and other parts of Germany and Europe, along with the associated human diseases in these areas, and another section on the effects of the weather on agriculture. These sections are not indexed. The rest of the issue is taken up with contributed reports or accounts of all aspects of natural, physical, or applied sciences, with a strong component of reports of medical interest.

Although many of the articles of medical interest are checked, there are numerous medical and biological contributions that are not. None of the articles that were not checked were found when they were searched for. The criteria for omission (brevity, out of scope) can be seen in many of the articles not checked, but cannot be as easily discerned in others. An article on the rupture of swine intestines, for instance, is indexed in an earlier volume of *Sammlung* (Crepatura 1719), although it is anonymous, while a human case in a later volume of *Sammlung* is not indexed even though it is attributed (Heinrich 1725).

The omitted articles cannot always be found in Ploucquet, as Billings had hoped. An omitted article on daytime somnambulism is easily found in Ploucquet along with two other articles from the same source, which he cites as *Sammlung.* He has references on the subject from many other sources, including *Hamburgischer Magazin*, but only the omitted one from *Sammlung* is "diurnal" (Muller 1725).

Medical Essays and Observations

Another important medical journal in the eighteenth century was *Medical Essays and Observations Published by a Society in Edinburgh.* The first edition apparently was published in six volumes in 1731, and it appeared in at least four other editions. An abridged edition was published as late as 1846. There were translations into French (*Essais*) and German (*Medicinische*) (1749).

Of the forty numbered articles in the first volume, only half are checked in the NLM copy for indexing. Many of the omissions can again be justified on the grounds of brevity, although many two- and three-page articles are indexed. Other omissions can be accounted for on the basis that they do not lend themselves to indexing, such as a description of Edinburgh, a meteorological register for the city that occupies over twenty pages, a mortality table for Edinburgh, and medical news items. Others, however, must represent judgments regarding the importance of the article or its relevance to contemporary medical practice.

Annales de Chimie

The last example in this brief survey of the indexing practices and policies of the *Index-Catalogue* is *Annales de Chimie*, which is one of the most important scientific journals in the last decade of the eighteenth century (Crosland 1994). It was proposed initially as a medium for translating Lorenz von Crell's German chemical journals into French in 1787, but met with resistance from the French authorities. The NLM copies were acquired well before indexing started for the *Index-Catalogue*, but there are no indications in the first four volumes that any of the articles were checked for that purpose.

Most of the articles in the journal can perhaps be considered outside the scope of the *Index-Catalogue*. In the first four volumes, however, there are at least four articles that merit some consideration. They are all by Antoine-Francois Fourcroy (1755–1809), the French physician who spent most of his life as a chemical teacher and investigator. He can in some ways be considered a forerunner of the physiological chemists. He was also an active member of the medical society as well as the Académie des Sciences, and one of the early supporters of Lavoisier.

SUMMARY AND CONCLUSIONS

It is clear that the *Index-Catalogue*, important as it is, is far from complete as a guide to the medical periodical literature of the seventeenth and eighteenth centuries. Many of the periodical titles covering this period were never acquired by the library. Some of the titles acquired by the library, moreover, appear not to have been indexed at all. Nor is it complete even for those titles that were selected for indexing. Billings apparently did make some policy decisions on indexing purely on pragmatic grounds. In fact, he was probably required to make these decisions in order to meet the financial and administrative problems of getting the *Index-Catalogue* published. Nevertheless, even if he did recognize usefulness to current medical practice as a criterion for selection, he fortunately did not adhere to this policy rigidly, or the *Index-Catalogue* would not be considered the invaluable tool among medical historians that it is today.

Questions may well be raised about the value of digging among these old bones, "dusty folios" as Billings called them, to seek out the antecedents of our current scientific beliefs. Ideas, however, are as much a product and a consequence of their history as people are of their ancestry. To understand some of our ideas and their place in contemporary thought, we must also find some of their genetic material. The bibliographer plays a distinct role here and makes an important contribution.

Despite the number of tools available to the medical historian to gain access to the scientific and medical periodicals of the seventeenth and eighteenth centuries, a scholar must still exercise great diligence to thoroughly survey the field. Numerous gaps still exist, in spite of the efforts of more recent bibliographers.

In the meantime, the *Index-Catalogue* remains one of the most useful tools available to medical historians. It encompasses the holdings of a library rich in the medical literature from the earliest period of printed writings to the latest. It is true that Billings found it necessary to omit a great deal of the early literature, both in his acquisitions for the library and in his selections for indexing. The remarkable thing, however, is not that he included so little in the *Index-Catalogue,* but that he included so much. It is a tribute both to his persuasive powers and his organizational ability that he was able to bring it into existence and sustain it. Our admiration still should not prevent us from recognizing the deficiencies of this great bibliography and perhaps not even from attempting to remedy some of them.

REFERENCES

Allgemeine Deutsche Bibliothek. 118 vols. Edited by Christoph Friedrich Nicolai. Berlin: Nicolair, 1768–1796.

Allgemeine Literatur-Zeitung. Edited by Christian Gottfried Schütz, Friedrich Justin, and Gottlob Hufeland. Leipzig: Sächsiche Zeitungsexpedition, 1785–1803.

Armytage, W. R. S. *The Rise and Fall of the Technocrats: A Social History.* London: Routledge and Paul, 1965.

Billings, John Shaw. *Selected Papers.* Chicago: Medical Library Association, 1965.

Brodman, Estelle. *The Development of Medical Bibliography.* Baltimore, Md.: Medical Library Association, 1954.

Callisen, Adolph Carl Peter. *Medicinisches Schriftsteller-lexicon der jezt lebenden Aertze, Wundaerzte, Geburtshelfelr, Apotheker, und Naturforscher aller gebildeten Völker.* 33 vols. Copenhagen, 1830–1845. Reprint: Nieuwkoop: B. de Graaf, 1964.

Chaumeton, François Pierre. "Bibliographie." In *Dictionaire des Sciences Medicale.* Paris: Pancoucke, 1812, vol. 3:104–8.

"Crepatura (De), Oder Zersprungung eines Darms in einem lebendigen Schweins." *Sammlung von Natur und Medicin* 7 (1719): 189–98.

Crosland, Maurice. *In the Shadow of Lavoisier: The Annales de Chemie and the Establishment of a New Science.* Oxford: British Society for the History of Science, 1994.

Debaux, M. "Lettre à M. de Condamine sur le Nombre des Inoculée de Provence." *Journal de Medécine* 15 (1761): 272–79.

Debaux, M. "Parallele de la Pétite Verole Naturelle avec l'Artificielle ou Inoculée." *Journal de Medécine* 16 (1762): 99–105.

Frankische Sammlungen von Anmerckungen aus der Naturlehre, Arztneygelahrheit, Ökonomie und damit verwandten Wissenschaften. 8 vols. Edited by Heinrich Friedrich Delius. Nürnberg: Monath, 1756–1768.

Garrison, Fielding H. "The Medical and Scientific Periodicals of the 17th and 18th Centuries." *Bulletin of the Johns Hopkins Institute of the History of Medicine* 2 (1934): 285–343.

Gentleman's Magazine. Edited by William Davis. London: Edward Cave, 1731–1907.

Gnothi Sauton, oder Magazin der Erfahrungsseelenkunde. 10 vols. Edited by Carl Philipp Moritz. Berlin: Mylliue, 1783–1793.

Heinrich, Ernst Friedrich. "De Crepatura Intestini oder Zerplatzung einer Darms mit Ausgebung der Speise." *Sammlung von Natur und Medicin* 33 (1725): 653–68.

Kastner, Abraham Gotthelf. "Anmerckung über die Aehnlichkeit des Auges mit einem verfinsterten Zimmer." *Hamburgischer Magazin* 8 (1751): 416–32.

Kronick, David A. *A History of Scientific and Technical Periodicals*. Metuchen, N.J.: Scarecrow Press, 1976.

Mayer, Claude. *New Terminology and the Index-Catalogue*. Washington, D.C.: published by the author, 1948. Mimeograph.

Muller, Johann Kaspar. "Somnabulo Diurni, oder von einer Mannes-Person, so bey heilen Tage da-jenige gethan, was ein Mondsichtigen ey Nacht." *Sammlung von Natur und Medicin* 33 (1725): 653–68.

Naturforscher, eine physikalische Wochenschrift. 3 vols. Edited by Christlob Mylius and Gotthold Ephraim Lessing. Leipzig: Johann Gottlieb Crull, 1747–1748.

Naturforscher. 30 vols. Edited by Johann Ernst Immanuel Walch and David Schreber. Halle: Gebauer, 1774–1804.

Neuberger, Albert. *The Technical Arts and Sciences of the Ancients*. Translated from the German by Henry L. Brose. New York: Barnes and Noble, [1930] 1969.

Plouquet, William Gottfried. *Initia Bibliothecaae Meduci-Practica et Chirurgica*. Tübingen: 1793–1797.

Reed, Charles B. *Albrecht von Haller; a Physician not without Honor*. Chicago: Literary Club, 1915.

Reuss, Jeremias David. *Repertorium Commentaationem a Societatis Litteratiis Editorum*. 16 vols. Göttingen: Dietrich, 1801–1821.

Rohlfe, Heinrich. "Eine literatische Legende." *Deutscher Archiv für Geschichte der Medizin* 1 (1880): 270–72.

Sammlung von Natur und Medicin wie auch, hierzu gehörigen Kunst-und Literatur-Geschichte. 9 vols. Edited by Johann Kanold, Johann Georg Braunschwitz, and Johann Christian Kundmann. Breslau: Hubert, 1717–1726.

Smeaton, William A. "*L'Avant-Coureur*: The Journal in which Some of Lavoisier's Earliest Research Was Reported." *Annals of Science* 3 (1956): 224–25.

Chapter 6

Authorship and Authority in the Scientific Periodicals of the Seventeenth and Eighteenth Centuries

ABSTRACT

The methods of establishing the credibility of the author as witness are explored in the periodical literature of science of the seventeenth and eighteenth centuries. The ambivalence of the early scientific societies and journal editors toward this subject can be explained on the basis of the need to maintain freedom of discourse because, although science is sustained by consensus, it also advances through its subversion. We can see in this early period of modern science the beginnings of the mechanisms that the scientific community has tried to develop to establish a balance between the two.

INTRODUCTION

The characteristic that distinguishes science from other disciplines, says Ziman (1968), is the fact that it is "public knowledge"—knowledge, that is, that is established through the formation of consensus among the workers in a particular field. The process by which consensus is formed is, therefore, an essential part of understanding science and scientific communication. It is, nevertheless, as Ziman suggests, full of ambivalence and ambiguity. In science, he says, "there is no authority but one's own intellect" (Ziman 1968:72), but still admits that few research results are validated unless they impinge directly on the investigator's own work. The scope of modern science requires the scientific worker to take much of his

Published in the *Library Quarterly* 48 (1987): 255–75

science on trust. We are coming more and more under the sway of a "new authoritarianism" that, like theological authoritarianism, requires the unquestioning acceptance of a body of beliefs that we cannot test.

An author reporting on his experience, whether in the laboratory or in the field, is serving essentially as a witness. The character of his testimony will depend on the verisimilitude of his perception, its correspondence to reality, and his ability to describe it accurately and persuasively. However, where his arguments are based on the object or the process witnessed, their acceptance is also related to his authority or credibility. In fact, Ziman also remarks on the close similarity in science and law of the concept of "evidence," which he defines as "information relevant to a disputed hypothesis" (1968:14). The scientific author in a sense then serves in the capacity of an "expert witness." His credibility and authority are established through a system of social and intellectual controls, which apparently have not yet been adequately analyzed or described. I would like, therefore, to explore some of these issues as they find expression in the scientific periodical as it developed in the seventeenth and eighteenth centuries.

Science in the seventeenth and eighteenth centuries, of course, differed in many ways—socially, intellectually, and economically—from the science of the twentieth century. There were, first of all, no well-established or recognized methods of training and certifying scientific investigators in any way similar to those of our graduate schools today. The social roles of the scientist and the reward systems associated with them were still in the early stages of evolution, along with the function of publication in that system. The canons and conventions of acceptable research practice had not been established. Science had not yet thoroughly separated or differentiated itself from humanistic studies or humanistic modes of thought.

Attention has been called to the high level of credulity in this period. Saul Jarcho asks the question: How were seventeenth-century readers able to swallow the stories of bizarre anomalies reported in such a respectable journal as the *Miscellanea curiosa*, which first appeared in 1670 under the auspices of the *Academia Naturae Curiosorum*? The learned physicians who made up the society appeared to give credence to such stories as that about a woman who vomited live kittens at a suitable interval after she had innocently swallowed some cat semen in well water. As Jarcho explains, the experimental tradition had not yet been established, no adequate criteria existed for testing truth, and philosophically the intervention of the supernatural was widely if not universally accepted (Jarcho 1972:66).

The processes associated with refereeing and review in scientific publication, however, were, as we shall see, in their early stages of development, and the issues involved in the formation of consensus in science can perhaps be more easily seen in the embryonic forms of these processes.

SOURCES OF AUTHORITY

One of the ways in which scientists tried to achieve the objectives of verification and consensus was by working together in groups on the same projects, as in the Accadèmia del Cimento or by bearing witness to others' work by reviewing it in societies organized for that purpose. The Accadèmia del Cimento only lasted for about ten years, from 1657 to 1667, and the evidence seems to be that the group experience did not work well (Middleton 1971). The Académie des Sciences of Paris, another highly subsidized seventeenth-century society, began with a similar commitment to anonymous group activity, but definitely abandoned the notion by the time of the 1699 reorganization of the society. One of the new regulations read: "Experience having shown that it is inconvenient for members of the Academy to work in common on a single project, each one will choose a particular object for his study, and by the account he gives of it in the assembly, he will try to enrich with his knowledge all those who make up the Academy and to profit from their remarks" (Hahn 1971:19ff.) Here the scientist achieved credibility through exposing his work to the scrutiny of his peers, but the authority of the work was derived not so much from him as an individual as from the society itself.

Antony van Leeuwenhoek provides a touching example of the need for group testimony or witness. Despite the fact that he had no obvious rivals as a microscopist in his lifetime, he had no credentials (credibility) in the scholarly world because he was not a physician, nor does he seem to have had any university education of any kind. In one of his communications to the Royal Society of London, he added the testimony of credible eyewitnesses to his observations. Among them, with a view perhaps to reassuring his English colleagues, he was careful to place the pastor of the English Congregation at Delft (Dobell 1958:175).

Authority and credibility in science, nevertheless, derived ultimately from the author or originator of the work. This will become clear as we examine the testimony of contemporary journal editors and society officers. The fact that the authority of the evidence is directly related to the reliability of the witness has been amply demonstrated by the care taken in modern times, for example, to establish the credentials of the expert in a court of law. Modern sociological studies have shown how the so-called prestige factor influences the effectiveness of a communication (Shapin 1994). There are doubtless also other factors at work, which derive from the nature of the discipline. It has been said, for instance, that physicists in general tend to give more credence to the writings of other physicists than biologists give to other biologists. This seems to suggest that credibility is related to the ability to quantify research results and to the number of variables involved in any experience. But these are more modern concepts that need exploring.

Credibility can also be conferred by sponsorship, as when a communication is submitted to a society or journal through someone who has already achieved a position of authority. This was a common practice in the seventeenth and eighteenth centuries, both for papers presented before scientific societies or submitted for publication in a periodical. The process has a kind of echo effect, as publication is itself a method of establishing one's credentials. Thus, the 1699 regulations of the Académie des Sciences specified "that no one can be proposed for a place as a pensioner or associate if he has not already published a significant work, or produced a machine or invention, or a particular discovery" (Fontenelle 1709:32).

Another way to achieve credibility is through establishing a professional reputation, although this can also be a circular and sometimes self-defeating process because the length of time involved may prove a disadvantage. An early example is Tommaso Cornelio, a physician at Naples and one of the founders of the Accademia degli Investiganti. He had completed a manuscript on anatomy and physiology in 1653 but, it is said, hesitated to publish it until he had established a sufficient reputation as a physician to support his views with authority. He finally published his views in his *Progymnasmata* in Venice in 1663, interestingly enough, in the same year that Accademia degli Investiganti was established. In the meantime, however, Glisson and Willis had published similar discoveries in England. It is, therefore, their names that are associated with these discoveries rather than Cornelio's (Fisch 1953:524).

The individual without reputation had to seek other sources for establishing his credibility. If he submitted his communication to an independent journal, as so many of them were in this period, his credibility then would have been based to a large extent on the reputation of the journal, journal editor, or sponsor. Identification with a recognized and established society, however, was one of the primary methods by which an author could establish his credentials.

THE SCIENTIFIC SOCIETY

One of the earliest of the scientific societies, the Accadèmia del Lincei, published works under its direct auspices, including Galileo's letters on sunspots and his *Il Saggiatore* (1624), but individual members also frequently identified themselves on their own publications as members of the society. With a characteristic sensitivity to the theological and political climate, the society was careful to specify in its statutes, or *Praescyiptiones*, which were drafted as early as 1604, that members could use the society's name only in connection with works on mathematics or physical sciences, which in their innocence they may have thought to be immune from polemics or controversy (Drake 1966).

Early in its history, the Académie des Sciences in Paris was assigned the responsibility of passing judgment on all projects and machines presented to the crown, primarily those for which the *privilége* had been requested. The expression "approuvé par l'académie" prominently displayed on a product or invention undoubtedly contributed greatly to its marketability. The academy's *privilége* was also extended to publications, for it was one of the few agencies under the Ancien Régime that could approve a work without having it first submitted to the process of official censorship. The importance of a society endorsement or affiliation is shown by the number of times these official designations appear on the title pages of this period. There may have been a belief that a direct relationship existed between the number of society memberships an author could list and the degree of credibility he could claim, because authors frequently listed all their society affiliations. One of many examples is that of Ignaz Edlen von Born, who edited *Abhandlungen* of Boehmischen Gesellschaft der Wissenschaften in 1775. He listed seven societies among his credentials on the title page.[1] He is outdistanced considerably, however, by the Abbé Pierre Bertholon, who had a reputation in the later half of the eighteenth century for contributions in the field of electricity. He also edited two scientific periodicals. On the title page of one of them he is identified as being a member of sixteen scientific societies: Montpellier, Lyon, Bordeaux, Béziers, Dijon, Marseilles, Nimes, Rouen, Toulouse, Valence, Madrid, Rome, Milan, Hesse-Hambourg, Lausanne, Florence, and others.[2]

Some of these societies, particularly those in the French provinces, certainly could not have conferred any great distinction. Of the society at Valence, for instance, little seems to be known except that it was formed as a literary and patriotic (that is, agricultural) society in 1784 and that it lasted until 1789, apparently without leaving many traces (Lasteyrie 1883,1). Although membership in such societies as the Académie des Sciences conferred great prestige as well as a state pension, there were many societies in which membership was largely a prerogative of position in the community. In the case of many of the so-called patriotic, improvement, or agricultural societies, membership was contingent upon the payment of an annual fee. It was probably also true, as it is today, that society memberships tended to accrue to those individuals who had already achieved recognition, a characteristic that Merton has referred to as the "Matthew effect" (Merton 1968), and that might be described as "the rich getting richer."

The importance of the society endorsement of a publication, in this period, is attested to by the number of times that independent journals claimed the sponsorship of fictitious organizations, or were the results of organizations formed solely for the purpose of publishing them. One of the most successful of the early journals that appealed to popular interest in science and included discussions on questions of science and medicine, as well as theology, history, and other subjects, was *Athenian Mercury* (1691–1700). To endow its contributions with greater credibility, the publisher, John Dunton, originated an Athenian Society along with a group of anonymous authorities. He even carried the pretense to the point of having one of his authors, Charles Gildon, write a history of the imaginary society (Cormick 1973:83). The case of *Museum Rusticum*, an agricultural journal that enjoyed a brief flurry of popularity in England between 1763 and 1766, was rather different; it claimed on its title page that its papers were "revised and digested by several members of the Society for the Encouragement of Arts, Manufactures and Commerce," when in fact they had no official connection with that association (Hudson 1954:17).

An example of an organization created primarily for the purpose of publishing a journal is the Society for Promoting Medical Knowledge. Its sole activity seems to have been the publication of a volume called *Medical Communications* in 1784 and once again in 1790. It differed from the professional or regulatory societies in that it included among its contributors apothecaries and surgeons as well as some of the highly reputable physicians of that day. The editors themselves indicated in the preface of the first volume (1784) that they were brought together for the purpose of "collecting and publishing such papers as they think worthy of being preserved." There is perhaps no special distinction to be inferred from the fact that the Germans translated the first volume of *Medical Communications* almost at once,[3] because they sometimes seemed rather undiscriminating in their zeal to convert English publications into their familiar tongue. The publication does, however, provide an example of an approach to therapeutic procedures, as well as sensitivity to the problems of credibility, which may be unusual for that period.

The first volume includes a communication on the treatment of venereal disease by opium, a paper that the editor identifies in the preface as the first publication in Great Britain on that subject. He adds the following comment:

> The author seems to be aware that much observation is yet wanting to determine the real value of this new remedy, and candidly acknowledges that it has not been always given with equal success. The Society are also convinced that the experiments hitherto made are by no means sufficient to establish a new method of treating the disease in question, and have published the paper rather with a view to excite inquiry than to satisfy doubt; hoping that those who have proper opportunities will exert themselves in the investigation of so important a subject.[4]

This passage is interesting from two points of view: first, it points to the society's awareness that it is somehow lending its authority to the communication by publishing it under its auspices, and second, the quotation suggests that the objective of publication is not so much to establish the credibility of a fact as to incite inquiry. This point of view is mirrored in modern times by J. D. Bernal when he says it is sometimes more useful when the research results reported are wrong than when they are right. He says: "In most cases we know that the material contained in a paper or scientific communication is wrong. It is probably more useful when it is wrong than when it is right, because when it is wrong, you can do something to make it better. When it is right, it finishes the whole thing" (Bernal 1939:432).

Societies were understandably extremely cautious about letting their names be attached to various publications and protested vigorously when it was done without their approval. When *Journal littéraire*, a French-language literary journal published in Berlin in 1772, implied an association with Akademie der Wissenschaften by merely stating on its title page that it was under the auspices of a "society of academicians," the society must have moved at once to ask them to clarify this identification because the second issue contains the following declaration: "Truth obliges us to repeat that the Royal Society of Sciences and Belles-Lettres has no part in this journal. It is edited by a separate society which several of the Academicians have formed among themselves for this purpose."[5]

As is well known, despite a much more direct association between the Royal Society of London and the highly respected *Philosophical Transactions*, the Society refused to legitimize the connection until 1753, almost a hundred years after its founding. As early as March 1, 1664, the minutes of the council had specifically set down "that the *Philosophical Transactions* to be composed by Mr. Oldenburg, be printed the first Monday of every month, if he has sufficient matter for it; and that tract be licensed by the council of the society, being first reviewed by some of the members of the same" (Andrade 1965:14). Yet in the issue for May 7, 1766, Oldenburg found it necessary to insert a notice that disavowed any formal connection between the society and the *Transactions* except through his membership in the Royal Society of London and as a result of his inserting, from time to time, material in the publication that had first been brought to its attention (Andrade 1965:14). When in 1682, after a hiatus of four years, Robert Hooke took over the publication of *Philosophical Collections*, as he called them, he too was quick to deny any formal association with the Royal Society of London by affirming that the publication "is not to be looked upon as the Business of the Royal Society" (Andrade 1965:14). Despite the fact, however, that the Royal Society of London did not take official responsibility for *Philosophical Transactions* until 1753, during the interval it maintained a "committee of correspondence" consisting of twenty members whose charge it was to assist the secretary in obtaining the communications that found their way into that publication (Adelmann 1966:670). However, as early as 1663, they had adopted an official procedure for its imprimatur for the works it licensed to be printed.

The Académie de Chirurgie in Paris, in their regulations, specifically forbade any member to use its name on their publication without its permission on pain of expulsion from the Académie.[6] A society that operated under governmental auspices, particularly when its members received stipends from the government, and when it exercised the privilege of publication without recourse to official censorship, such as the Académie des Sciences of Paris, could, of course, apply these rules more vigorously than others who did not have these advantages. Another example is the Sociedad de Medicina y Demás Ciencias of Seville, which had its origins in 1700 but did not receive official sanction until 1729. Under a royal decree, the Royal Society of London was awarded an annual income derived from shipping fees, which was to be distributed among its members. The members at the same time were also constrained from publishing anything without the prior approval of the society.[7]

A provincial society such as that at Toulouse had stringent rules about publication by its members. Statute no. 24 of its regulations reads: "No academician may place his title at the head of a work he wishes to have printed, unless it has been approved by the Society, and after this approval has been given him subsequent to a reading before the Assembly, or at least after it has been examined by the Commissioner named for that purpose."[8] This was particularly true of publications that appeared under the auspices of such societies, but that were not by their members. The Académie des Sciences, Arts et Belles-Lettres of Dijon, which made Rousseau famous (or vice versa) for his prize essay, awarded the physician Godard the prize in 1764 for his work on the subject "Determine the nature of the so-called antispasmodics, explain the way in which they operate, distinguish the various kinds, and indicate their use in various diseases." After Godard had published the essay, Maret, the Society's perpetual secretary, wrote and asked him to revise it by making changes on specific pages on which he had discussed exorcism and witchcraft, or at least to insert a note that this material had been added to the essay after the society had rendered its judgment. The society apparently had no confidence that Godard would comply with this request because it decided also to send a letter to several periodicals indicating which pages it disavowed and informing that it had no knowledge of these additions until after Godard had printed them in Paris (Tisserand 1936:583).

Societies developed various other techniques to guard and protect their credibility. The regulations established at the reorganization of the Académie des Sciences in 1699 specified that "all the experiments reported by any academicians will be verified by them in the assembly, or at least in the presence of several academicians" (Fontenelle 1709:36). The Académie Royale de Chirurgie set up an official procedure for confirming the accuracy of its records—approval of the minutes, so to speak—in one of its regulations, which stated: "The deliberations of the Academy will not be registered until after they have been approved; it suffices to have them signed by the President and the Secretary."[9] The "registers" kept by many of the societies were, of course, not public records and served more to fulfill a historical and contractual requirement rather than one relating to the recording and dissemination of information. They did, however, frequently provide the basis for the retrospective publication of the society's *Mémoires* and sometimes provide the only surviving records of a society's activities.

Another protective measure used by a society was not to publish everything that was presented at its meetings, even as a part of its official proceedings. The registers and manuscripts relating to the activities of the Académie des Sciences, Belles Lettres et Arts of Besançon in the eighteenth century, which are deposited in the municipal library of that city, are said to comprise more than a hundred thousand pages. Among them are four volumes in folio containing its statutes, *lettres de cachet*, *procès verbaux des shances*, summaries of the discourses given before it, and *résumés* of the works presented *au concours*, as well as six volumes of memoirs contributed by its members and one volume by foreign authors (Cousin 1954:19). The only official publication that seems to have resulted from all of these activities and records is a series of short accounts of its public sessions, which was published fairly regularly from 1754.[10] At that time, the society also decided to publish those works of its members that it regarded as of sufficient merit to bring to the public's attention. The society's historian describes the process as follows:

> A choice of this kind is very delicate, the modesty of some, the scruples of others was such that on November 24, 1755, they decided to read in a chronological order in assembly the articles which had been transcribed on their registers, to choose by vote those that were to be printed, and to entrust certain members with the responsibility of meeting with the authors themselves on the texts selected for publication. (Fisch 1953: 29)

The readings began in 1755 and continued until 1758 without any choice being made. The task was resumed in 1768, but in 1776 Turgot was still expressing astonishment that nothing had been printed, despite the subsidies the society has received from the state for that purpose (Cousin 1954:2)

The Academy at Dijon was founded in 1725 but did not receive its official charter until 1740. It was not until 1769 that it published the first volume of its *Mémoires.*[11] The publication was first proposed in 1764, but the commission, which was charged with selecting the contents, did not meet until 1769 (Tisserand 1936:519). They decided to follow the model established by the Académie des Sciences of Paris, which was widely copied in eighteenth-century Europe. First the society briefly summarized in the *Histoire* the papers that had been read before its assemblies, including those that had already been printed separately. It was only from 1761 onward that the society systematically kept records of the papers presented, except by title and date. Consequently, a great many of them were printed in the independent periodicals of the time, among them the celebrated *Journal de Physique* edited by Rozier, but also in such places as *Mercure de France, L'Avant-Coureur*, and even *Almanach des muses*. Tisserand, the society's historian, indicates that great precautions were taken before any member was permitted to publish anything that identified him with the society, but does not specify what these precautions were (Tisserand 1936:95). In 1774, *Journal de Physique* printed a memoir that had been presented that same year by the physician Morveau, a promptness that was not unusual for that journal. The article carried a statement signed by Maret, the society's secretary, to the effect that Morveau had been authorized to have it printed,[12] but this practice does not seem to have been followed consistently. Some societies developed techniques of selection and approval similar to the refereeing methods used by modern journals. In the preface to his translation of *Medical Essays, Revised and Published by a Society in Edinburgh*, Demours, the translator, described the process as follows:

> Memoirs sent by correspondence are distributed according to the subject matter to those members who are most versed in these matters. The report of their identity is not made known to the author. Nothing is printed in this review which is not stamped with the mark of utility and if the number of good memoirs is not sufficient to make a proper volume, they wait for a new harvest until it can be achieved.

The preface of the first volume of the original edition of *Medical Essays* spells out the editorial policies carefully; that is to say,

> The Manner in which each Subject expected from our Correspondents should be treated in our judgement. (1) In historical Essays, the proper Vouchers are to be particularly quoted. (2) The Descriptions and Virtues of simple Drugs are to be clearly and succinctly told, without enlarging or Arguments a priori, which are too liable to lead into Error.[13]

The listing of policies goes on through nine additional rules and precepts. Some years later, when the organization had ascended to the dignity of becoming the Royal Society of Medicine of Edinburgh, these rules were elaborated and extended. Papers were required to be distributed to members for review and selection well in advance of their presentation. Until 1780, each week the society appointed three "impugners" whose function was to be prepared to provide a critique of the papers scheduled to be read the following week. Attempts were made to revive the practice in 1857 and again in 1912, but apparently they did not succeed (Gray 1952:66–69).

It is interesting to see the editors of *Memoirs of the Literary and Philosophical Society of Manchester* express the sensitivity of the process of editorial selection in terms very similar to those used by the historian of the Académie des Sciences, Belles-Lettres et Arts de Besançon, echoing the sentiments of the society at Manchester:

> The selection of the papers has been made with as much impartiality, and as strict attention to their comparative merits, as could be expected, in decisions of so delicate a nature. Yet the committee are sensible that a majority of votes, delivered by ballot, is not an infallible test of excellence in literary or philosophical productions. (36)

The method referred to was one in which an appointed "Committee of Papers" was charged with voting on the selection of articles. In spite of these precautions, the editors take pains to include the following disclaimer:

> The sanction which the Society gives to the work, now published under its auspices, extends only to the novelty, ingenuity, or importance of the several memoirs which it contains. Responsibility concerning the truth of facts, the soundness of reasoning, or the accuracy of calculation, is wholly disclaimed; and must rest alone, on the knowledge, judgement, or ability of the authors who have respectfully furnished such communications.[14]

Disclaimers of this kind are to be found frequently in the literature associated with the scientific societies of this period. In one of the early prize essays published by the Académie des Sciences of Paris, the society makes it clear that appearance of an essay under its imprint does not represent an endorsement of the ideas expressed therein, but merely indicates that it was preferred over the others submitted on the subject.[15] The editor of *Medical Essays and Observations*, cited above, after going into detail on the difficulty of avoiding error and misrepresentation in publications, warns that "if we ought, therefore, to be cautious in giving implicit Credit to original Authors, we must be no less on our Guard in consulting the voluminous Compilers, who indifferently collect from all Books every Case relating to the Title of their Sections or Chapters." The editor of a collection of papers issued by a medical society in Hamburg in 1776 makes it clear that, as secretary of the society, he was charged simply with the task of "permitting" some of these papers to be printed: "Beyond this, however, he takes no part in the essays included (except for his own) for which each author is himself answerable."[16] The Royal Irish Academy, in the first volume of its *Transactions*, published in 1787, makes the statement in stronger terms: "The Academy desire it to be understood that, as a body, they are not answerable for any opinion, representation of facts, or train of reasoning which may appear in the following papers. The authors of the several papers are alone responsible for their contents."[17] The American Philosophical Society in Philadelphia gave evidence of even greater insecurity, as may have been proper for a provincial organization:

> As soon, therefore, as the Society judged that they had received a sufficient Number of Communications for a Volume, they appointed a Committee, to assist the Secretaries, in selecting out of them such as might be most proper for the public view. And for their direction in the execution of this trust, the following Rules were given, viz. First: that the grounds of the Committee's Choice of Papers for the Press, should always be, the importance or singularity of the subjects, or the advantageous Manner of treating them, without pretending to answer, or to make the Society answerable, for the certainty of the facts, or propriety of the reasonings, Contained in the several Papers so published, which must still rest on the credit or judgement of their respective authors. *Secondly:* That neither the Society, or the Committee of the Press, do ever give their opinion as a body, upon any paper they may publish, or upon any subject of Art or Nature that comes before them.[18]

THE INDEPENDENT JOURNAL

Publication by a scientific society conveyed with it all the authority that had been conferred upon it by official sanction, whether the society wished to acknowledge it or not. There was no similar source of authority for periodicals, which were started by individuals. The credibility or authority here derived from the issuer or editor who, as we have seen, sometimes sought to establish his credentials by associating himself with an existing society or by creating a society that would serve that purpose. In this respect, he was not much different from an individual author seeking a publisher for his work. In fact, in this early period, journals were frequently largely written by the editor himself, who in many cases could be considered the author of a series of essays. There are numerous examples of learned and scientific journals inaugurated in this period by editors with the hope of attracting other contributors. They frequently ended their efforts without any other content but that supplied by the editor himself. Many of the independent journals of the seventeenth and eighteenth centuries were secondary publications that served the purpose of transmitting information that had theoretically been authenticated or validated elsewhere. Even in the case of Rozier's *Journal de Physique*, which in the middle of the eighteenth century was one of the outstanding vehicles for scientific information, many of the contributions consisted of preliminary reports of papers that had been presented to various scientific societies in France and Europe. They frequently were published here much earlier than they appeared in the society's *Memoires.*

The learned and scientific journals of this period cover a wide range of types with differing objectives and variations in formats. From the earliest prototype, however, like *Journal des Sçavans*, which was widely copied throughout Europe, they were to a large extent news media rather than repositories of primary scientific information. In this respect they took their characteristics from the contemporary newspaper from which they derived their inspiration. This is consistent with the fact that the news was repeated from one journal to another, as one newspaper reprints a story that originated elsewhere in order to disseminate it to its own readership. Newspapers seem to be more concerned with the reliability of the report and how accurately it conveys what has been said or what has happened, rather than with the reliability of the data reported. The emphasis is perhaps more on the skill of the reporter than on the methods by which the data were derived. This is borne out in the introductory comment to the first issue of *Journal des Sçavans*, which may have been written by the editor, de Sallo, himself. The purpose of the journal, he says, is that of informing the reader about "that which is new in the Republic of Letters." He also warns that he takes no responsibility for the ideas expressed by the authors who are included, "neither guaranteeing their validity nor offering criticism of them" (Brown 1972:369).

For the most part, therefore, the learned or scientific journal did not adopt many of the techniques for review and evaluation that the scientific society had developed. Its credibility rested upon the editor and, in turn, on the authors whose papers he accepted, and thus on the editorial process, the editor's policies, in accepting or rejecting material. This made the task a very responsible one, and a difficult one, as Albrecht von Haller, even with all his extraordinary talents, had to confess when he was editor of *Göttingsche Gelehrte Zeitung*. The editor, said Haller, must be a master of his subject. He must understand as many languages and as many sciences as human frailty permits. If this were not possible, he added, then it would be better if the task were assigned to a group of scholars (Haller 1772:121).

We do not know how many seventeenth- and eighteenth-century editors met these standards, as no collective study of the scientific editors of this period seems to exist. One comment that we can make is that, in contrast to the practice today, editors tended to be self-selecting, a natural consequence of the fact that most of the periodicals were individual enterprises rather than the group or organizational undertakings they are today. We also do not know a great deal about the process of editorial review in this period, or whether it resembled at all the modern practice called "refereeing." Although editors in those days were more frequently inspired to write lengthy prefaces even to single issues, and especially to the collected volumes of their journals, they were not always revealing about their editorial policies. They were much closer to the tradition of the scientific "intelligencers" such as Oldenburg, Haak, Collins, Peiresc, and others who tended to regard themselves as transmitters rather than evaluators of information in the sense of the modern journal referee.

One of the ways the early journals tried to improve the editorial process was by the introduction of collective review and editorial boards. This is revealed in the history of *Journal des Sçavans,* the inner workings of which we perhaps know better than any other early journal. As an official or semiofficial organ of the state, its history has been well documented. It is said to have been a collective enterprise from its very beginnings in 1665, in that individuals were recruited in France and abroad to send the editor books published in their geographical regions and to report the interesting events in the scholarly world that occurred there (Brown 1972:376). This can hardly be considered joint editorial responsibility, however. It is another example of the similarity to the newspaper of the early learned journal to which these far-flung contributors served as correspondents. In fact, under La Roque's editorial direction, a cabal was formed in 1684 to attempt to convince the chancellor, Louis Boucherat, that a single editor could not possibly keep himself *an courant* with respect to so many different sciences (Trenard 1969:133). It was not until the Abbé Jean Paul Bignon became editor in 1701 that the journal was put in the hands of an editorial board under his direction. The scholars appointed were made pensioners of the state and included Audry for medicine, Fontenelle for mathematics, Vertot for history, and Pouchard for philology and literature (Trenard 1969:136). They met weekly at Bignon's home so that "the work could be assembled under their eyes and be perfected by their councils" (Barnes 1934:255).

Journal de Trévoux, which was inaugurated as a Jesuit alternative to *Journal des Sçavans* in the same year that Bignon took over the latter as editor, was under the direction of an editorial board of five from the very beginning. Diderot confirmed the need for this kind of multiple editorship when he wrote in his article in the *Encyclopédie*:

> A journal embraces such a large variety of matters that it is impossible for a single editor to issue even a mediocre journal. One cannot at the same time be a great geometer, a great orator, a great historian, or a great philosopher; no one has universal erudition. A journal must be the work of a society of scholars. (Dumas 1931:221)

Diderot's emphasis, as in most of the other arguments for collective editorial responsibility in this period, is on the need for specialization, rather than on the need for consensus, a need we find expressed in the publications of the scientific societies. For example, *Journal Oeconomique*, an outstanding agricultural periodical published in France between 1751 and 1772, covered industry and commerce. It was under an editorial committee made up of specialists (Bourde 1967:1507). The prospectus for *Bibliothek der neuesten medicinisch-chirurgischen Literatur* announced that the editorial staff would be composed of twenty assistants, partly German and partly foreign.[19] However, when the special journal, *Annales de chimie*, finally began publishing in 1789, after overcoming official resistance, the editorial board was also made up of specialists, but this time all were involved in a single specialty, chemistry.

For the most part, however, the independent learned and scientific journals of the seventeenth and eighteenth centuries were the responsibility of individual editors. It is not often that they took the trouble to state their prerogatives. When they did they could be found at the extremes of the spectrum of subjects and goals. The statement of a prolific medical editor of the period, Ernest Gottfried Baldinger, who was also a well-known German physician, represents perhaps the barest minimum as far as the responsibilities of the editor are concerned: "I will not compose any regulations for anyone who honors us with his reports. The only rule which I will permit myself is this, to be concerned that nothing unworthy is printed, and merely that. I extend to myself the unreserved right, however, to see every essay before it is printed."[20] The editor of the natural history journal *Neues Magazin für die Liebhaber der Entomologie*, Johann Caspar Fuessli, is, on the other hand, much more forthright. In the preface to the first volume of the series, he states unequivocally that the complete direction and responsibility of the journal is his: "If I should find among the submitted papers any such of which I can make no use I will send them back to their authors."[21] He could perhaps afford to make this statement more than many of the editors, as he was one of the few who offered to pay honoraria to his contributors.

AUTHORSHIP

In the final analysis, it is the author who bears the burden of responsibility for the communication that finds its way into print. The corporate bodies or societies, and the individuals such as editors who intervene, function to a large extent only to regulate the flow of traffic. It is the author's name that lends the communication its authority, and it is he, therefore, it is now generally agreed, who should reap the appropriate social and fiscal rewards. In view of this observation, it is difficult to understand the curious eighteenth-century ambiguity regarding attribution of authorship. Is it because of some fundamental philosophical differences between then and now about the nature and derivation of knowledge? Or is it an outgrowth of the great changes that have taken place in the social role of the scientist? Some of the scholarly amenities regarding the practice of literature citation had not yet been well established, and scholars were sometimes careless in assigning authorship. This attitude is also reflected in the almost complete freedom with which scholarly journals borrowed from each other, often without indicating their sources. The best demonstration of these differences, however, lies in the ambivalent attitude in this period toward anonymity.

On the one hand, one finds anonymity firmly rejected. An example is *Almanach Véterinaire*, one of the first truly specialized veterinary periodicals. It was published in Paris in 1782 and, as its title suggests, still had many features of the traditional almanac. However, it also solicited and printed original papers that were within its scope and compatible with its objectives. In the preface to the first volume, the editors are careful to add the statement that "we warn that we will accept only those papers that are signed by their author since we cannot be responsible for the errors which those who are not expert in the art may send to us."[22] For an opposite position, we can cite an English agricultural journal published in 1764 that recognized fully, although perhaps with some reluctance, the right of their contributors to withhold their names. While *Museum Rusticum*[23] had no official association with the Society for the Encouragement of Arts, Manufactures, and Commerce, many of the distinguished and noble members of that organization who were interested in agriculture did contribute papers to the publication with the proviso that their name be withheld. As the editor explains:

> It is, however, to be lamented, that we are not permitted to disclose the names of many of our benefactors, some of whom stood so high in the learned world that their sanction would carry into practice many excellent precepts which had now nothing to support and recommend them beyond their own intrinsic merit. But while we thus notice the reserve of our anonymous correspondents, we hope none will think we undervalue the great favour conferred on us by the gentlemen whose names do honour to the pages of this volume. (531)

Arthur Young, a member of the society and editor of the most important agricultural journal of the eighteenth century, disagreed vigorously with this attitude, and in fact apologizes when on rare occasions he is forced to print a communication without a signature. "Not inserting the name of my correspondents," he says, "is directly contrary to my engagements."[24]

The whole question of anonymity is a complex one as it relates to authorship and authority in the scientific periodicals of this period and in the history of science in general. In modern times, the use of anonymity in authorship has been argued for in terms of withholding the identity of referees from authors whose papers they are reviewing. It has been suggested, and practiced in some instances, that authors' names be withheld from reviewers to avoid any undue influence upon them by the reputation or lack of reputation of the author. In the eighteenth century, withholding the name of the author was a standard practice in submitting essays for the *prix* that proliferated among the scholarly, scientific, and technical societies of that period. A standard technique evolved of shielding the author's name by means of a Latin motto that was affixed to the contribution and also deposited with the society's secretary in a sealed envelope. Contemporary scientists who wished to maintain secrecy for a period or to protect priority used similar devices and other techniques. Christiaan Huygens, for example, embodied his discovery of Saturn's rings in the form of a cryptogram that he sent to Oldenburg (Thorndike 1941, vol. 6:629). Over a hundred years later, Lavoisier, for similar reasons, deposited a sealed note on combustion with the secretary of the Académie des Sciences that was not to be opened for six months (Duveen 1954:92).

Many of Lavoisier's papers, particularly those associated with his official duties for the Académie and letters that he addressed to various periodicals, were published unsigned. Lavoisier's publication practices have been extensively documented, and some of his motivations for publishing anonymously have been explored (Scheler 1960, 1961; Smeaton 1936, 1957; Storrs 1968). The publication behavior of other scientists in this period has not been studied in any detail. Studies of such behavior may lead us to some new insights into the psychology and sociology of the communications process in science. To take one example, Huygens had the option of having some of his unpublished papers printed under the Royal imprimatur in France when De la Hire invited him to include them along with the papers of Frenicle, Roberval, and Picard in a collection published by the Académie des Sciences, of which Huygens had served as a member.[25] Huygens chose instead to publish them in Holland in his own *Traité de la Lumière,* which appeared earlier than the Parisian publication. Lavoisier, on the other hand, held back the printing of many of his early memoirs even after the Académie had approved them for publication. Many of them consequently were not published until many years after his death. His bibliographers insist that in these instances, Lavoisier was exercising a strict control on the quality of his output. "He considered these papers to be of little value," they say, "and therefore restrained his output with fastidious judgement" (Duveen 1954:9).

But here we are beginning to get away from the problems of authorship as they relate to authority and are starting to concern ourselves with other aspects of authorship. The comparison between the publication behavior of Huygens and Lavoisier goes beyond perception of the author's credibility and impinges on his motivations for publishing and also on the publisher's incentives for printing. The author's motivations involve us with all the philosophical and psychological aspects of authorship, as well as the constraints that are imposed on him by the scientific community.

Economics intrudes in more direct ways in publishers' incentives, because printed communications can also be marketable commodities.

J. D. Bernal in his *Social Function of Science* called for a drastic revision of the system of scientific communications. His arguments seem to be based on dramatically opposed points of view. First, he pleaded for a more open system of publication without the constraint of referees on the basis that erroneous reports sometime stimulate more productive work than correct ones (21). On the other hand, he repeats the common complaint about the absolute bulk of scientific communication: "It is of unequal value," he says, "a large proportion of it, possibly as much as three quarters does not deserve to be published at all" (Bernal 1939:118). He thus points to, however, the delicacy of the balance between the need for freedom of expression and exchange of ideas in science and the desire on the part of the scientific establishment to manage the system of scientific communication in as efficient a manner as possible. The important thing about consensus in science is that it is not fixed and immutable and that scientific communication exists not only as an instrument for the development and establishment of a consensus but also as a force for changing it.

NOTES

1. *Abhandlungen eine Privatgesellschaft in Böhmen zur Aufnahme der Mathematik, der vaterlandischen Geschichte und der Naturgeschichte* (Prague), vol. 1, 1975.
2. *La Nature Considerée Sous ses Diferens Aspects* (Paris), vol. 1, 1787.
3. *Medical Communications of the Society for Promoting Medical Knowledge*, vol. 1 (1784), preface.
4. *Medicinische Beiträge*, edited by Christian Friedrich Michaelis. Göttingen: Dieterich, 1785.
5. *Journal Littéraire Dedie an Roi Par un Société d'Académiciens*, vol. 1 (November–December 1772), preface to the 2d issue.
6. Académie de Chirurgie, Paris. *Règlement pour une Académie de Chirurgie.* British Museum, 557d2 (13), regulation 32. n.d., 1732(?).
7. *Memorias Academicas de la Real Sociedad de Medicina Demás Ciencias de Sevilla*, vol. 1 (1776), preface.
8. *Histoire et Mémoires de l'Académie Royale des Sciences, Inscriptions et Belles Lettres de Toulouse* 1 (1782): 12.
9. *Mémoires de l'Académie Royale de Chirurgie* 1 (1743): Regulation XI.
10. *Séance Publique de l'Académie des Sciences Belles-Lettres et Arts de Besançon.* 1754–1783.
11. *Mémoires de l'Académie de Dijon* 2 (1774): xxxvii–xi.
12. *Introduction aux Oberations sur la Physique* (1774): 416–25.
13. *Medical Essays, Revised and Published by a Society in Edinburgh*, vol. 1 (1733), preface.

14. *Memoirs* of *the Literary and Philosophical Society* of *Manchester*, vol. 1 (1785), preface.
15. Académie des Sciences, Paris. *Pièces qui ont Remportée les Deux Prix*. Paris: Claude Joubert, 1721.
16. *Abhandlungen und Beobachtungen aus der Arzneigelahrtheit*, vol. 1. Hamburg: Meyn, 1776.
17. *Transactions of the Royal Irish Academy*, vol. 1 (1787), preface.
18. *Transactions of the American Philosophical Society*, vol. 1 (1771), preface.
19. *Bibliothek der neuesten medicinisch-chirurgischen Literatur*, vol. 1 (1789), preface.
20. *Magazin vor Aerzte*, vol. 1 (1775), preface.
21. *Neues Magazin für die Liebhaber der Entomologie*, vol. 1 (1782), preface.
22. *Almanach Véterinaire*, vol. 1 (1782), preface.
23. *Museum Rusticum*, vol. 1 (1766), preface.
24. *Annals of Agriculture*, vol. 2 (1784): 9.
25. *Divers Ouvrages de Mathématiques et de Physique par Messieurs de l'Academie Royale des Sciences*. Paris: Imprimerie Royale, 1693.

REFERENCES

Adelmann, Howard B. *Marcello Malphigi and the Evolution of Embryology*. Ithaca, N.Y: Cornell University Press, 1966.

Andrade, E. N. da C. "The Birth and Early Days of the *Philosophical Transactions*." *Notes and Records of the Royal Society* 20 (1965): 9–27.

Barnes, Sherman B. "The Beginnings of Learned Journalism." Ph.D. diss. Cornell University, 1934.

Bernal, John D. *The Social Function of Science*. London: Routledge & Sons, 1939.

Bourde, André. *Agronomie et Agronomes en France au XIIIe Siècle*. 3 vols. Paris: SEVPEM, 1967.

Brown, Harcourt. "History and the Learned Journal." *Journal of the History of Ideas* 33 (1972): 365–78.

Cormick, Jean. "Medical Advice in Seventeenth-Century Journalism." *Journal of the American Medical Association* 224 (1973): 83–86.

Court, Susan. "The *Annales de Chimie*, 1789 1815." *Ambix* 19 (1972): 13–28.

Cousin, Jean. *L'Académie des Sciences, Belles Lettres et Arts de Besançon*. Besançon: LeDoux, 1954.

Dobell, Clifford. *Antony van Leeuwenhoek and His "Little Animals."* New York: Russell & Russell, [1932] 1958.

Drake, Stilllman. "The Accademia dei Lincei." *Science* 151 (1966): 199–20.

Dumas, Gustav. *Histoire du Journal du Trevoux depuis 1702 jusqu'en 1762*. Paris: Boivin, 1931.

Duveen, Denis L., and Herbert S. A. Klickstein. *Bibliography of the Works of Antoine Laurent Lavoisier, 1743–1794*. London: Dawson, 1954.

Fisch, Max H. "The Academy of Investigators." In *Science, Medicine, and History*. Edited by E. A. Underwood. London: Oxford University Press, 1953, 521–63.

Fontenelle, Bernard le Bovier de. *Histoire du Renouvellement de l'Académie en 1699*. Amsterdam: 1709. Facsimile Reprint, Brussels: Culture et Civilisation, 1969.

Gray, J. *History of the Royal Medical Society, 1737–1937*. Edinburgh: University of Edinburgh, 1952.
Hahn, Roger. *The Anatomy of a Scientific Institution: The Paris Academy of Sciences, 1666–1803*. Berkeley: University of California Press, 1971.
Haller, Albrecht von. *Sammlung kleines Hallischen Schriften*. 2d ed. Bern: E. Haller, 1772.
Hudson, Derek, and Kenneth W. Luckhurst. *The Royal Society of Arts, 1754–1954*. London: John Murray, 1954.
Jarcho, Saul. "Seventeenth-Century Medical Journalism." *Journal of the American Medical Association* 220 (1972): 64–8.
Lasteyrie, Robert de. *Bibliographie Generale de Travaux Historique*. Paris: Imprimerie Nationale, 1883.
Merton, Robert K. "The Matthew Effect in Science." *Science* 159 (1968): 56–63.
Middleton, W. E. Knowles. *The Experimenters: A Story of the Accademia del Cimento*. Baltimore: Johns Hopkins University Press, 1971.
Scheler, Lucien. "Antoine Laurent Lavoisier et le *Journal d'Histoire Naturelle*." *Révue d'Histoire des Sciences* 14 (1960): 1–9.
Scheler, Lucien. "Antoine Laurent et Michael Adanson, Redacteur de programms et Prix à l'Académie des Sciences." *Révue d'Histoire des Sciences* 14 (1961): 257–64.
Shapin, Steven. *A Social History of Truth: Civility and Science in Seventeenth-Century England*. Chicago: University of Chicago Press, 1994.
Smeaton W. A. "Two Unrecorded Publications of the Régie de Poudre Salpetrê, Probably Written by Lavoisier." *Annals of Science* 12 (1936): 179–97.
Smeaton W. A. "*L'Avant-Coureur*: The Journal in which Some of Lavoisier's Earliest Research Was Published." *Annals of Science* 13 (1957): 224–25.
Storrs, F. C. "Lavoisier's Techncial Reports, 1768–94." *Annals of Science* 22 (1966): 251–75; 24 (1968): 179–97.
Thorndike, Lynn. *A History of Magic and Experimental Science*. 8 vols. New York: Macmillan, 1923–1958.
Tisserand, Roger. *L'Académie de Dijon au Temps de l'Encyclopedie de 1740 à 1793*. Paris: Boivin, 1936.
Trenard, Louis. "La Press Française des Origens à 1788." In *Histoire Général de la Press Française*, vol. 1. Edited by Claude Bellanger, et al. Paris: Presse Universitaire de France, 1969, 22–404.
Ziman, John M. *Public Knowledge: An Essay Concerning the Social Dimension of Science*. Cambridge: Cambridge University Press, 1968.

Chapter 7

Literature of the Life Sciences

The Historical Background

To understand any phenomenon, it is useful to know its antecedents. Physicians apply this principle when they take a history from their patients to establish a diagnosis. This approach may also be relevant to understanding the "body" or "bodies" of literature in the life sciences, as it is used in such expressions as the Hippocratic Corpus when referring to the literary works of the legendary and historic Greek physician who, in another allusion, is called the "father of medicine." Such expressions imply an organic nature to the literature and suggest that, like all organic forms, it has gone through a process of evolution, which can be described historically. It also implies that by studying the evolution of the medium, we can more easily understand the more complex forms we see today.

In the development of scientific communication, some of the species, as in all evolutionary processes, survive and continue to function, although in some cases with permutations attributed to changes in the social and scientific environment. Keele argued for knowing the history of a phenomenon in another similar manner. He compared history to a Bach fugue, because history, he said, "possesses a unique power of showing how the complexities of the present have evolved from relatively simple roots. It displays the logic of this growth, much in the same way as Bach presents us with a simple theme before confounding us with the contrapuntal wizardry of his fugue" (Keele 1967:256).

In the history of scientific communication and of communication in general, no communications technique ever seems to be completely obliterated.

Published in the *Bulletin of the New York Academy of Medicine*, 60 (1984): 858–75

It may even be enhanced by the introduction of new techniques. In the historical antecedents of our present communication systems, scientists and practitioners will easily recognize prototypes of media and communications practices that they use today. Oral transmission of information, our most ancient medium, is still one of the most used today, especially as enhanced and facilitated by electronic means and by the ease and speed of travel. In fact, historians of science lament the lack of written records, which in the past were available in the form of correspondence and diaries. As a result, our perceptions of how research is conducted and how new discoveries are made may be distorted. Historians of contemporary science must therefore depend on what they call oral histories—that is, interviews with individuals who have been involved in particular scientific events.

Communications systems result from interactions between the perceived needs of its users and the available technology. There are, therefore, two lines of development in their evolution, extrinsic ones that relate to technical changes, such as improvements in roads and modes of travel, changes in means of producing documents through printing, photography, or electronics, and intrinsic ones that relate to the changing philosophical basis of science, the roles of the scientist in society, and his modes of organization, education, and interaction. No comprehensive history of scientific communication has been written in these terms, as far as I know. In this chapter, we can consider only the major developments beginning during the long period that precedes the new approach to describing natural phenomena that emerged during the sixteenth and seventeenth centuries, characterized as the "scientific revolution," to the present, when science has become a highly developed and institutionalized part of our social and industrial systems.

The invention of writing is probably the most important technological advance that has occurred in communication. It very likely has not existed for more than five or six thousand years, and before this, society depended entirely on oral transmission from one generation to another to retain its historical and technical heritage. In spite of this limitation, elaborate belief systems and impressive technology developed involving agriculture, astronomy, and the ability to manufacture both domestic utensils and instruments of war. Oral transmission of ideas as a primary medium has persisted through the ages even after the invention of writing and, millennia later, of printing. This is particularly true of the knowledge of crafts, passed on through apprenticeship systems and seldom made a part of the written record. The Académie des Sciences in Paris recognized this in the eighteenth century when it started a massive effort to collect and publish information on the various crafts of the time, from wig making to coal mining. From this effort they published an impressive series of monographs in some 103 numbers under the title *Descriptions des Arts et Metiers* (Paris, 1761–1788).

Oral transmission of information continued long after the invention of printing, most likely due to the scarcity of books. It was the tradition of professors in medieval universities to read to their students from established texts and transcribe the lectures, a system that persists even today in our lecture halls. It is difficult to say how much of our knowledge is still transmitted orally and never committed to paper. This is certainly true of some of our social beliefs, and it may also be true even of some of the technical aspects of the laboratory. It has been suggested that the scholarship of preliterate societies, based on oral transmission, tended to be nonanalytical because only when ideas are embodied in a static physical form can they be organized, analyzed, and explored.

Although writing introduced the ability to fix ideas in a static, physical form, some of the problems of oral transmission remained because manuscripts are disseminated by copying and errors of transmission easily occur. Now as in the past, original authors may also fall victim to the editorial zeal of copyists who may wish to "correct," amend, or even change the text.

One of the earliest scientific manuscripts we possess is the Edwin Smith papyrus, named after the man who acquired it in Egypt in 1862. Translated by the famous American Egyptologist, James Henry Breasted, and published in facsimile with a hieratic and English text in 1930, it dates from about 1500 BC but is probably based on older texts going back to 3000 BC. It includes, for example, explanations of words whose meaning had already become obscure when it was compiled. The Smith papyrus, now in the Library of the New York Academy of Medicine, is one of eight principal Egyptian medical papyri so far discovered and studied. Containing a series of surgical cases with descriptions of the injuries and prescriptions for treatment, it is obviously only a fragment of the whole text because cases are arranged in the traditional head-to-foot sequence and it stops about in the middle of the body. Each case description is accompanied by a statement about whether the physician will or will not treat it. At first glance, this may seem a precaution against charges of malpractice, particularly considering the severe penalties we find in some texts of this period (Harper 1904). It has been explained, however, as an option traditionally exercised by physicians and recognized as late as the eighteenth century.

Another Egyptian medical text of about the same period, the Ebers papyrus, is concerned with nonsurgical problems. One chapter is headed: "Eye diseases treated according to the priest physician as revealed by a Semite of Kepni." This indicates that medicine, even in this early period, was international in scope. A prominent American referred to the Smith papyrus almost with awe: "Conceived in ancient Egypt about 5000 years ago, it is the most remarkable book in the entire history of surgery. Compiled by an unknown writer at a time when medicine was magico-religious, when the vocabulary of science had not yet been created and when the first groping steps in inductive reasoning were being taken, this volume is as logical as a modern textbook in surgery" (Zimmerman 1957).

It is difficult to know the relationships between any of the texts of the manuscript period unless earlier versions exist. These are lacking in many cases. It has been estimated that of all the literary writings produced by the classical Greeks, only about 10 percent have survived. For instance, the oldest manuscript we have of the work of Hippocrates was copied during the ninth century AD, more than a thousand years after his era. The whole corpus of Hippocrates' writing consists of some one hundred treatises. Many were written by his students and their descendents, but are considered to be of the school of Hippocrates rather than by the great physician himself. Others apparently were notes taken by his students during lectures, a practice that did not stop even with the introduction of printing. Many student notebooks transcribed from lecture notes taken as late as the eighteenth and nineteenth centuries have found their way into library collections, and some even formed the basis for textbooks.

The deficiencies of the manuscript period exist in the process of making copies. The reader has no assurance that his text is exactly that produced by the author or a copy in the possession of another reader. Texts were transformed in the process of making successive copies and reinterpreted through successive commentaries. Each copy, in a sense, constitutes a new edition in which the copyist can exercise and perhaps abuse his editorial prerogatives of correction, expurgation, and emendation without the author's approval. Manuscript texts thus were subjected to endless alterations, additions, and abridgements. They were frequently brought together in collections of diverse and sometimes disparate items. Transmission of scientific ideas was largely through a process of synthesis, and many encyclopedic works on science and medicine were produced in this way from the Roman through the medieval period. Medieval writers borrowed freely and without acknowledgment from other compilers who in turn had borrowed from other sources until remote origins are lost in antiquity. In fact, the ideas of some of the classical writers survive only in the form in which transmitted by these encyclopedic compilers. These practices have not necessarily disappeared in modern times, and some of our scientific ideas may be transmitted in the same way through textbooks. Excerpts from published texts paraphrased or quoted out of context may be subject to the same abuses as the copied manuscript.

An established trade in books existed during Greek and Roman times. Manuscript copies could be turned out quite cheaply, as they were copied in large shops staffed by slave labor. In some ways this method of production had advantages over our system of publishing. A "publisher" could turn out a book almost at a moment's notice. The author handed his manuscript to the publisher, who turned it over to his staff of slave readers and transcribers. If a comparatively short work, it could be ready for distribution in twenty-four hours without the expense of typesetting or printer's corrections. Estimates have been made that some books in the manuscript period were produced in editions of five hundred to one thousand copies, which is larger than many of the editions produced by the early printing presses (Mumby 1974:24). Texts were often borrowed verbatim without any acknowledgment of the source. Other texts maintained their identity, if not their integrity, for a long time. There is, for example, a text of the physician John of Bordeaux of which the British Museum alone has twenty-two manuscript copies, seventeen in English, four in Latin, and one in Dutch, the earliest dated 1365, and one copied as late as the sixteenth century (Singer 1916).

Use of manuscript books persisted long after the introduction of printing around 1450, and reflects the cultural lag that often accompanies technical innovations. Objections were also raised to printed books on scholarly and aesthetic grounds, although some equaled and even surpassed in beauty many of the manuscript books on which they were modeled. The great advantage claimed for the printed book—that it could replicate identical copies of the same text—was valid only if the original text was properly edited and composed. In their haste to meet market demands, the early printers often did not take care to produce accurate editions. Manuscript copies were sometimes made from printed books that may still have been the cheapest way to produce a limited number of copies before the advent of photocopying.

The introduction of printing, however, ultimately radically transformed the nature of scholarship and the transmission of information. Texts available in identical copies could be disseminated more widely than had been possible with manuscript books. It was not only a matter of being able to reproduce texts more easily—authorship could be more readily established and ascertained. A recent writer aptly sums up the impact of printing on scholarship:

> First, it enabled priority of discovery to be established by referring to any copy of a printed text. But this can be done only if the text is accurately dated, and if the author is clearly named. Hence, the accurate establishment of priority of discovery depends on the development of conventions of presentation, which record the authorship and date of published works. . . . The possibility of achieving a definitive version of a text, through its publication in a uniform edition of identical copies, also enables research results to be criticized, validated by replication, and incorporated into an accepted body of knowledge by citation. (Katzen 1980:179)

These changes in scholarship were as much due to changes in attitudes toward the writings which had been inherited from the past as toward changes in technology. It is said that up to the Renaissance no sense of personal property in a piece of knowledge existed. The scholastic method consisted largely of the analysis and citation of authorities. In fact, many genuinely original works were published under the names of earlier well-known writers because an established name conferred an authority that it might otherwise not have (Ravetz 1971:179). Although scholars are finding more and more expressions of original ideas in the Middle Ages, it was a period of faith in which the ancient authorities were for the most part accepted unquestioningly. The great German physiologist Helmholtz recognized that this phenomenon still existed during the nineteenth century when he remarked that a large part of our knowledge is still "accepted uncritically and without examination, indeed unconsciously from the past" (Helmholtz 1893).

It was not only the introduction of printing that ushered in an age of improved communication. Improved transportation and travel were also involved, as well as the growth of centralized governments that brought about such changes as better postal services. It is indeed difficult to say whether improvements in communications or the radical changes in attitudes toward the authoritative teachings of the past brought about the "scientific revolution." Travel in ancient times was one of the primary new methods of acquiring new knowledge. There was a certain amount of mobility in the ancient world, but the political unity imposed by Roman rule facilitated greatly the exchange of ideas among scholars. With the breakdown of the Roman Empire, the medieval church took over this function until the growth of the great national states.

Improved transportation and postal services greatly helped to increase communication among scholars. Scholarship recognized no political boundaries and the only qualification was dedication to the pursuit of knowledge. Vesalius, whose book on anatomy (published in 1543) is one of the landmarks of modern science, was born in Brussels. No country in Europe, however, can really claim him as its own. He studied medicine in Paris and taught anatomy in the universities of Italy. When it came time to print his great book, he sent the engravings, painstakingly executed in Padua, by caravan across the Alps to Switzerland, and then followed himself to see it through the press. He served in the latter part of his life as a physician of the Imperial Court in Spain, and died in 1564 while on pilgrimage to the Holy Land.

Vesalius exemplifies another important change, which took place in the Renaissance, the beginning of the coming together of scholars who worked with ideas and words, and of craftsmen who worked with things and with their hands. This change can readily be seen by comparing the title page of the *Anathomia* of Mundinus, completed in 1316, although not printed until 1487, with the celebrated title page of Vesalius's *De Fabrica*

Humani Corporis, printed in Basle in 1543. In Mundinus, the physician sits on a platform high above an amphitheater where an assistant performs the dissection, whereas in the *Fabrica*, Vesalius stands at the table performing the dissection himself.

Medical books were well-represented among the first printed books because, aside from the clergy and lawyers, physicians were one of the largest literate classes in Europe. Medicine also was a part of the academic background of educated men, and even nonmedical readers were interested in medical books as a part of their general interests. Medical books, therefore, were a fairly good investment for early printers who then, even as now, had to show a profit or close shop. Many early books printed popular texts that had existed in manuscript form for centuries, but contemporary authors also began to publish their works.

Before the seventeenth century, the literature of scholarship was comparatively static. Represented by a relatively fixed body of knowledge, although constantly modified by commentaries and recast into new encyclopedic collections, it remained of relatively the same order of size as preceding bodies of knowledge. With the emphasis on experiment and personal observation, introduced by such men as the philosopher Francis Bacon in the sixteenth century, a new element was introduced. The need to report each observation as a separate entity began to take precedence over the tendency to incorporate them into larger syntheses, especially as the means to do so became available through improvements in communications. There was a new sense of urgency that did not characterize the earlier period, where scholarship was regarded as the celebration of old knowledge rather than the heralding of the new. This was exemplified by a new pioneering spirit in science in which each scholarly adventurer sought to stake a claim in the new territory that he was exploring. In this environment, the ideas of progress and change went hand in hand with the changes in the nature and transmission of information.

Personal correspondence was the first medium available to scholars to disseminate new information before the revolutionary emergence of scientific journals during the latter half of the seventeenth century. Personal in that it was usually addressed to an individual, it was also impersonal in that it conveyed news about observations and experiments in science rather than personal events. It introduced a new genre, the erudite letter, a letter intended frequently for a broader audience than the individual to whom it was addressed. It forms one of the important influences and precedents in the development of scientific journals, which today retains some of these functions in letters addressed to the editor, as well as in the research and other papers that can be regarded as letters addressed to a relatively undifferentiated and untargeted audience.

Sigerist suggests that the erudite letter continued to serve a purpose even after scientific journals appeared:

> When a scientist made a discovery in the eighteenth century, he did not publish it immediately but described it in a letter written in Latin that was sent to some friends abroad. They in turn would discuss these letters with their students and colleagues, would repeat the experiments described and report what their experience had been. After a discovery had been tested in such a way, it might then be published in a monograph, or in the transactions of an academy. (Sigerist 1947:12)

The volume of such correspondence during the seventeenth and eighteenth centuries (not all of which took place in Latin, although that was still the preferred language for international communication) is indicated by the bulk of some of the collections of letters of individual scientists of this period. The Danish physician, Thomas Bartholin, who also issued one of the early medical periodicals from 1673 to 1680 (*Acta Medica* 1671), published five volumes of his correspondence during his lifetime and was preparing to publish three more when his manuscripts were accidentally burned in 1670. Albrecht von Haller's letters fill seven large volumes and include some thirteen thousand letters to more than sixteen hundred correspondents in French, German, English, Italian, or Latin, all languages he used with ease.

To these erudite gentlemen, international conflicts imposed no barriers. The Dutch microscopist Anton van Leeuwenhoek continued to send letters to the Royal Society of London during the seventeenth century, although Holland and England were at war. Much of this correspondence was in fact addressed to the new scientific and learned societies, which were beginning to form during the latter half of the seventeenth century. In some cases, both the society and the correspondence were managed by the same individuals who became the editors of the early scientific journals. Henry Oldenburg, a preeminent example of this class, was first secretary of the Royal Society of London as well as the founder of *Philosophical Transactions*, which appeared in 1665, three years after the Royal Society received its Royal charter. His published correspondence filled thirteen volumes published from 1966 to 1986 (Oldenburg 1966). They contain letters to and from the leading scientists of Great Britain and the continent, including Robert Boyle, Edmund Halley, Robert Hooke, Martin Lister, Christopher Wren, Christian Huygens, Anton van Leeuwenhoek, Gottfried Wilhelm Leibniz, Marcello Malphigi, Francesco Redi, and Baruch Spinoza. These letters were to a large extent intended for members of the Royal Society with whom he shared them. They provide fascinating insights into the scientific life of the seventeenth century. The letters presented to the Royal Society were entered in the Record Books of the Society, but many also found their way into Oldenburg's new *Philosophical Transactions*, where he shared them with a wider readership.

Another technical innovation that preceded the first scientific journals was the media, which established predictable channels of communications and created "periodicity in print." These came in various forms such as annual book catalogs published for the various European book fairs soon after the introduction of printing. Manuscript and printed newsletters were issued for a select clientele on a regular basis. Among the paramount forms were the calendars and almanacs that provided annual vehicles not only for lists of phases of the moon and saints' days, but included such things as agricultural and health information and advice. Printed newsletters could also serve as vehicles of scientific information. It is sometimes suggested that in 1608 Galileo first learned of a device to see objects at a distance from a twelve-page Dutch newsletter a little larger than a library catalog card: included in an account of the visit to Holland of an embassy from Siam was a short note about the use of a combination of lenses that made distant objects appear closer (Drake 1976).

Newspapers were widespread by the middle of the seventeenth century. The first French newspaper was issued by the crusading physician Theophraste Renaudot (1586–1653), who issued the *Gazette* under the auspices of the government in 1631. His sensitivity to the conditions of the poor led to other innovations, including a public health clinic and other activities that provoked the bitter enmity of the medical establishment represented by Guy Patin and the Faculty of Medicine of the University of Paris.

The early newspapers, even as they do today, carried news of interest to the scientific community, but it was not until 1665 that the first two journals, which can be regarded as true organs for the dissemination of scientific ideas, appeared. *Journal des Sçavans*, as it was then called, appeared in Paris. The Académie des Sciences had been formed shortly after in 1665 in the same city and although it had no formal association with *Journal des Sçavans*, news of the Académie's activities appeared in its pages. Although *Journal des Sçavans* did report scientific discoveries such as Jean Denis' curious experiments with the transfusion of blood and other scientific news gleaned from their correspondents, it was more a general literary journal than a scientific one. It was primarily concerned with reviewing books in every field of knowledge, except for those on religion and politics, which the editors treated very gingerly. *Journal des Sçavans* was soon reprinted in Amsterdam and other places and became a model for literary journals in Holland, Italy, Germany, and the rest of Europe.

The Académie des Sciences pursued another course by publishing its proceedings in quite a different format, which in turn became a prototype for similar publications by other learned societies on the continent and later in America. Both *Philosophical Transactions* and *Journal des Sçavans* celebrated their three hundredth anniversary in 1965, although in quite altered form. *Journal des Savants* today is a general literary journal, and the Royal Society now uses *Philosophical Transactions* to publish two series of highly specialized monographs, one in mathematics and physics and the other in the biological sciences.

Modern commentators assign a high importance to the appearance of the journal in the history of science. Ziman puts it most emphatically: "The invention of a mechanism for the systematic publication of fragments of scientific work may well have been the key event in the history of modern science" (Ziman 1969:218). Cole and Eales say pretty much the same: "It may, in fact, be claimed that science could not have made the advance that it has but for the recognition of the periodical as the most convenient and efficient method of encouraging research" (Cole and Eales 1917:588).

In many ways the scientific and learned journals of the seventeenth and eighteenth centuries did not resemble very closely the research journals we know today. Borrowing from one journal to another was a common practice. For the most part, sources were acknowledged but frequently no indication of origin was given. Sometimes the borrowed information is paraphrased and commentary added, and in this respect they were very much like the newspapers that provided their models. They did not worry about redundancy because they regarded themselves principally as news media. Duplication of information becomes a problem only when its recipient has access to both sources. This poses an interesting difference between the newspaper and the scientific journal in that the quality of a newspaper is improved to the extent that it includes information from other media because it is often a unique source for a particular reader. A scientific journal, on the other hand, particularly one publishing original research, is proscribed from this practice. In the seventeenth and eighteenth centuries, journals were not as closely differentiated from monographs as today and were frequently reviewed in the same manner. This reflected that in some instances a large part of the content of a journal issue could be attributed to the editor, who sometimes referred to himself as the author. They were also characterized by short life spans, and many did not last beyond the second or third year, perhaps because many appeared under the auspices of a single individual.

Specialized journals began to appear very early in the history of scientific journals, although much scientific material continued to be reported in general learned journals. Among the first specialized journals was one edited or written by another controversial French physician, Nicolas Blegny (1652–1722), regarded by some historians as rather unsavory although also credited with outstanding work on hernias in the seventeenth century and other interesting social and medical innovations. His *Nouvelles Descouvertes sur Toutes les Parties de la Médecine* appeared in 1679 in the form of letters to a provincial physician. It can also be considered an example of how scientific journalism can be used for self-aggrandizement as well as to disseminate information. Blegny took care in some of his discussions of new drugs to indicate where they could be obtained, and was not constrained from speaking well of a new kind of hospital he had opened (Kronick 1960).

The earliest specialized journals were medical because physicians were the largest organized group interested in the new science. Journals also fostered the intellectual and social status of a specialty. The Académie de Chirurgie organized in Paris in 1731 to support the surgeons' claim to academic standing began to publish *Memoires Pour les Chirurgiens* in 1736. Specialized journals in other disciplines did not begin until later in the century. The German chemist Lorenz von Crell (1744–1816) published his *Chemisches Journal* in Lemgo from 1778–1781 and followed it with other journals in chemistry. In London, William Curtis (1746–1799) began his *Botanical Magazine* in 1787, which is still avidly sought by collectors today for its beautiful colored plates. Before the end of the century there was even a journal in psychiatry, what the Germans called "knowledge of the soul," *Magazin der Erfahrungsseelenkunde,* edited by Carl Philipp Moritz from 1783 to 1793 in Berlin.

The predominance of independent journals, those not sponsored by scientific or learned societies, continued throughout the middle of the nineteenth century, although society sponsorship became more prevalent as time went on. Today the situation is quite the reverse in that most journals are sponsored in some way by societies. The famous Danish physician Thomas Bartholin (1616–1680) published his *Acta Medica et Philosophica* from 1671 to 1680 while a member of the faculty of the University of Copenhagen, but the journal had no official connection with the school. Early scientific societies nevertheless had a strong influence on the development of scientific journalism, and both responded to the same intellectual and social needs.

It is therefore more than a coincidence that the earliest journals arose in close association with scientific societies. The Royal Society of London contented itself with *Philosophical Transactions* to disseminate news of their activities. The Académie des Sciences in Paris, however, published their proceedings in a series called *Histoires et Mémoires.* These, in a sense, formed an archival record of the activities of the society as well as a repository of the most important papers delivered in person or communicated to the society. They were often, however, published sporadically or irregularly, or long after the papers were presented and sometimes not at all. *Histoires et Memoires* of the French Academy formed a model for the scientific and learned societies that proliferated in French provincial cities, throughout small and large German towns, and elsewhere in Europe. To a large extent they are more accurate prototypes of modern scientific journals than the independent journals published at the same time. Some current editorial practices, such as peer review, began in the methods these early societies devised for accepting communications for publication.

Booth argues that the Royal Society of London first "introduced the concept of refereeing" (Booth 1982:105) in the middle of the eighteenth century by setting up a committee to review all papers before they were published in *Philosophical Transactions.* There were, however, many antecedents to this practice. Oldenburg screened communications for presentation to the Royal Society, but after the papers were read, they were "ordered to be reviewed by several of the Fellows" (Hall 1975:70). Early in its history, the Académie des Sciences in Paris established select committees to determine whether a member could or could not publish under its auspices. The peer review process almost as we know it today is described in the preface to the French edition of *Medical Essays and Observations* published by a "Society in Edinburgh" in 1731. Papers submitted, it informs us, were distributed according to their subject content to those members of the society who are more versed in these matters for their review. It also specifies that the identity of the reviewer is not made known to the author, an early example of the controversial anonymous reviewer (*Essais* 1740). The Société Royale de Médecine, soon after its institution in 1776, inaugurated a system by which two members examined each paper submitted to the society and provided the other members with a summary and critique (Hannaway 1974:121). Validation of scientific work through review and discussion was in fact a major function of early scientific societies.

Reasons for the formation of scientific societies may be sought in general in a gregarious human nature, but another important reason was failure of contemporary universities to respond to the scientific revolution. These organizations of individuals interested in the new science were sometimes called "invisible colleges" because they carried on some of the traditional functions of the university outside their precincts.

Some early scientific societies required contributions of papers as a condition of membership. Many also gave prizes for essays submitted on questions proposed by the society and widely publicized through general and special periodicals. These prize essays in many ways can be considered closer precursors of modern scientific papers than any other eighteenth-century publication format. They can to some extent be regarded as a form of sponsored research, clearly evident in prizes awarded by such organizations as the Royal Society of Arts (originally the Society for the Encouragement of Arts, Manufactures, and Commerce), which from its origins in 1754 offered prizes for essays on such subjects as the improvement of the production of wine and the development of a more efficient ship's pump. The Society of Arts and Science in Utrecht proposed a question in 1783 with a decidedly modern inflection: What are the causes for the increase in nervous illnesses, and which causes lie in nature and which in the mode of life? They offered a prize of thirty ducats and expressed their willingness to receive papers in Dutch, French, or Latin (Almanach 1783:43). The venerable Faculté de Médecine in Paris was stimulated at last by the activities of the newly organized Société de Médecine to publish an account of their public meetings. In 1778 they announced that awards had been made for essays submitted on the subject of the treatment of military fever in pregnant women. Five essays were regarded as of special merit, but only two were elected to share in the prize (*Séances* 1789:13–7).

The role of the prize essay in the history of science perhaps has not yet been fully explored. Prizes are still awarded today for essays or scientific papers, but they are usually on subjects of the author's own choice, while the eighteenth-century prize was on a subject announced in advance and open to international competition. Contestants submitted their papers anonymously to the society's secretary with some kind of device to identify them later, usually a Latin motto. They resemble to some extent the graduation dissertation, a centuries old academic tradition. In fact, the Akademie der Wissenschaften in Berlin published a collection of their prize essays in 1748 under the title *Dissertations qui a remporté le prix*. Societies were careful to preserve the anonymity of the author in order not to be influenced by his reputation. The mathematician Daniel Bernoulli nevertheless was awarded the prize by the Académie des Sciences ten times in the eighteenth century. The attempt to preserve anonymity was in keeping with their efforts to eliminate personal bias not only in judgment of research results but in the observation of nature. It is interesting that this period, characterized as one of the depersonalization of science, was accompanied by a greater emphasis on priority, which gave rise to a number of acrimonious disputes in this period. It reminds us of Merton's discussion of the clash between the social norms of science and the personal needs of the scientist (Merton 1963), and Kuhn's comments about the "essential tension" in science (Kuhn 1977).

Almost half of the learned and scientific journals that appeared in the eighteenth century were published in German, a phenomenon attributed to a German quality that has been described as *schreiblustigkeit*, or joy in writing, and to the highly decentralized nature of the German speaking states at that time, in which each provincial state created its own journals. It also reflects the parochial nature of many of these publications because German was a language not widely used outside of its sphere of dominance. Latin dominated scholarly writing until about 1600, when such vernacular languages as French and English began to be used more frequently. French assumed ascendancy in the eighteenth century, but this came from social and political factors as much as scientific ones. French retained its dominance in the first part of the nineteenth century because of the outstanding contributions of its scientists and physicians, but gradually gave way to German until at the end of the nineteenth century it was almost mandatory for any serious scientific worker to read German (Ben-David 1960).

This phenomenon has been said to be produced by the great increase in productivity among German scientists. One of the strong reasons for this phenomenon, says sociologist Ben-David, was the decentralization of the German universities, which, like the proliferation of German journals, was a consequence of the political decentralization that existed when they were created. The existence of a large number of academic institutions made possible a high degree of mobility among students and teachers. This in turn produced a scientific competition not duplicated elsewhere. It was accompanied, Ben-David says, by a change from the dominance of *Naturphilosophie*, a philosophic way to look at natural phenomena, to greater emphasis on experimental science. This in turn led to recognition of the university as a seat of scientific research and to the creation of scientific institutes, all of which figured in the competition for faculty. American universities with graduate programs in the sciences tended to model themselves on the German universities during this period. This change, Ben-David concludes, created some of the conditions under which American science assumed ascendancy in the twentieth century, until English, rather than German, is now the predominant language of the scientific world (Ben-David 1960).

Many journals that began during the latter half of the nineteenth century bear the names of outstanding scientists of this period, such as *Wilhelm Roux' Archiv für Entwicklungsmechanick der Organismen*, which began in 1865. These were not merely honorific titles because, in most cases, they were inaugurated by scientists who took personal editorial responsibility for them. This tradition of concern for the literature by eminent scientists has an old tradition and is indeed a hallmark of professional status. The Swiss physician, physiologist, botanist, and poet Albrecht von Haller (1708–1777) has been called by William Osler "the greatest bibliographer in our ranks." Early in his life, Haller developed the practice of systematically reading and abstracting the literature. This activity culminated in his eight-volume textbook on physiology (Haller 1771) and in his monumental bibliographies in anatomy, surgery, medicine, and botany. One biographer credits him with having written twelve thousand book reviews for *Göttinger Gelehrte Anzeiger* during his tenure at the university from 1745 to 1777. This would average out to about four hundred reviews a year, a formidable achievement even for a genius like Haller, considering the other things he was doing, such as developing a theory of tissue irritability. Fortunately for our sense of credulity, a later scholar found that his biographer had been overcome by admiration and added an extra zero to the twelve hundred reviews with which he had originally been credited (Rohlfe 1880).

Many leaders in biology and medicine began journals in this period to create outlets for their own work and for that of their colleagues and students. The German anatomist and physiologist Rudolf Albert von Kölliker (1817–1905) founded *Zeitschrift für wissenschaftliche Zoologie* in 1848 and edited it for half a century. Rudolf Carl Virchow (1821–1902), probably the best known German physician of his time, who, as the founder of cellular pathology, contributed greatly to the scientific basis of modern medicine, began *Archiv für pathologische Anatomie und Physiologie* in 1845 and remained its editor until his death. After his death, it became known as *Virchow's Archiv.* This is also true for other outstanding journals of the period, such as that of Albrecht von Graefe (1828–1870), who made numerous contributions in ophthalmology and edited *Archiv für klinische und experimentelle Ophthalmologie*; Felix Hoppe-Seyler (1825–1895), who occupied the first chair of physiological chemistry in Germany and edited the *Zeitschrift für physiologische Chemie*; and Carl Gegenbaur (1826-1903), the anatomist who was responsible for *Morphologischen Jahrbuch.* Today these journals publish largely in English, and some have even acquired English titles.

It has been estimated that the number of journals increased from about nine hundred in 1800 to almost sixty thousand in 1901 (Brodman 1954:102). During this period, the scientific journal changed from primarily a news and book review medium to a vehicle and repository for scientific research. It did not lose any of its old functions while adding new ones, but did undergo specialization and diversification, which produced the complex and cumbersome system we know today.

One of the consequences of this growth has been the increase in secondary media, that is, publications based on antecedent publications and presenting guides or digests (another biological term) to enable scientists better access to or control of the burgeoning literature. Efforts to present summaries or syntheses of the current state of knowledge ("state of the art reviews," as we say today) go back to the beginnings of recorded knowledge. The great encyclopedic compendia of the classical and medieval periods, such as Roman Pliny's *Natural History*, Isidore of Seville's *Etymologiae*, and the Franciscan known as Bartholomew the Englishman's (ca. 1260) *On the Properties of Things*, represent efforts of this kind.

With the proliferation of journals during the eighteenth century, anxieties began to be expressed about the difficulty of gaining access to the literature.

One new technique introduced at this time was the abstract journal. A journal published in Mannheim in 1760 describes the function of this new medium in its title: *Journal des Journaux; ou Précis des Principaux Ouvrages Périodiques de l'Europe* (Journal of Journals; or a Summary of the Periodical Works of Europe). The new medium met a real need because the number of active abstract journals in all fields increased from 9 in 1790 to 249 in 1920 (Manzer 1977:111–36). Germany dominated these services in the sciences until World War I, when many of these publications were suspended and did not resume. The *Chemisches Zentralblatt*, begun in 1830, established a model for a series of *Zentralblatter* in various disciplines, some of which lasted for only a few years but many of which continued for decades.

Other efforts to organize and synthesize the literature took the form of progress reports and annual reviews, which took place such as the *Jahresbericht* (annual report) *über die Leistungen und Fortschritte im Gebiete der Ophthalmologie* that began in 1870. They also began with such terms as *Jahrbuch* (yearbook), *Ergebnisse* (results), and *Fortschritte* (advances), all with cognates in English and other languages.

The first efforts to index the journal literature took place almost immediately after the first journals appeared. An obscure Flemish bookseller named Cornelius a Beughem published an index to *Journal des Sçavans* in 1683 under the title *La France Sçavante*. Like all Gaul, it was divided into three parts: a chronological part in which the contents of each issue are listed by date of publication, an author index, and finally a classified subject index. It is interesting that the *Current List of Medical Literature*, published by the National Library of Medicine from 1941 to 1959, was in almost the same format. Beughern followed his single journal index with an index to a larger number of journals, which included *Philosophical Transactions* in its Latin edition, *Acta Eruditorum*, a Latin equivalent of *Journal des Sçavans* published in Germany, and seven other journals. Many other indices, either on a current or retrospective basis, appeared during the eighteenth and nineteenth century. They culminated in a large retrospective index under the title *Catalogue of Scientific Papers* that the Royal Society of London began to publish in 1858 to cover the scientific literature in all disciplines published since 1800. Later in the century, the National Library of Medicine (then the Library of the Surgeon General's Office) began to publish the monumental *Index-Catalogue of the Library of the Surgeon-General's Office*, which, although primarily devoted to medicine, is of interest to all the life sciences. It appeared in five series from 1896 to 1961 when it was discontinued because it could not keep pace with the growth of the medical literature. These publications are still of great use for historians of science or investigators looking for early antecedents of scientific ideas. They represent precursors and prototypes of a host of publications in all disciplines in the life sciences, many of which are now mediated by a powerful new tool that has only just recently become available, and that provides a new technology for dealing with some of the problems of control and access to the literature—the computer. The social and intellectual factors that produced them, however, cannot be ignored in considering the direction in which the evolution of the media of scientific communication is taking us today.

REFERENCES

Acta Medica et Philosphica Hafniensis. 5 vols. Copenhagen: 1673–1680.

Almanach für Aerzte und Nichtaerzte für das Jahre 1783.

Ben-David, Joseph. "Scientific Productivity and Academic Organization in Nineteenth-Century Medicine." *American Sociological Review* 25 (1960): 828–43.

Booth, C. C. "Medical Communication: The Old and the New." *British Medical Journal* 285 (1982): 1055–8.

Brodman, Estelle. *The Development of Medical Bibliography.* Baltimore: Medical Library Association, 1954.

Cole, F. J., and N. B. Eales. "The History of Comparative Anatomy: A Statistical Analysis of the Literature." *Science Progress* 11 (1917): 578–96.

Drake, Stillman. *The Unsung Journalist and the Origins of the Telescope.* Los Angeles: Zeitlin & Ver Brugge, 1976.

Essais et Observations de Médecine de la Société d'Edinbourg. 1740, vi.

Hall, Marie Boas. "Science in the Early Royal Society." In *The Emergence of Science in Western Europe.* Edited by Maurice Crossland. New York: Science History Press, 1975, 57–77.

Haller, Albrecht von. *Primae Lineae Physiologicae.* Lausanne: Grosset, 1771.

Hannaway, Caroline. *Medicine, Public Welfare, and the State in Eighteenth-Century France. The Société Royale de Médecine of Paris.* Ph.D. Diss. Johns Hopkins University, 1974.

Harper, R. F. *The Code of Hammurabi, King of Babylon.* 2d ed. Chicago University Press, 1904.

Helmholtz, Hermann von. "On Thought in Medicine." In his *Popular Lectures on Scientific Subjects.* Vol 2. London: Longmans Green, 1893, 199–216.

Katzen, M. F. "The Changing Appearance of Research Journals in Science and Technology." In *Development of Science Publishing in Europe.* Edited by A. J. Meadows. Amsterdam: Elsevier, 1980.

Keele, K. D. "Cybernetic Aspects of Medical History." *Scottish Medical Journal* 12 (1967): 250–59.

Kronick, David A. "Nicolas Blegny: Medical Journalist." *Bulletin of the Cleveland Medical Library* 7 (1960): 47–56.

Kuhn, Thomas. *The Essential Tension.* Chicago: University of Chicago Press, 1977.

Manzer, Bruce M. *The Abstract Journal, 1790–1920.* Metchun, N.J.: Scarecrow Press, 1977.

Merton, Robert. "The Ambivalence of Scientists." *Bulletin of John Hopkins University Press* 112 (1963): 77–97.

Mumby, Frank A. *Publishing and Bookselling*. 5th ed. London: Jonathan Cape, 1974.
Oldenburg, Henry. *Correspondence*. Edited and translated by A. Rupert Hall and Marie Boas Hall. Madison: University of Wisconsin Press, vols. 1–9, 1966–1973; London: Mansell, vols. 10–11, 1975–1977; London: Taylor and Francis, vols. 12–13, 1986.
Ravetz, J. R. *Scientific Knowledge and the Social Problems*. Oxford: Clarendon Press, 1971.
Rohlfe, Heinrich. "Eine literarische Legende." *Deutscher Archiv für Geschichte der Medizin* 1 (1880): 270–72.
Séances Publique Tenue par la Faculté de Médecine. Paris: Université de Paris, 1789.
Sigerist, Henry E. "Nationalism and Internationalism in Medicine." *Bulletin of the History of Medicine* 21 (1947): 5–16.
Singer, Dorothy W. "Some Plague Tracts (14th and 15th Centuries)." *Proceedings of the Royal Society of Medicine*, History Section 9 (1916): 159–212.
Ziman, John M. "Information, Communication, Knowledge." *Nature* 244 (1969): 328–24.
Zimmerman, L. M. "The Beginnings of Surgery and the Edwin Smith Papyrus." *Journal of the International College of Surgery* 21 (1957): section 11, 14–21.

Chapter 8

Bibliographic Dispersion of Early Periodicals

ABSTRACT

As scholars are well aware, the scientific and technical periodicals of the seventeenth and eighteenth centuries are difficult to identify and locate. To test this observation, 640 scientific and technical periodicals published in the period 1791 to 1800 were drawn from four well-known source lists and searched in three major national union lists of serials. The results revealed that not only were the titles widely scattered in the source lists, but that only 57 percent of all the titles could be located in any of the national union lists.

The scientific and technical periodicals of the seventeenth and eighteenth centuries provide an important resource for historians. Yet anyone who has worked with early scientific periodicals can heartily endorse the comments of the editor of *British Union Catalogue of Periodicals* who classifies them "among the most elusive of all forms of printed matter" (*British Union-Catalogue* 1955:preface).

An opportunity to attempt a quantitative assessment of these sentiments came recently with an effort to update a study of scientific and technical periodicals, which covered the period 1665 to 1790 (Kronick 1976). In order to cover the time span used, Garrison's list of titles covering the period 1791 to 1800 were added for review (Garrison 1934). The following three additional source lists were also checked for titles within the scope of the study: Bolton's bibliography (Bolton 1897), Scudder's list

Published in the *Serials Librarian* 12 (1987): 55–59

(1879), and Joachim Kirchner's *Bibliography of German Periodicals from Their Beginning to 1830* (Kirchner 1969).

These source lists vary considerably in scope and, therefore, in coverage. Most of the problems confronting catalogers of contemporary serials are compounded in relation to serials of the seventeenth and eighteenth centuries—problems of name changes, volume numbering gaps in publication, and so on. In earlier periods, editors do not seem to have been quite as compulsive about the periodicity that marks the contemporary periodical. In many cases, they judiciously waited until they had enough material for an issue. No sharp lines were drawn between newspapers and periodicals, and modern compilers of periodical bibliographies frequently honor this practice by including seventeenth- and eighteenth-century titles that may more properly be considered newspapers. There is also greater difficulty in the period in distinguishing between periodicals and "subscription books" or "books-in-parts." These factors, as well as variations in scope and coverage and changes in bibliographic conventions, create wide variations in the titles included in the source lists and in the way they are listed, particularly because in many instances the compilers were not able to see or even locate the titles for examination.

Scudder's list, which is the earliest of the four, explicitly omits periodicals dealing exclusively with the applied sciences (medicine, agriculture, technology, etc.). In addition to research libraries, primarily in Northeastern United States, he drew from a large number of printed sources but did not indicate how many of the titles he was able to personally examine. "Years of labor and prolonged visits to a hundred European libraries," he says "would not suffice barely to verify at first hand all the titles collected" (Scudder 1879, viii), a sentiment that other investigators would surely endorse. The titles are arranged first by country, then alphabetically by title under cities. There is an index of towns in which the periodicals were published, an alphabetical index of titles in abbreviated form, and an index of minor subjects that oddly, in view of the work's scope, includes such subjects as architecture, military science, and telegraphy. There are 439 numbered entries but many of them, particularly those concerning societies, include several titles.

Bolton elected to include all the "principal independent periodicals of every branch of pure and applied science, published in all countries," but with a few exceptions chose to omit transactions of learned societies. Like Scudder, he drew his data from available bibliographies, as well as the printed and manuscript catalogs of libraries in the United States, England, France, Belgium, Germany, and Italy. He circularized 200 American libraries to get information on holdings and received 133 returns. He was able, however, to provide locations for only 3,160 titles of the 8,600 he listed. An appendix provides a chronological table for a selected list of titles through which it can be determined which title was being published in any particular year, but only 400 or so of the 8,600 titles are covered, and of these only 76 were published in the seventeenth or eighteenth centuries.

Garrison's list is the most comprehensive of the four in terms of coverage of the scientific and technical periodicals of the seventeenth and eighteenth centuries. However, it includes a number of titles that the other compilers would consider out of scope, such as dissertation collections and multivolume works published serially. Garrison had the rich collections of what was then the Library of the Surgeon General's Office (now the National Library of Medicine) to draw on, but he also used a number of printed sources. The most important source he used was the monumental nineteenth-century bibliography of Adolph Carl Peter Callisen (1786–1866), which in volumes 23 to 25 provides us with what Garrison describes as "a definitive catalogue, not only of the medical periodicals of 1780–1833, but also of secular and scientific journals of the period containing matter of medical interest" (Garrison 1934:286). He also characterizes it as "chaotic to an exasperating degree." It is nevertheless a very useful repository of information about the scientific literature of this period. It includes not only brief biographies and citations to reviews of an author's books, but also reviews of journal issues, an indication that the two forms of publication were not as sharply differentiated as they are today. Counting all the "continuations, translations, abridgements, and separate publications of scientific societies," the Garrison list includes 1,199 titles, of which 107 were published in the seventeenth century.

Kirchner's bibliography is an update of the list he published in his *Die Grundlagen des deutschen Zeitschfiftenwesens* (Leipzig, Hiersemann, 2 vols. 1928–1931), which also covered the literature only to 1790. The new list covers all periodicals published in German and in Germany to 1830. It is arranged in broad subject groups, from which the titles of scientific and technical interest can easily be extracted. Kirchner also used a number of printed sources. His bibliography of sources includes 31 titles published from 1747 to 1962 (xii–xiii). Kirchner was able to provide locations for about 75 percent of the titles, although, as he indicates, some of them may have been destroyed in World War II or moved to other locations.

Of the 640 titles that met the criteria of the review (published between 1791 to 1800 and of interest to the history of science and technology), the number derived from each source list were:

Bolton, 198
Garrison, 345
Kirchner, 274
Scudder, 213

The overlap between the lists varied from 67 for Kirchner/Scudder to 135 for Bolton/Scudder. While it is probably not wise to insist too much on the precision of any of these figures—the data was hand-tallied in a large number of categories and there were undoubtedly instances in which the tallier ticked rather than tocked—they do help to underline some of the problems that scholars have in identifying and locating relevant titles.

To try to determine the availability of any of the selected titles, they were searched in three national union lists (*British Union-Catalogue*, *Catalogue Colléctif*, and the *Union List of Serials*). These lists are all of reasonably recent vintage, and although one is sometimes disappointed on visiting a designated library to discover that a title cannot be found or more lamentably has been "destroyed in the war," they are reasonably reliable. They are all also reasonably similar in scope and coverage, although they vary in arrangement, filing rules, and other conventions, as they were compiled at different times and under different cataloging policies and codes. They tend to be stronger in titles published in the country in which the list was compiled, although French titles not in the French list appear in the British and U.S. list and vice versa. As an example, only 17 of the 29 unique titles in the French list have French imprints. They vary also in bibliographic detail because entries are based on the holdings actually on hand in a particular library.

This is not the place to rehearse all the variations in form of entry, bibliographic convention, and other features of each of these lists. The user, however, must be aware of these conventions in each case in order to be successful in searching them. We are concerned here with the extent to which these lists reveal the availability of the 640 titles in our sample list. The tally results are as follows:

Titles found in British list	248 (39 percent)
Overlap with U.S. list	29
Overlap with French list	54
Overlap with both	129
Unique titles	36
Titles found in French list	213 (33 percent)
Overlap with British list	30
Overlap with U.S. list	26
Overlap with both	129
Unique titles	20
Titles found in U.S. list	278 (43 percent)
Overlap with British list	54
Overlap with French list	26
Overlap with both	129
Unique titles	64

Of the 640 titles derived from the four source lists (Scudder, Bolton, Garrison, and Kirchner), 278 or 43 percent were not located in any of the three union lists searched. Some of the faults to find titles may be based on variations in scope and policies between the lists and the source lists or on searching errors, but the results reveal clearly how widely dispersed early periodicals are and that more than one list needs to be searched in order to have some reasonable assurance about the availability of any particular title. The *Deutsche Zeitschriften Datenbank,* an online database recently initiated in west Germany (it is now also available in microfiche updated twice a year) and other union lists may increase the percentage of titles that can be located, but finding early periodicals will still require diligence, patience, and, frequently, good luck.

REFERENCES

Bolton, Henry C. *A Catalogue of Scientific and Technical Periodicals (1665–1895) Together with Chronological Tables and a Library Checklist.* 2d ed. Washington, D.C.: Smithsonian Institution, 1897.

British Union-Catalogue of Periodicals: A Record of the Periodicals of the World, from the Seventeenth Century to the Present Day in British Libraries. 5 vols. Edited by James D. Stewart, Muriel E. Hammond, and Erwin Saengcr. London: Butterworth, [1955–1958] 1968, 1970.

Callisen, Adolf Carl Peter. *Medizinischer Schriftsteller-lexicon der jetzt lebenden Aertze, Wundaerzte, Geburtshelfer, Apotheker, und Naturforscher aller gebildeten Völker.* 33 vols. Copenhagen, 1830–1845. Reprint, Niewkoop: B. De Graaf, 1964.

Catalogue Collectif des Périodiques du Debut du XVII Siècle à 1939 dans les Bibliotheques Universitaires des Départiments. 4 vols. Paris: Bibliothèque Nationale, 1967–1977.

Garrison, Fielding H. "The Medical and Scientific Periodicals of the 17th and 18th Centuries." *Bulletin of the Institute for the History of Medicine* 2 (1934): 285–334.

Kirchner, Joachim. *Die Zeitschriften des deutschen Sprachgebietes von den Anfängen bis 1830.* 4 vols. Stuttgart: Hiersemann, 1969–1977.

Kronick. David A. *A History of Scientific and Technical Periodicals: The Origins and Development of the Scientific and Technical Press 1665–1790.* 2d ed. Metuchen, N.J., Scarecrow Press, 1976.

Scudder, Samuel H. *Catalogue of Scientific Serials of All Countries, Including the Transactions of Learned Societies in the Natural, Physical, and Mathematical Sciences: 1663–1876.* Cambridge, Mass.: Library of Harvard University, 1879.

Union List of Serials in Libraries of the United States and Canada. 5 vols. Edited by Edna Brown Titus. 3d edition. New York: Wilson Co. 1965.

Chapter 9

Anonymity and Identity

Editorial Policy in the Early Scientific Journal

ABSTRACT

A significant change in the publication behavior of scientists can be revealed by studying the policies of editors of seventeenth- and eighteenth-century journals toward anonymity. Although a great deal more anonymity is to be found in this period, there is also evidence of ambivalence and even rejection of the practice. It reflects the Baconian ideal of collective research and a belief that the role of the scientist was to uncover rather than to discover the truth, an attitude that is reflected in scientific writing today, where the author seeks to cast himself in the role of a passive observer. The explicit and implicit reasons revealed for the assumption of anonymity have some relevance even today, when it is rarely to be found. The study of anonymity in scientific writing can serve to illuminate the changes in the social role of science and the functions of scientific communications.

"Although scientific knowledge is impersonal in the sense that its claim to truth must be assessed entirely apart from its source, the historian of science is called on to prevent scientific knowledge from sinking (or rising) into anonymity, to preserve the collective memory of its origins. Anonymous givers have no place in this scheme of things. Eponymity, not anonymity, is the standard" (Merton 1973:302).

Published in *Library Quarterly* 58 (1988): 221–37.

THE BACKGROUND

Among the basic social norms accepted by the scientific community is that of a communality of ideas, that is, that scientific ideas are a shared commodity that belongs to everyone. This notion is contravened by the many disputes about priority that have occurred in the history of science and by the sensitivity displayed by authors when they feel they have not been given adequate recognition for their ideas (Merton 1973). One reflection of the acceptance or rejection of this norm can be found in the changes in attitudes and practices regarding anonymity in scientific writing. Today, when such a high premium is placed on publication for scientific advancement, there seems to be little ambivalence about the need for authors to identify themselves clearly, and criticism is leveled rather at those who attach their names to communications in which they have not adequately participated (Relman 1983).

In the early history of modern science, the anonymity of the scientist was more likely to be condoned and even praised. In the Baconian model, which was widely acclaimed, the scientist worked with his colleagues in a collective enterprise and his product had significance only as the work of a group. This was regarded as one of the points of contention in the so-called war between the ancients and the moderns in which the ancient philosophic systems were being assaulted as the works of isolated individuals, while Baconian science was to be the result of the collaboration of many (Jones 1982: xii). The early scientific academies tried faithfully to implement this model. The proceedings of the famed Accademia del Cimento founded by the Medicis in Florence were published as a collective volume without citing the names of the individual researchers. Among them were some of the most distinguished names in seventeenth-century science. Hahn characterizes the point of view as follows: "In reporting the eternal truth of nature, would it not be foolish and vain to record the names of mere mortals who happened to have stumbled across a right answer?" (Hahn 1971:25). In a favorite phrase of the period, scientists were said to be "uncovering the secrets of nature," and therefore should not take credit for what was a fortuitous happening. Science was the result not of a flight of fancy but a voyage of discovery. Although the social environment of the scientist today differs greatly from that of the seventeenth and eighteenth centuries, this attitude still persists to some extent in the impersonal style that is usually affected in scientific writing, in which the scientist tries to convey the impression that he is a passive observer (Meredith 1966:131).

There is evidence that there may not have been complete acquiescence among the celebrated members of the Cimento on these sentiments. Middleton suggests, in fact, that the submersion of their individuality in the scientific enterprise may have been one of the reasons the society broke up so early (Middleton 1971:328). The names of the authors were actually used in the first draft of *Saggi*, as the proceedings were entitled, but they were removed after a long delay in their publication. It is not clear whether this was the result of a decision to accept collective responsibility for the work or because of a sense of prudence (Middleton 1971:54). Leopold, their patron, had in the interval been elevated to the position of cardinal, and philosophical speculations and experimentation were already beginning to be regarded as matters of risk in Italy at the time, as Galileo had helped to demonstrate. The evidence is that the papers were not the result of collective enterprise. The names of many of the authors of the original papers can be determined from manuscript copies of contemporary diaries that have been preserved (Middleton 1971:54). There is also evidence that the individual scientists were not particularly pleased with the arrangement. Borelli, one of the most distinguished of the group, revolted against the imposition of anonymity and published many of his papers outside *Saggi* (Hanlon, 1970:21).

The pattern was repeated in some of the other early scientific societies of the period. The transactions of the Collegium Experimentale Sive Curiosorum, published in 1676, used the same anonymous style as the Cimento. The early works of the Académie des Sciences in Paris in the seventeenth century also played down the individual author, assuming the responsibility and credit for the society. In these early days, the Académie also tried to follow Bacon's model and to pursue science as a collective activity. As it moved away from the role of a research institute and toward one in which it served more to validate and evaluate work performed outside the society, it tended more toward acknowledging individual authorship. In fact, in 1699, when the Académie was reorganized, it definitely abandoned by regulation the ideal of collective activity and anonymous publication and recognized, as Hahn concluded, that scientific discovery was usually an individual act (Hahn 1971:291).

Anonymity did not completely disappear in the society publications of the eighteenth century. A learned society founded in Bohemia in 1746 proclaimed its devotion to anonymity by taking the name Societas Eruditorum Incognitarum, or Gesellschaft der Unbekannten (Müller 1975:29). Out of seventy-three papers that appeared in a collection of prize essays of the Académie des Sciences in Paris, ten were published without names. The authors of eight are still unknown, but two were fairly transparent. One of the two prizewinners in 1738 is politely referred to as a "young lady of high rank," revealed later as Madame du Chastelet, a "friend" of Voltaire who won the other prize in that year (Jaeggli 1975:4081). It may be a matter of interest that all the other anonymous papers deal in one way or another with problems of navigation or shipping, which may provide historians with some clues or perhaps lead to some interesting speculations. None of the papers that were included in *Mémoires Pour l'Histoire Général du Diocese de Bèziers Tiré des Registres de l'Académie des Sciences et Belles-Lettres de la Ville de Bèziers,* published in Bèziers in 1728, is attributed or signed.

Anonymity was even more prevalent when a publication was issued under the auspices of a state institution, such as *Mémoires de Mathematiques et de Physique Rédiges à l'Observatoire de Marseille,* published in Avignon in 1756. Most of the articles, which cover the wide range of subjects reflected in the title, are not signed. Anonymity of this kind sometimes grows out of the fact that authors are acting as surrogates of the society or organization that they represent in the same way that some documents today are assigned by catalogers to "corporate authorship." The "programmes" written to announce the prizes offered by scientific and other societies in the eighteenth century fall into this category. They have historical interest because the "author(s)," in posing the question to be answered for the prize, frequently attempt to define the subject and to set standards that reflect contemporary points of view. The Académie des Sciences in Paris generally assigned this task to one of its members most knowledgeable about the subject. Several of these programs have been assigned to Lavoisier, for example, by Dumas, Grimaux, and others (Scheler 1964).

In his monumental bibliography covering the medical and related literatures from 1780 to 1820, Callisen (1830) devotes two volumes to "Die anonymische Schriften." Many of these are the works of multiple authors in the sense that they are collections of the separate works of individual authors, but they also contain works of what we would call "corporate authorship," such as pharmacopoeias and dispensatories where no individual is singled out for recognition. Multiple authorship, as we know it today, also occurs when many authors sign their names to a single document. As Merton points out, this also confers a kind of anonymity on the document, as the contribution of the individual is difficult to discern (Merton 1973:332). Little of this kind of authorship is noted in Callisen. However, there are a large number of individual authors (certainly large compared with today's practices, although there seems to be little quantitative data) who did not sign their papers. Callisen is able to identify some from extraneous evidence or later writings, but many remain unknown.

MOTIVATIONS

Many reasons have been adduced for why authors chose to maintain their anonymity. Often these are not apparent from the publication in hand, nor is evidence always available to support the suspected reasons. The easiest, perhaps, to speculate on are those in which identification of the author would have put him in some jeopardy. There is unfortunately a long history to support the statement that scientific as well as religious and political opinions can bring severe repercussions against the author. That this kind of danger did not disappear with the passing of the Renaissance is shown by Barber's study of the resistance to scientific ideas (Barber 1961). These dangers occurred most frequently when religious beliefs seemed to be challenged by scientific ideas, as they were during the Enlightenment. D'Holbach, for example, wrote over fifty books, including his famous *Système de la Nature*, and four hundred articles and published them all anonymously (Kors 1975:14).

Other authors chose anonymity even when identification might not have threatened them with any direct repercussions, but might have had an adverse effect on their careers and livelihood. Thomas Young, a successful practicing physician, published papers on such subjects as the wave theory of light, the Rosetta stone, and physical optics under a pseudonym because he was afraid his medical practice might suffer if it were known that he was devoting time to these other pursuits (Lagemann 1956:132). Erasmus Darwin, the physician who anticipated his illustrious grandson in some of his discoveries, published the first edition of his celebrated *Botanic Garden* anonymously because, as he wrote his publisher: "I would not have my name affixed to this work on any account as I think it would be injurious to me in my medical practice, as it has been to all other physicians who have published poetry" (Schofield 1963:204–51).

Arthur Young, who took such a strong stand against anonymity in his agricultural journals (see below), did not use his name when publishing in another context. During the period in which he was establishing himself as England's outstanding agricultural writer, he wrote a novel under the title *Adventures of Lucy Watson*, which he published anonymously in 1768 (Gazely 1973:351). We can only speculate that he thought that his credibility as a farm expert would be threatened by being identified with such a popular novel. In other cases, the motivations are more apparent. Callisen lists an anonymous pamphlet published in 1810 in which a German physician pleads with his colleagues not to impoverish their patients by excessive fees (Callisen 1830, vol. 22:no. 557). This solicitude may not have been received with any great gratitude or approval and may even have cost him some consultations had he revealed his name.

Another instance in which an author may have had reason to conceal his identity is a poetical lampoon on the proceedings of the Royal Medical Society of Edinburgh. It was said to describe "at considerable length, in ludicrous coulors and rather libelous terms, one of the Society's most exciting debates. The members were thereby thrown into a great ferment, and much curiosity was felt as to the authorship. All acknowledged the cleverness of the production, and this induced the author in an unlucky hour to print and publish it." The Society took action against this affront to its dignity in a curious way. It applied to the courts for an interdict against the publisher on the grounds that it had the sole right to publish its proceedings without consent. The courts granted the appeal and all the unsold copies of the lampoon were confiscated and destroyed (Gray 1952:122).

Sometimes the wish to avoid anticipated controversy is the sole motivation for anonymity. Charles Bonnet (1720–1793), the Swiss naturalist, chose to remain anonymous when he first published his revolutionary *Essai de Psychologie* in 1754. As he explained in a letter to his countryman and fellow naturalist Abraham Trembly in 1779, "the subject matter was very delicate, very contentious, on which I had expressed myself with too little reserve." His anonymity was quickly exposed in Geneva, but he persisted in denying his authorship because he did not "want to be exposed to a multitude of questions and objections." Bonnet was torn between preserving his anonymity and defending his work against his critics. He compromised by citing it in his other writings as the work of another author and did not actually acknowledge it until his collected works were published toward the end of his life (Starabinski 1975).

Another motivation for anonymity that is frequently alleged in this period is the diffidence or modesty of the author. "Diffidence" has been defined as a form of timidity and hesitation to expose oneself. It can be seen clearly in the case of William Thomas (later Lord Kelvin), who submitted some papers on advanced mathematics for publication when he was only sixteen and still a student at Cambridge. In other cases, the diffidence is not always possible to confirm. Johann Kanold (1679–1729), who edited the monthly *Sammlung von Natur- und Medizin,* maintained the anonymity not only of the authors who wrote for the journal, but also—for the first six months of the publication—of the editors. The editors' names were revealed in an issue of *Neuen Zeitung von gelehrten Sachen.* In the third quarterly issue of *Sammlung,* Kanold confessed:

> It is true that we preferred not to present our names in the first issue, not because of modesty or fear but only to show the entire world that our intentions were based on public need rather than an empty desire for glory. However, despite all our efforts, our names have been made publicly known by other means. Therefore, in order that we may not be accused by anyone of being ashamed of our defects, or of making our work appear like a bastard child, the anomaly of whose birth is established by the concealment of its parents' name, and in part also that our learned and well-disposed friends may be more assured that it is truly the work of our hands, we do not scruple in announcing (our names) publicly. (Kronick 1976:83)

When Jenner published his first paper in the form of an anonymous pamphlet in 1783 (Jenner 1783), age was not a factor; he was thirty-four at the time. An intimation of his motivation is afforded by the word "cursory" in the title. John Hunter persuaded him to communicate it to the Society for the Improvement of Medical and Chirurgical Knowledge, where it was read in 1784. It was not published until 1793 in the society's *Transactions* where it finally appeared under his name (LeFanu 1957). Benjamin Franklin also chose when he did and did not wish to have his name attached to one of his scientific communications. In a letter to his English correspondent Peter Collinson, he requested "that you would not expose these letters, or if you communicate them to any Friends, you would at least conceal my name" (Cohen 1941:63).

Motivations for concealing identity cannot always be considered meritorious. In fact, concealment may sometimes seem downright dishonest and even malicious. A notorious example is that of Jean-Paul Marat, who had made a promising beginning as a scientist before he gave himself over completely and violently to the excesses of the French Revolution. To his chagrin and frustration, Marat was thoroughly rejected by the French scientific establishment, apparently for good reasons (Fayet 1960:27–44). He found some of the acceptance and praise he sought through his scientific articles in the Parisian periodicals, but primarily for articles he wrote anonymously. In one anonymous article, he described his work on optics in the following words: "The Revolution which M. Marat brought about in optics made such a strong impression among the physicists who cultivate this science that they still have not recovered from their astonishment." This duplicity did not prevent him later from accusing the Academy of Sciences of abusing its press privileges to suppress scientific ideas that they did not favor (Fayet 1960:39). We may find it easy to believe these things about Marat, who may have had some "bad press," but even such a scientific lion as Linnaeus has been accused of similar behavior. "He had a habit," says one of his biographers, "of writing anonymous reviews of his own books and getting them printed in appropriate places." It is also claimed that some of the work that has been attributed to him was the result of his failure to give credit to other authors (Goerke 1973:65).

The Parisian periodical press fell into other traps as a result of its cavalier attitudes toward anonymity. The general literary journal, *L'Avant-Coureur*, which even Lavoisier used as an outlet for some of his scientific announcements, appeared under the imprint of the well-known Parisian publishers Pancoucke and Lacombe. Many of the contributions that appeared in it were unsigned, including an article that accused merchants in Paris of selling adulterated tobacco. The Académie des Sciences became involved because of its interest in maintaining standards in the tobacco trade. An investigating team found that the author of the article was attempting falsely to discredit the tobacco retailers because he had recently been refused a monopoly on the right to grate tobacco, a process in preparing it for sale (Smeaton 1957:221). Examples of this kind of use of the press are, of course, not confined to the eighteenth century. Nicholas Blegny, a seventeenth-century French physician who published the first medical journal in the vernacular in 1679, was a master of this kind of personal aggrandizement. One of his many enterprises was a directory or guide to services and facilities in Paris and in France, which was published under the title *Livre Commode des Addresses*. It is a veritable gold mine of information about Paris in the late seventeenth century. He published it under a pseudonym, taking this opportunity to extol the virtues of his wife as the best midwife in Paris. He also used his *Journal des Nouvelles Déscouvertes* to advertise and promote his own remedies (Kronick 1960).

The positive values of anonymity are seen most clearly in the practices that prevailed in the competition for prizes offered by many of the scientific societies of the eighteenth century. In addition to the prestige that accrued from winning, many of these prizes were quite substantial in value and the competition for them was often very keen. Almost all of them provided for withholding the identity of the author until the prize was awarded. In a few cases, suspicions have been expressed about the care taken to guard the author's identity. It is alleged, for example, that when both Bernouilli and d'Alembert submitted papers to the Berlin Academy of Sciences on the question of the cause of winds, Bernouilli took the precaution of telling Leonhard Euler, who favored his candidacy, what the secret motto was by which his paper could be identified. D'Alembert nevertheless was awarded the prize, but this has been attributed in part to the influence of Maupertius, so it is possible that he too had some means of secretly revealing his identity. Euler was a rival of d'Alembert and was able to block d'Alembert's contributions for some time. The situation changed when Maupertius came to Berlin as president (Hanlon 1970:45). The rule of maintaining secrecy of the candidates for the prizes seems for the most part seem to have been rigorously observed. This can be seen in the number of cases in which the prizes were repeatedly awarded to the same person, so that in some instances a limit was set to the number of times the prize could be awarded to the same individual.

Other positive reasons for anonymity were also raised from time to time in this period. Theophraste Renaudot, the controversial physician who touched so many areas of French life in the seventeenth century, did not include any of the names of those who participated in his conferences, which were first published in Paris in 1634 under the title *Premiére Centuries des Questions Traitées ez Conferences du Bureau d'Adresse* (Solomon 1972:256). The reasons are given in the preface to an English translation that appeared in London in 1664:

> The Publisher craves Pardon of the Readers that he hath forborn to divulge the Names of the Persons of Honour, who held these Conferences weekly on

> *Mondays* at *Paris*, it being the principal condition which they requir'd of him; Some, that the judgement of their opinion might be left free to everyone, which the knowledge of the Authors commonly prepossesses; and Others, out of desire secretly to discover what Sentiment the publick would have of theirs (like the Knights Errant of old, who fought under borrow'd Armes) leaving their Names to be conjectur'd by such as found any thing in them for which to desire it; but All, through a Modesty as commendable in respect of themselves, as injurious to the Publick. (*General Collection* 1664:viii)

Isaac Walton is even more disingenuous in explaining why he omitted his name from the first edition of his *Compleat Angler.* He did so, he said, because he did not want to distract the reader from the subject matter.

EDITORIAL POLICIES

Wide variations in attitudes toward anonymity can be seen in the editorial policies and practices of the editors of scientific journals of the seventeenth and eighteenth centuries. There do not seem to be any quantitative data about the differences between anonymous publication in either scientific or nonscientific writing, in this period and the present, but even casual acquaintance with the literature of these two periods makes it clear that anonymous publication was much more widespread in the earlier period. In reviewing the periodical literature of the eighteenth century, Baesel concludes: "Eighteenth century authors seemed to prefer writing anonymous articles. The writer by not revealing his name was free to make honest comments and evaluation of potentially sensitive subjects" (Baesel 1974:80).

The early issues of *Philosophical Transactions* are full of communications signed only as "Extract of a letter written from Paris" or "Extract of a letter written from Rome." Sometimes the editor, Oldenburg, prefaces the communication with such a comment as, "I have received an account from very good hands." The articles and reviews in *Acta Eruditorum*, published soon after on the continent, were also generally unsigned, although it appears that the authors' names were known to the editor, Otto Mencken, because there are issues of *Acta* in the university library in Leipzig that have the authors' names written in the margin in Mencken's hand (Kirchner 1928:83). In another contemporary learned journal, *Nova Literaria Maris Balthici et Septentionis*, issued in Lübeck from 1698 to 1768, not even the editors are named, and their identity is still in doubt today. The contributors were also anonymous (Kirchner 1970:200-206). This is only one of many journals in the period published without the names of their editors. Examples include a general natural history journal, *Erlauterte Merckwürdig-keiten der Natur* (Danzig, 1736) (Harff 1940:24),

Physikalisch-oeconomische Realzeitung (Stuttgart, 1755–1757) (Harff 1940:38), and *Algemein nützlicher chemisch-physikalisches Mancherlei* (Berlin, 1781) (Harff 1940:59).

Many of the contributions to *Journal des Trévoux*, the Jesuit answer to *Journal des Sçavans*, were also published anonymously. The names of many of the contributors, who include Leibniz, the sculptor François Girardin, and the English "oculiste" Woolhouse, were not known until archival files were discovered containing article manuscripts with the authors' names written in the margins (Dumas 1931:192). Sometimes this policy is editorially defended. The editor of *Bibliothéque Raisonée* says that, in this way, reviewers can more easily retain their independent judgment and impartiality without the need to praise their friends or castigate their enemies.[1] The practice, however, may also have led to abuses. Johann David Michaelis, when he took over as editor of *Göttingsche Anzeigen von gelehrten Sachen* in 1753, admitted that not all of the reviews were submitted by members of the Göttingen University faculty, as had been assumed, and that some of them were indeed received unsigned, despite the fact that this permitted some scholars to praise their own work.[2]

Similar opinions are expressed in other editorial statements. Many of the articles in Giseke's *Abhandlungen und Beobachtungen aus der Arzneygelahrtheit*, published in Hamburg in 1776, are signed only with an initial. This, says the editor, is so that the reader will not be prejudiced either for or against the author (Kronick 1976:129). *Beyträge zur Naturgeschichte* was published in Altenburg from 1774, and, as the editor indicates, was derived "from the unprinted letters of learned naturalists and observant friends of nature," most of whom did not sign their contributions. Some of the contributors had in fact expressly asked him not to use their names. He, therefore, decided not to use any names in the publication, although in some cases he adds the geographic origins of the author as, he says, it may have some significance in articles dealing with natural history. It is questionable whether "authorship" can be ascribed to some articles in any case, since, as he admits, he has edited many of the letters to improve their style and even brought several together into a single essay, which often included his own comments.[3]

Editorial opinion is by no means uniform on this matter. As early as 1693 the editor of *Memoirs for the Ingenious* says clearly: "I desire to be excused from receiving, answering or publishing any relation whose Author will not be known or stand by it."[4] Arthur Young expressed himself just as strongly almost a century later when a correspondent asked him to print a communication in which he would not be identified. In the history of agricultural journalism between his own *Annals of Agriculture* and the agricultural journal edited by Houghton, he says in an accounting at the end of the first year that anonymity had been one of the principal reasons "they dropped into obscurity almost as soon as they started for public favor."[5] "My idea," he adds, "was to admit no communication without the names and places of abode of the authors." He does not seem to provide a rationale for this position, but he is obviously convinced that the policy led to better results: "If the reader will take the trouble of turning over the *Museum Rusticum, De Re Rustica,* and *Foreign Essays* I believe he will find more original and authentic communication in this one volume, than in those ten."[6] An indirect corroboration of these sentiments is provided by the editor of *Sammlung medicinischer und chirurgischer Original-Abhandlungen aus dem Hannoverischen Magazine von 1750 bis 1786,* published in Hannover in 1786. As the title indicates, this is a collection of original papers on medicine and surgery, many of them written for a medical audience, although they were first published in general regional literary journals. The editor states clearly in the preface that he intends to omit all papers that have been "issued without the real name of the author."[7] There are other indirect examples of this kind of editorial policy. *Naturforscher,* edited by Walch from 1774 to 1778, is one of the outstanding natural history journals of the eighteenth century.[8] In a recent index to its articles published in 1971, only one entry is listed under "anonymous"; two are signed with initials and not identified. There are 478 articles contributed by 107 authors (Walch himself signed forty during this period) (Geus 1971). In some instances, readers are given access to the author in another way. One contributor to Rozier's *Observations sur la Physique* is identified only as "D. T., Correspondant de l'Académie Royale des Sciences." A footnote indicates that those who are curious about verifying the experiments discussed should address themselves to the editor.[9]

The fact that authors also had some ambivalence about revealing their names in scientific publications in the eighteenth century is indicated by the frequency with which editors give their contributors the option of withholding their identities. Once given, the option is usually honored. In *Medical Essays and Observations,* the editor of a paper entitled "An Anomalous Tumour of the Leg Unsuccessfully Treated" notes: "the author of this paper having claimed the performance of the promise we made in our proposals, we have complied with the desire of his letter, by suppressing his name."[10] Is it possible that the author's choice would have been different if his title ended with the words "successfully treated"? This option is made available to the author by many other editors. The editor of *Gesellschaftliche Erzehlungen für die Liebhaber der Naturlehre,* a weekly periodical published in Hamburg in 1753 devoted to general science and domestic affairs, invites contributors to submit their articles either with or without their signatures.[11] The editor of *Sammlung von Beobachtungen aus der Arzneygelahrheit und Naturkunde* instructs his potential contributors that they are free to sign their names if they wish to do so.[12] The first two volumes are primarily devoted to reviews and contain only a few articles, which are all unsigned. Beginning with the third volume, more of the articles tend to be signed at least with an identifying initial, which may or may not have revealed the author to his audience. The fifth volume is occupied primarily with an account of the epidemic in Nordling in the winter of 1771–1772 and is signed "J. A. P. G.," a combination of initials that can belong only to the editor himself.[13]

Even such a prestigious periodical as Rozier's *Observations sur la Physique* contains articles that are signed only with an initial, such as the paper "On Elementary Composition of Bodies," which is signed M. M.[14] but also lists the author's academic affiliations in order, one suspects, to establish a credibility that may be lost when a full name is not provided. In many of these cases it is difficult to discern any reason for anonymity, at least without extensive research, such as the reasons that have been adduced for a paper of Lavoisier that appeared in this journal without his signature (Meldrum 1933:409). Numerous unsigned case reports, reviews, announcements of prize essays, and so forth were probably written by the editor or his staff. In most cases, it seems the author's name is known at least to Rozier, such as in a lengthy memoir on the composition of light that appears in the May 1773 issue with the author identified only as a "correspondent of the *Académie*"—who, says the editor, has asked him to withhold his name despite the editor's reluctance to do so.[15]

In the first issue of the *Annales de Chimie*, Baron de Dietrich, the Commissaire du Roi a la Visite des Mines, reported on the founding of the Société de l'Art de l'Exploitation des Mines at Schemnitz in Hungary in 1786. Among the French members were Hassenfratz, Morveau, Lavoisier, Malesherbes, and the Baron himself. The membership also included Gmelin and Goethe from Germany, and Kirwan, Withering, and Watts from England. Article 9, section 3 of the society's regulation specified that members could submit memoirs anonymously if they wished.[16] In the first issue of his *Frankisches Sammlungen*, Delius invites all those "who love and esteem nature, all inquisitive physicians, as well as all those devoted to industry" to submit their work to him. They will be printed, he says, either with or without their names as they wish (Delius 1755). All the articles in the first issue are apparently written by Delius himself. Although he continued to write a substantial part of the subsequent issues, he also included many contributed and signed papers. Many of the apparent anonymous or unsigned articles in this period may in fact result from the lack of distinction between the author and editor in periodical publication. Many of the essay journals of the period, such as the famous *Tatler* and *Spectator,* and *Moralische Wochenschriften* issued in Germany in their image, were written largely by their editors. This practice was extended to essay journals devoted to scientific subjects. An example is

Dresdnisches Magazin, a general natural history journal edited by Michael Groll. It includes articles on health and natural phenomena, as well as occasional poetry. Sometimes these journals were started with expectations of receiving contributions but ended up as collections largely of the editor's own essays, in a way similar to the experiences of the *Minerva, oder Abhandlungen über Gegenstände der Natur und Kunst*, a monthly edited in Osnabruck by C. L. Reinhold in 1778.

All the articles in the first issue of *Magazin der Bergbaukunde* issued in Dresden in 1783 are neither signed nor attributed. The editor, Johann Friedrich Lempe, who describes himself as a mathematician at the Churfurstlich-Sachs, Bergakademie, apparently wrote the entire first issue himself. He invites contributions in the preface: "If experts desire to enrich this mining periodical they need only take the opportunity to send their contributions to me or to the Walther Hof-Buchhandlung in Dresden." There do not seem to have been any immediate responses to his invitation, because the entire second and third issues also consist of unsigned articles, with the exception of a translation of a paper by De Lisle on fossils, which appears with a commentary by Dietrich Ludwig Gustav Karsten. Karsten seems to be the principal other contributor beside the editor himself.

The failure in some cases to identify an author in this period may be a result of the general indifference to attributing texts. This to some extent reflects the hereditary relation of the learned periodical to the newspaper, where journalistic anonymity was and is an accepted part of reportage. A reporter, at least theoretically, acts as a disengaged, unbiased, and neutral transmitter of information, and his name in a byline is often more an exception than the rule, except that a good reporter is expected to identify sources of information and validate them. It was a common practice in the early days of newspaper journalism, particularly in the provinces, to take copy from other newspapers. Barnes points out that the dearth of copy in the seventeenth and eighteenth centuries made it necessary for editors to plagiarize reviews and articles from other journals (Barnes 1936:53). The journal sources are named in some cases, but the authors of the original articles are much less frequently cited. *Encyclopådisches Kalendar*, edited by Johann Christoph Heppe, is made up primarily of excerpts from English and French scientific and technical journals, including *Gazette ſsalutaire*, *Universal Museum*, and *Gentleman's Magazine*, and in one case simply "from the French." Frequently, however, not even the journal source is named, as Trieloff discovered in his study of the reviews in *Frankfurter Gelehrte Anzeiger*, which took many of their articles directly from *Monthly Review*, *London Magazine*, *Gentleman's Magazine*, and even from *Mercure de France* (Trieloff 1908:59).

Not all journals adhered to this practice. *Hanauisches Magazin*, which covered a wide range of subjects, including natural history, affirms in the preface to the third volume for 1780: "We completely decline essays and reports which have been previously communicated to other journals since we are not issuing a newspaper and do not need to be an echo." Most of the articles, nevertheless, are unsigned or signed only with an initial, and the editors' ambivalence becomes clear when they ask their correspondents to "at least have the discretion or modesty" to indicate the source of their material when they have taken it substantially from some other source.[17]

SUMMARY AND CONCLUSIONS

The purpose of this brief survey of anonymity and identity in the scientific journalism of the seventeenth and eighteenth centuries has not been to make a case either for or against anonymity in scientific writing. While many authors chose the prerogative of remaining anonymous in the earlier period, the practice rarely occurs today, and when it does, it is not as easily condoned. Although there are not enough data to make any quantitative assessments, it is clear that there are significant qualitative differences between the editorial practices and the science of that period and those of today. Although specialization among scientific journals occurred as early as the seventeenth century, most of them could still be read easily by general audiences throughout the eighteenth century. The social environment of science and scientific journalism has also undergone great changes, in the professionalization of the scientist, the premium placed on publication, the greater emphasis on experimental than on observational data, and so forth.

In some ways, however, the nature of the controversy still has not changed. This is particularly true in the maintenance of anonymity in the review process (Shils 1975). The role of the scientist as a witness raises quite a different issue. Scientific evidence, like other kinds of evidence, is based essentially on two factors: the cogency of the argument and the credibility of the witness. While the anonymity of the witness does not affect the cogency of the argument, it contributes nothing to the credibility of the witness, unless other kinds of bona fides can be established (Kronick 1978). The anonymous or "masked" witness is sometimes accepted in a court of law and sometimes for the same reasons that might be presented for the scientific witness. There are, of course, differences between scientific and legal evidence. One apparent difference, which might weigh in favor of anonymity in science versus anonymity in law, is that legal evidence is to a large extent historical, whereas evidence in science is based primarily on experiences (experiments) that can be replicated. Our ability to assess the biases, the predilections, the background, and the training of the observer presenting the evidence provides us with a means of calibrating the instrument. Essentially, however, it is the cogency of the argument that must prevail.

The editors of *Berlintsche Monauschrift* confront the issue squarely in a long editorial on the subject published in 1788. They were responding to criticism that they had published anonymously unfavorable reviews of Johann Caspar Lavater, the author of a popular theory of the time on the relation between character and facial expression. "In the final analysis," they conclude, "it is what is said, and not who said it." The editors footnote the comment with one word: "Vortreflich" (excellent).[18]

NOTES

1. "Avertissement des Libraires." *Bibliothèque Raisonée des Ouvrages des Savants de l'Europe* 1 (1728): v–xvii.
2. Michaelis, Johann D. "Vorrede." *Göttingsche Anzeigen von gelehrten Sachen* 1 (1753): i–xiv.
3. "Vorrede." *Beyträge zur Naturgeschichte sonderlich des Mineralreichs* 1 (1774).
4. "Preface." *Memoirs for the Ingenious Containing Several Curious Observations in Philosophy*, 1693 (1, no. 8).
5. Young, Arthur. "Preface." *Annals of Agriculture* 1 (1784): 5–8.
6. Young, Arthur. "To the Reader." *Annals of Agriculture* 1 (1784): 461–67.
7. "Vorbericht des Herausgebers." *Sammlung medicinischer und chirurgischer Original-Abhandlungen aus dem Hannoverischen Magazine von 1750 bis 1786* 1 (1786): i–x.
8. *Naturforscher.* Edited by Ernest I. Walch, 1774–1778, and J. C. D. Schreber, 1779–1804. Nos. 1–30. Halle: Gebauer, 1774–1804.
9. *Observations sur la physique* 2 (July 1773): 11.
10. "An Anomalous Tumour of the Leg Unsuccessfully Treated." *Medical Essays and Observations* 21 (1733): 234–38.
11. Unzer, Johann A. "Vorbericht." *Gesellschaftliche Erzehlungen für die Liebhaber der Naturlehre, Haushaltungs-wissenschaft, Arzney-Kunst und Sitten*, vol. 1 (1753).
12. "Vorbericht." *Sammlung von Beobachtungen aus der Arzneygelahrheit und Naturkunde*, 1 (1769).
13. Gesner, J. A. P. "Beobachtungen über das epidemischen Fieber in Nordlingen." *Sammlung von Beobachtungen aus der Arzneygelahrheit und Naturkunde* 5 (1776): 87–302.
14. *Observations sur la Physique* 10 (1777): 286.
15. *Observations sur la Physique* 1 (1773): 339–75.
16. *Annals de Chimie* 1 (1789): 118.
17. "Vorbericht." *Hanauisches Magazin*, vol. 3 (1789).
18. "Ueber die Anonymität der Schriftsteller." *Berlinische Monatsschrift* 11 (1788): 137–51.

REFERENCES

Baesel, Don R. "Natural History in the British Periodicals in the Eighteenth Century." Ph D. diss. Ohio State University, 1974.

Barber, Bernard. "Resistance by Scientists to Scientific Discovery." *Science* 134 (1961): 596–602.

Barnes, Sherman B. "The Editing of Early Learned Journals." *Osiris* 1 (1936): 155–72.

Callisen, Adolph Carl Peter. *Medicinisches Schriftsteller-lexicon der jetzt lebenden Aerzte, Wundaerzte, Geburtshelfer, Apotheker und Naturforscher aller gebildeten Völker*. 33 vols. Copenhagen, 1830–1845. Reprint, Nieuwkoop: B de Graaf, 1964.

Cohen, I. Bernard, ed. "Benjamin Franklin's Experiments." Cambridge: Harvard University Press, 1941.

Delius, Heinrick F. "Vorrede." *Fränkische Sammlungen und Anmerckungen aus der Naturlehre, Arzneygelahrtheit, Oekonomie und denen damit verwandten Wissenschaften* 1 (1755): i–xii.

Dumas, Gustav. *Histoire du Journal de Trévoux depuis 1701 jusqu'en 1762*. Paris: Bouvier, 1931.

Fayet, Joseph. *La Revolution Française et la Science, 1789–1795*. Paris: Riviere, 1960.

Gazely, John G. *The Life of Arthur Young, 1741–1820*. Philadelphia, Pa.: American Philosophical Society, 1973.

General Collection of Discourses of the Virtuosi of France. Render'd Into English by G. Havers. London: Thomas Dring & John Starkey, 1664.

Geus, Armin, ed. *Indices naturwissenschafatlich-medizinischer Periodica bis 1850*. Stuttgart: Hiersemann, 1971.

Goerke, Heinz. *Linnaeus*. Translated from the German by Denver Lindley. New York: Charles Scribner's Sons, 1973.

Gray, J. *History of the Royal Medical Society, 1737–1937*. Edinburgh: University of Edinburgh, 1952.

Hahn, Roger. *The Anatomy of a Scientific Institution: The Paris Academy of Sciences, 1666–1803*. Berkeley: University of California Press, 1971.

Hanlon, Thomas L. *Jean d'Alembert: Science and the Enlightenment*. Oxford: Clarendon Press, 1970.

Harff, Horst. "Die Entwicklung der deutschen chemischen Fachzeitschrift bis zum Beginn des neunzehnten Jahrhunderts." Ph D diss. Friedrich Wilhelms-Universitäts zu Berlin, 1940.

Jaeggli, Alvin E. "Recueil des Pièces qui ont Remporté le Prix de *l'Académie de Sciences* Depuis 1720 jusq'en 1772." *Gesnerus* 32 (1975): 408–14.

Jenner, Edward. *Cursory Observations on Emetic Tartar*. Wooten-Under-Edge: Bence, 1783.

Jones, Richard F. *Ancients and Moderns: A Study of the Rise of the Scientific Movement in Seventeenth-Century England*. St. Louis: Washington University Press, 1961. 2d ed. Reprinted by New York: Dover, 1982.

Kirchner, Joachim. "Zur Entstehung und Redaktiongeschichte der *Acta Eruditorum*." *Archiv für Buchgewerbe und Gebrauchsgraphik* 65 Sonderheft 4 (1928): 78–88.

Kirchner, Joachim. "Die älteste Monatsschrift der Ostseeländer." In *Aufgewhälte Aufsätze*. Stuttgart: Hiersemann, 1970, pp. 200–206.

Kors, Alan. *D'Holbach's Coterie: Enlightenment in Paris*. Princeton: Princeton University Press, 1975.

Kronick, David A. "Nicolas de Blegny, Medical Journalist." *Bulletin of the Cleveland Medical Library Association*, 1960: 47–56.

Kronick, David A. *A History of Scientific and Technical Periodicals: The Origins and Development of the Scientific and Technical Press, 1665–1790*. Metuchen N.J.: Scarecrow Press, 1976.
Kronick, David A. "Authorship and Authority in the Scientific Periodicals of the Seventeenth and Eighteenth Centuries." *Library Quarterly* 48 (1978): 255–75.
Lagemann, Robert "Pseudonyms of Physicists." *Scientific Monthly* September (1956): 130–34.
LeFanu, William R. *A Bio-bibliography of Edward Jennet, 1748–1813*. London: Harvey & Blithe, 1957.
Meldrum, Andrew N. "Lavoisier's Early Work in Science." *Isis* 19 (1933): 396–425.
Meredith, Patrick. *Instruments of Communication: An Essay on Scientific Writing*. Oxford: Pergamon Press, 1966.
Merton, Robert K. *The Sociology of Science: Theoretical and Empirical Investigations*. Edited by Norman W. Storer. Chicago: University of Chicago Press, 1973.
Middleton, W. E. Knowles. *The Experimenters: A Study of the Accademia del Cimento*. Baltimore: Johns Hopkins University Press, 1971.
Müller, Hans Heinrich. *Akademie und Wirtschaft*. Berlin: Akademie Verlag, 1975.
Relman, Arnold S. "Lessons from the Darsee Affair." *New England Journal of Medicine* 308 (1983): 415–17.
Scheler, Lucien. "Antoine Laurent Lavoisier et Michael Adanson, Redacteurs de Programme de Prix à l'Académie des Sciences." *Révue d'Histoire de Sciences* 14 (1964): 257–84.
Schofield, Robert F. *The Lunar Society of Birmingham*. Oxford: Clarendon Press, 1963.
Shils, Edward. "The Confidentiality of Anonymity of Assessment." *Minerva* 13 (1975): 135–51.
Smeaton, W. A. "*L'Avant-Coureur*: The Journal in which Some of Lavoisier's Earliest Research Was Published." *Annals of Science* 13 (1957): 224–25.
Solomon, Howard M. *Public Welfare, Science, and Propaganda in the Seventeenth Century: The Innovations of Theophraste Renaudot*. Princeton: Princeton University Press, 1972.
Starabinski, Jean. "L'Essai de Psychologie de Charles Bonnet: Un Version Corrigée Inédité." *Gesnerus* 32 (1975): 1–15.
Trieloff, Otto P. "Die Entstehung der Rezensionen in den *Frankfurter Gelehrten Anzeigen* von Jahre 1772." Ph.D. Diss. Universität Münster, 1908.

Chapter 10

Notes on the Printing History of the Early *Philosophical Transactions*

ABSTRACT

Henry Oldenburg, one of the first secretaries of the Royal Society of London, established a new medium of scientific communication when he began the first scientific journal, *Philosophical Transactions*, in 1665 in London. It has been characterized as one of the most important contributions to the scientific revolution of that period. A study of its early printing history, relationship with the Royal Society, distribution mechanisms, editorial policies, and procedures provides some insights not only into the history of science of that period, but also into the evolution of the scientific journal as the most important medium of scientific communication in the modern world. The checklist of editions, translations, and abridgments appended is one indication of the impact it had in the seventeenth and eighteenth centuries.

Henry Oldenburg, one of the first secretaries of the Royal Society of London, by inaugurating the *Philosophical Transactions*,[1] created a new medium for disseminating scientific information that provided momentum to the scientific movement that still continues today. Although *Journal des Sçavans* preceded it by two months, appearing on January 5, 1665, *Philosophical Transactions*, which appeared on March 6 of the same year, fully deserves to be called the first scientific periodical. *Journal des Sçavans* also served in its early years as a vehicle for scientific information, but from its beginning it continued to be a general literary periodical, while *Philosophical Transactions* was dedicated from the start to the dissemination of scientific knowledge. It

Published in *Libraries and Culture* 25 (1990): 253–67

quickly became the preeminent scientific journal of the seventeenth century and maintained that position throughout the following century. The Oldenburg correspondence provides sufficient evidence, as the editors say, that "everyone wanted it . . . everyone was gratified to see his own contribution printed in it."[2] By way of complementing the large number of articles about *Philosophical Transactions*,[3] this chapter suggests the complexity of its editing, printing, and distribution history as summarized in the appended checklist.

The desire for a periodical publication to meet the needs of a constantly broadening audience for news of the world of scholarship and science became apparent to both the French and British community of scholars at about the same time. As early as September 1661 Robert Moray indicated in a letter to Christiaan Huygens that the Royal Society intended "from time to time . . . [to] . . . print what passes among ourselves, at least everything that may be published."[4] A note from Robert Hooke, which originally was thought to have been written in 1663, records the intention of the Royal Society to publish its activities in "a Paper of advertisements once every week, or fortnight at furtherest, wherein will be contained the heads or substance of the inquiries they are most solicitous about." Some scholars, however, have attributed the statement to the period after Oldenburg's death in 1677 when the society was considering an alternative form of communication.[5] In France, François Eudes de Mezeray drew up a draft for a "journal littéraire," dated from circumstantial evidence to 1663, which is regarded to have provided De Sallo with the inspiration to begin *Journal des Sçavans*.[6] In any case, Oldenburg was aware of it as early as November 24, 1664, when he reported to Robert Boyle on the interest in France "to publish from time to time a journal of all what passeth in Europe in matters of knowledge both Philosophical and Politicall."[7]

There were, however, significant differences between the two publications. In its first issue *Journal des Sçavans* indicated broadly that its mission was "de faire navoir ce qui se passé de nouveau dans la République des lettres," while Oldenburg made it clear in his title that it was to give "some accompt of the present undertakings, studies, and labours of the Ingenious in many considerable parts of the world."[8] Although *Journal des Sçavans* included science and mathematics among the subjects it covered, theology also played a large role in its contents.[9] Justel, in his letter to Oldenburg on January 27, 1666, asking him to send *Philosophical Transac*tions, complains that "those who work at our Journal are rather Historians than Philosophers, that is why you see nothing in it concerning physics."[10] After the appearance of the *Observations sur la Physique* of the Abbé Rozier in 1773, it completely abandoned any efforts to cover the sciences.[11] On the other hand, Weld, in his classification of papers published in *Philosophical Transactions* from its origin in 1665 to 1848, reveals that only about two percent of the contributions are assigned to "miscellaneous" and presumably nonscientific subjects.[12]

The differences between the two periodicals in their relationship to the scientific community are also significant. Although *Journal des Sçavans* had a tenuous connection with the Académie des Sciences, its orientation was directed more toward the Académie des Inscriptions et Belles Lettres. It was not until later in the eighteenth century that it is said to have become "a quasi-official publication" of the Académie des Sciences by virtue of the fact that its secretary, Fontenelle, was selected as the official book reviewer for the sciences.[13] Most of the associate editors, however, were chosen from the Académie des Inscriptions et Belles- Lettres.[14] Both the English and the French scientific societies had been granted the privilege by the state of licensing books under their imprimatur. The French society chose to exercise it by officially publishing annually, or even less frequently, its *Histoire et Mémoires.* The Royal Society, on the other hand, maintained an ambivalent and even contradictory relationship with *Philosophical Transactions* up to 1753, when it first made public avowal of its custodianship of the journal.

Although *Transactions* was published under the Royal Society's license granted in their first charter, both Oldenburg and the society agreed from the beginning that they were his sole responsibility. In the "advertisement" in the twelfth issue (May, 7 1666) he insists that "whereas 'tis taken notice of, that several persuade themselves, that these Philosophical Transactions are publish't by the Royal Society," they are nevertheless published "upon his Private account." He hastens to add, however, that "though he denies not, but that, having the honour and advantage of being a Fellow . . . he inserts at times some of the Particulars that are presented to them."[15] The issue also comes up often in his correspondence. In a letter to Palmer he refers to *Philosophical Transactions* as "my owne private diversion in broken hours, "and again on November 21, 1670 as "my Transactions."[16] In 1674, writing to Von de Becke, he requests him not to ascribe *Philosophical Transactions* to the Royal Society "since they are edited and published by me alone, its secretary."[17]

The Royal Society maintained this position after Oldenburg's death. After a lapse in publication of several years that was partially filled by *Philosophical Collections* (checklist no. 2) edited by Hooke, the unofficial nature of *Philosophical Transactions* is again pointed out on the first page of issue no. 143 for January 1682–1683 where it is stated that they were "not to be looked upon as the Business of the Royal Society." As late as 1735–1736 in volume 39, dedicated to Hermann Boerhaave, the editor, Cromwell suggests the maintenance of independence for the journal when he asks him for his "patronage," which would not have been necessary, he says, "were these Volumes published in the Name of the Royal Society." Yet there are numerous contradictions in this period. First, *Philosophical Transactions* was published under the society's sponsorship by officially designated "Printers to the Royal Society." Halley, in the dedication in volume 29 for

1714–1716, states that one of the objectives of *Philosophical Transactions* is "to give an exact account of such Experiments as having been made before the illustrious Royal Society, they shall please to order or permit the Publication of." In addition to "ordering" the printing and publishing of *Philosophical Transactions*, the society also took the prerogative of prohibiting the publication of papers, as it did in 1665 with William Petty's report on the double-bottomed boat, which might have given comfort to the enemy.[18] The close connection between the society and *Philosophical Transactions* is also apparent in the frequency with which *Transactions* was suspended during the summer months when the society was not in session. The council of the Royal Society also took direct action whenever the printing of *Philosophical Transactions* stopped, as it did after the cessation of *Philosophical Collections*. In its meeting of January 28, 1690–1691 it was "resolved, that there shall be *Transactions* printed, and that the Society will consider of the means for effectually doing so."[19]

Some of the reasons the society chose not to take direct responsibility for *Philosophical Transactions* may have been fiscal, as it received no direct subsidy from the state as its French counterparts did. This may also have been the reason it did not publish *Histoires* or *Mémoires* as did the French academies. This gap was filled to some extent by the retrospective histories by Sprat and Birch and later by Thomson and Weld, which serve to record the administrative history of the society as well as to print papers that did not find a place in *Philosophical Transactions*.[20] The society finally found a way to establish a fiscal base for *Philosophical Transactions* when it took over full responsibility with the forty-seventh volume for 1751–1752. The change, however, has not been attributed to fiscal problems, but to the bad repute *Philosophical Transactions* had acquired, as reflected in John Hill's *Review of the Works of the Royal Society of London*, published in London in 1751. The review, according to Thomson,[21] called attention to the "much greater proportion of trifling and puerile papers" to be found there. Nevertheless, before this change was effected, the statutes of 1752 were amended to make provision for the gratis distribution of *Philosophical Transactions* to Fellows and for raising the "admission money" from three guineas to five "in order to defray the additional expenses incurred by the free issue of the *Philosophical Transactions*,"[22] thus establishing a basis on which scientific and other organizations have subsidized their publications ever since.

PRINTING, EDITING, DISTRIBUTION

The roles of the editor, publisher, and printer in seventeenth-century periodicals were not clearly differentiated. Theoretically these functions are divided among an editor, who acquires and prepares copy for a publisher, who arranges for printing and distribution, and a printer, who is responsible for producing the final product. In the period we are discussing, printing, publishing, and distribution (bookselling) could all be carried out by a single firm, by two firms in various combinations, by three different firms, or by all possible combinations and permutations of partnerships. Oldenburg refers to himself variously, as we have seen, stating that *Philosophical Transactions* was published for him and also that *Transactions* was "edited and published by me alone."[23] Halley, in the preface to volume 17, for which he served as editor, refers to himself as "the publisher" and clears the Royal Society, "which is no way concerned therein," from any responsibility.[24]

It has been suggested that the financial responsibility was entirely Oldenburg's from the beginning and that "he probably started the journal as much for profit as from altruistic motives."[25] It is nevertheless clear that the publication of *Philosophical Transactions* did not confer great fiscal advantage upon him but rather at times caused him some financial difficulties.[26] It has been reported that Oldenburg hoped to make £150 a year from *Philosophical Transactions.* After two years' experience he wrote Boyle on December 17, 1667, that he had never received more than £40 a year from that source "and that is little more, than my house-rent."[27]

It is not clear what financial arrangements Oldenburg had with the publishers but, in his letter to Boyle on September 18, 1666, complaining of his difficulties in getting *Philosophical Transactions* printed, he suggests that he was being paid by the printer for the copy he provided, either as an outright payment or on the basis of sales. "Meanwhile," he writes, "I have given them the *Transactions* of this month freely, if yt will doe them any kindness or at least be an encouragement to ym to continue."[28] From a letter that the printer Davies wrote to Oldenburg on December 16, 1665, when *Philosophical Transactions* was being printed in Oxford because of the plague in London, it is also suggested that the fiscal responsibility may not have been solely Oldenburg's. Davies indicates he has not sold over three hundred copies, expresses his fears that he will not sell enough to cover the costs of paper and printing, and suggests he be "secured against loss."[29] Some of these arrangements may have changed after Oldenburg's death. Thomson reports that, when Robert Plot became secretary in 1683, he agreed to revive *Philosophical Transactions* after the cessation of *Philosophical Collections*, but only "on condition that the Society would bind themselves to purchase 60 copies of each number."[30] Weld cites a resolution of the council passed on February 15, 1692–1693, when Halley was editor, "that Dr. Plot shall print the *Transactions*, and that, for his encouragement therein, he leave the 60 books agreed by the bookseller to be allowed for the copy, and that the Society will make up to the Dr. what the value of the said books shall fall short of £40 per annum, and they will take the said sixty books as for money, and allow him 12 of each sort to present his chief correspondents: to which the Dr. agreed."[31] It is also reported that the society permitted Halley, when he was editor, to take sixty copies of *Philosophical Transactions* as soon as they were printed and that he sold them to the society for £12 in one instance and £13 12s in another.[32]

Until 1743, when the society assumed official responsibility for *Philosophical Transactions*, the "editors" were clearly identified with the several successive secretaries, but it is not clear what precisely their roles and responsibilities were. How much did a change in editorship affect publication policies? Robert Merton has called attention to the significant change in emphasis in *Philosophical Transactions* published between 1684 and 1687, when Halley was editor. Papers in the physical sciences accounted for 46.3 percent of all the papers published and the biological sciences for 34.7 percent. From 1691 to 1693, when Waller was editor, papers in the physical sciences accounted for 28.9 percent against 48.9 percent in the biological sciences. The ratios changed even more under the editorship of Sloane between the years 1700 and 1702, when the percentages were respectively 17.7 and 57.9.[33] After the society assumed official responsibility for *Philosophical Transactions* in 1752, the editorial responsibilities were assigned to a "Committee on Papers." Here also we need to know more about how the editorial responsibilities were exercised. Although it undoubtedly occurred earlier, the earliest mention in the society records of a paper being technically "referred" is May 25, 1780. No similar record is found until 1831, although it is reported that by 1832, when the custom of calling upon "two or more Fellows" as referees began, the practice seems to have become common.[34]

PRINTERS AND PUBLISHERS

Although accounts have been published of the early printing history of *Philosophical Transactions*[35] little attention has been paid to such matters as its distribution, frequency of publication, and availability, as well as the printing variants that may have appeared. The six printing variants reported here were found in the first volume and may offer some evidence as to the availability of *Philosophical Transactions* and the number of times issues were reprinted. One variant is well known. It is the celebrated issue 27, which first appeared on July 22, 1667, when Oldenburg was confined in the Tower of London on suspicion of espionage. It has been suggested that Wilkins, who also served as the society's secretary, was responsible for this issue, although there appears to be no record of it being licensed by the council in the record book.[36] It contained a translation of a letter of the French physician Denis dealing with his blood transfusion experiments. Oldenburg, on his release at the end of August, convinced of the priority of the English investigators in this area, attempted to suppress the issue by publishing a substitute issue 27 dated September 23, 1667, for the months of July, August, and September. The earlier issue is easily recognizable because it has no masthead. It also shows no imprint, which raises the question of whether it was really intended as an issue of *Philosophical Transactions.* The paging, nevertheless, falls into place in the volume.[37]

I do not believe the other variants have been reported, as this kind of bibliographic study seems to have been undertaken rarely with scientific periodicals even of the stature of *Philosophical Transactions,* although there are several such studies relating to literary periodicals.[38] The findings listed below are based on a simple examination of the colophons in the first volume. At the end of most of the issues, a colophon appears that indicates the publisher and/or printer responsible for the issue. The information in the colophons sometimes differs from that in the imprint on the title page for the volume, which was generally printed after the last issue for the volume had appeared. This is because the printer may have changed during the course of printing the volume, and it was the last printer who was responsible for printing the title page and prefatory matter.

1. Issue no. 6 (November 6, 1665): One copy had no colophon (University of Chicago) while the others showed: Oxford: Printed by Leonard Lichfield for Richard Davis.
2. Issue no. 7 (December 4, 1665–1666): One copy had: Oxford: Printed by W. Hall for Ric. Davis (New York University) while the others showed: Oxford, Printed by A. & L. Lichfield for Richard Davis.
3. Issue no. 8 (January 8, 1665–1666): One copy had: Oxford, Printed by W. Hall for Ric Davis, 1669 [sic],[39] while the others showed: Oxford, Printed by A. & L. Lichfield for Richard Davis.
4. Issue no. 18 (October 22, 1666): Half of the copies showed: London, Printed by T. R. for John Martin,[40] and the other half: London, Printed, for John Crook.
5. Issue no. 19 (November 19, 1666): Half of the copies showed: London, Printed by T. R. for John Martyn, and the other half: London, Printed for John Crook and Moses Pitt.[41]
6. Issue no. 20 (December 17, 1666–1667): The copies were equally divided among those that showed: London, Printed for John Martin, and those that showed, London Printed for Moses Pitt.

There are also a number of other variants indicated in the literature. Rivington, for instance, cites issue no. 9 as being printed at Oxford for

Richard Davis.[42] The *National Union Catalog*, in a note on *Philosophical Transactions*, indicates variations in later volumes: "Some of the volumes were issued with new title pages and minor changes in text and type: thus the title page of v. 12 (1677–79) and v. 13 (1683), which in some copies are incorrectly numbered, are correctly numbered in other copies."[43] The editors of the Oldenburg correspondence, with some justification, refer to the page numbering in the early *Philosophical Transactions* as "utterly bizarre."[44] In some cases, errors in numbering were corrected in other copies.[45] There are some discrepancies in the collations of the early volumes of *Philosophical Transactions* in Callisen and in Banks.[46] Some of the errors are obviously typographical, but others are likely discrepancies that resulted from the way the issues were brought together in volumes. This was probably the case with Christiaan Huygens, who complained to Oldenburg that their volume numbers did not match, as might easily be the case when pages are numbered consecutively from year to year as they were in the first twelve volumes.[47]

What is the significance of all these variations? Since there do not seem to be any differences in the imprint dates in some of the variants, it is possible that only the colophons vary.[48] In some cases, this may represent a change made for distribution by different booksellers, as was the case with publishing partnerships or "congers" in this period, although this problem was usually solved by including the names of all the publishers or booksellers in the imprint or colophon. Additional evidence, however, may be provided by an examination of what we can find out about the frequency, availability, and distribution of *Transactions* in this period, which may indicate whether a need existed for reprinting the individual issues.

FREQUENCY

While the proceedings of learned societies may be irregular in their schedules, the scientific periodical is by definition expected to appear at regular intervals. Oldenburg started off bravely to fulfill the society's resolution that the *Philosophical Transactions* "be printed the first Monday of every Month,"[49] until the plague caused the removal of the printing to Oxford and a gap of three months between the fifth and the sixth issue occurred. After this, the issues were infrequently dated in the first week and more often in the second and even the third week. Of the first 100 issues, only 16 were dated in the first week, 24 in the second week, 45 in the third week, and 15 even in the fourth week of the month. The day of issue continued to be listed after Oldenburg's death, but the practice was finally discontinued with the 177th issue in January 1685.

There does not seem to have been a close relationship between the date in the imprint and the time when *Philosophical Transactions* was available for distribution. It has been noted that issue number 61 was "dated 18 July 1670, but (as always) was actually on sale a week or so later."[50] There is frequently a difference between the issue date on the masthead and the date in the colophon. Issue 128 has an issue date of September 25, 1676, and an imprint date of October 3, 1676. There are other similar examples, although it is not clear whether the difference is significant. Similarly, the date of the society's "imprimatur," which appears on many of the issues, also differs from the issue date, often being dated up to a week or more after the issue date. In issue 179 for January–February 1686, the imprimatur signed by Pepys as president is dated May 15, 1686.

There also does not seem to have been any uniformity in the number of issues that made up a volume and the number of issues that appeared in any given year. Between 1665 and 1750, the number of issues per volume ranged from 8 to 15, and the number of issues per year from 5 to 10.[51] This was a result of the practice of publishing issues covering several months, which occurred for many of the issues beginning with the end of the seventeenth century, and also because of the gaps in publication that occurred at various times during the period. Some of these gaps are readily explained, such as the gap between issue 26 (June 3, 1667) and issue 27 (September 23, 1667) when Oldenburg was imprisoned, and the gap after Oldenburg's death between issue 136 (June 25, 1667) and 137 (February 10, 1667–1668). Others seem to be without explanation except for such notices as the one that appeared in an advertisement in issue 86 (August 19, 1672): "It was thought fit, herewith to give the Reader notice, that the next of these Tracts will not come abroad till the end of October next."[52]

The practice of not publishing during the Royal Society's recess began well before the society took official responsibility for *Philosophical Transactions*. There were no August issues in volume 2 (1667) and volumes 8–11 (1673 1676). In his letter to Huygens on November 3, 1673, Oldenburg explains that there were no issues for August and September because of the "vacation,"[53] which presumably refers to the recess. However, writing to Leibniz on July 12, 1677, he gives quite a different reason for the omission of the July and August issues, explaining that it was "because in this summer holiday season my bookseller hesitates to print these *Transactions*."[54] The policy of not publishing during the summer months was made official by the society in its statement of responsibility for *Philosophical Transactions* in volume 65, when the notice was added that beginning with volume 63, *Transactions* would be published in two parts, twice a year, the first part to cover November to February, and the second "the remaining months to the recess."

There was a gap of three years between issue 191 (December 1687) and issue 192 (January/February 1691) that is between the landing of William of Orange and the flight of James I.[55] Thomson echoes the idea in saying that it was "owing obviously to the deficiency of material."[56] There were a number of other gaps in publication that have not been explored, but it is apparent that, in spite of them, timeliness of publication was an important consideration. Sometimes this occurred with exemplary promptness, as when Flamsteed's astronomical observations made between March 16 and 19 were printed in the same year in number 82 of *Philosophical Transactions* dated April 22, 1672. In volume 58 for the year 1768, the practice began of indicating when communications were received by or read in the society, reflecting sensitivity to the questions of priorities. This note, however, did not occur regularly. Even when *Philosophical Transactions* appeared only twice a year, Weld took pride in their promptness by stating that "in no case have papers, read before the Society and ordered for publication, been delayed more than a year (generally only a few months) between their being read and appearance in the *Transactions.* In this respect the Society is far in advance of the French Institutes, where papers frequently remain unpublished for two, three and even four years."[57]

AVAILABILITY AND DISTRIBUTION

There seems to be little information on the size of the press runs of *Transactions* at any time in the seventeenth century. Moray, in a letter to Oldenburg from Oxford dated October 11, 1665, speaks of negotiating with the Oxford printer Davies to print *Philosophical Transactions* during the London plague.[58] Moray had bargained for 1,200 copies, but Davies agreed to only 1,000. Even if Davies printed only a thousand copies, it surely would have been adequate to take care of the needs of the Fellows and probably have provided for any sale or distribution among the relatively small number of other interested readers.[59] It also has been indicated that the print run for the first version of number 27 was 500.[60] Weld reports that, on the assumption of responsibility for *Philosophical Transactions* by the Society in 1752, "orders were given to print 750 copies of the *Transactions*, for the benefit of the Fellows, and that all copies exceeding the number of Fellows were sold to the Society's bookseller at 25 per cent under the selling price to the public."[61]

Nevertheless, there is some evidence that *Philosophical Transactions* may have been in short supply from time to time or difficult to obtain. Rivington cites a note in Birch that indicates that all the unsold copies of *Philosophical Transactions* were destroyed in the Great Fire in London in 1666 and that Oldenburg had difficulties in getting outstanding numbers printed. He quotes Oldenburg as writing on October 23, 1666: "To this very houre, I have got none yet to print them."[62] There is additional evidence in the Oldenburg correspondence on the difficulty of acquiring the early issues. Thus, Colepresse, in his letter to Oldenburg of February 1, 1666–1667, expresses his desire to complete his set of *Philosophical Transactions*, "which since London's fiery Calamity I cannot (otherwise) possibly procure" and asks Oldenburg's help in doing so.[63] He wrote to Oldenburg again on October 28, 1667, when 28 issues of *Philosophical Transactions* had appeared, saying that he had "only 4 copies of Trans. viz. 11. 13.14.19, three of which I must againe acknowledge as your favour, ye other Mr. Boyle's."[64] Even Wallis apparently had difficulties procuring copies because he writes to Oldenburg from Oxford on October 6, 1668, that he sought the August and September issues "at our Book-sellers several times in vain"[65] and again on November 1, 1675, thanking him for the August and September issues and acknowledging the receipt of the March and May issues, which Oldenburg sent previously. At the same time he informs him that he has been unable to procure the April, June, and July issues.[66]

In his letter to Hevelius on August 2, 1669, Oldenburg explains why the run of issues 16 to 49 he is sending him is incomplete and informs him: "No. 27 of the *Transactions* is lacking and the bookseller has no more copies of it; he is to have that number and others reprinted in the near future and when it is done it shall certainly be sent to you."[67] On April 24, 1671, Oldenburg is able to inform Leibniz that he has been able to procure *Philosophical Transactions* for 1668, 1669, and 1670 to send him, although it is not clear whether these were complete volumes or not.[68]

Although there is little direct evidence, we can assume that the experienced booksellers who were involved with the distribution of *Philosophical Transactions* used the normal book trade channels of the time, including those that involved the international market. Oldenburg also used whatever channels were open to him, including diplomatic couriers and visiting notables, to relay copies to selected correspondents on the Continent. For instance, on October 15, 1673, he informs Nazari, the editor of the *Giornale de'Literati*, that he is sending him some of the issues he has requested by "a friend who is setting out for Venice hence."[69] To Hevelius on February 11, 1675–1676, he writes that he is sending issues 113–115 "with a certain young German names Getz who was making for Danzig,"[70] and later to Huygens that he is sending the latest issue (September 26, 1675) with the "only son of Mr. Evelyn who is coming to Paris in the suite of the King's Ambassador."[71] Back issues for 1668, 1669, and 1670 were sent for Leibniz in April 1671 by the bookseller Martin to a bookseller in Frankfurt. On April 24, 1671, Oldenburg writes to Leibniz, "I am fully confident that you will arrange for Zunner to satisfy Schulz as regards the price, then Schulz can satisfy Martin; if this is not done our Mr. Martin will make difficulties next time."[72] In 1677, arrangements appear to have been made for Martin to send issues on a regular basis to the bookseller Schulz in Hamburg, but the service apparently was not very reliable.

Barnes suggests that "the circulation of the *Transactions* was confined almost entirely to Great Britain,"[73] but is obviously in error, as the Oldenburg correspondence attests. Circulation on the Continent would necessarily have been restricted in any case because *Philosophical Transactions* was published primarily in English, a language not accessible to many scholars in Europe. In Great Britain, it is reported *Philosophical Transactions* "was on sale at the book stalls at various places in London, Cambridge, Exeter, Bristol and perhaps in some towns in Ireland and Scotland."[74] Nevertheless, in 1665 we find the Oxford printer Davies suggesting that copies should be sent to three or four booksellers in London as well as to Cambridge and other places.[75] Of course, all these conditions might have prevailed at one time or another during a period of over a hundred years.

TRANSLATIONS

The deficiencies in accessibility of *Philosophical Transactions* were compensated for in a large degree because its contents were available in other forms such as translations, reprints, abridgments, and through abstracts and reprints in other periodical publications. In a letter to René François Sluse on April 2, 1669, Oldenburg justifies his publishing *Philosophical Transactions* in English "because they are intended to be for the benefit of such Englishmen as are drawn to curious things, yet perhaps do not know Latin."[76] Selected papers such as some of Leeuwenhoek's contributions, nevertheless, were published in Latin, which was still the language of the scholarly world. This was also so when papers were submitted in Latin, as was the case even with some of the English authors, particularly in mathematics or astronomy. It also occurred when the editor particularly wanted to address a non-English-speaking audience, as Oldenburg did in issue 106 (September 21, 1674), when he comments on the Latin Version made by C. S. of the *Philosophical Transactions* of 1665, 1666, 1667, 1668, printed at Amsterdam by Henry and Theodor Boom: "Which reflextions since they chiefly concern Forreignors, it was thought most proper to give them in Latin."[77] This practice persisted through the eighteenth century. In volume 38 (1733), for instance, the contributions in Latin were published as submitted, whereas all others not in English were translated. Nevertheless, Oldenburg received many complaints from his foreign readers about the inaccessibility of the contributions in English, among them the plea made by Hevelius in his letter of November 19, 1669.[78] Even by the time the first volume of a Latin translation appeared, Erasmus Bartholin is still expressing "a great longing for the Latin edition of your *Philosophical Transactions* in order that everyone may read them."[79]

The story of the Latin edition of the early volumes is a complex one and not easy to reconstruct. Oldenburg early expressed his intentions to publish *Philosophical Transactions* in Latin as well as English. Robert Moray writes to Christiaan Huygens on February 3, 1665, that Oldenburg planned to publish a Latin version every three months.[80] On December 3, 1667, Oldenburg writes to Boyle that he has "discussed with M. Martyn about principall matters of ye *Transactions* in Latin with a small letter, for a pocket, yt ye Dutch may not be able to print upon him with advantage. This he seems to like well off, and ready to undertake. I want for Sr. R. Moray to deal with him about ye consideration I think, I deserve for ye labor, ye version will cost me."[81]

The project never came to fruition. Oldenburg was a skilled and busy translator, particularly of Boyle's works from English into Latin, and may have been reluctant to devote time to a task that might prove unremunerative. Nevertheless, he writes to Wallis on December 24, 1667, that he is hoping to issue a Latin translation by Whitsuntide, some six months hence, in order that it "be kept from usurpation."[82] Rumors of a Latin edition on the Continent were apparently rife, because on November 26, 1668, Wallis asks Oldenburg to forward the Latin translation of his paper that Oldenburg asked him to prepare "to those who are putting ye *Transactions* into Latine in Holland."[83] On November 15, 1670, Oldenburg is still insisting "that I am translating the *Transactions* into Latin,"[84] but less than a month later, in his letter to Leibniz on December 8, 1670, he informs him that a Latin translation is being printed at Hamburg and suggests it may serve him better than the English edition.[85] As late as January 30, 1676–1677, Wallis is still encouraging Oldenburg to prepare his own translation. "I find, you do ever & anon find fault with some imperfect Translations of your Transactions abroad: & seem to promise a more perfect yourself. I think it not amiss if you did so do."[86] Can this be construed as a suggestion to stop complaining and to remedy the situation by doing the work himself?

The first we hear of a Latin translation on the Continent is in a letter of Colepresse to Oldenburg on November 10, 1668, in which he informs him: "Tis said, a Professor here, welwisher to our Nation, & Philosopher, is now translating your *Transactions* [underlined in text] into Latine, resolving to alter ye Method somewhat by reducing ye severall accounts belonging to ye same subject in to one place."[87] This interesting proposal presages the abridgments that appeared much later and were arranged in subject order. It does not seem to have materialized. The professor is later identified in one of Colepresse's letters as one Van Horne who wrote a book on reproduction that also seems not to have been published.[88] We hear next in a letter from Vogel to Oldenburg dated February 1670–1671 that "the printing of the English *Transactions* has begun at Copenhagen by Daniel Pauli, translated by a Scotsman, who was, it seems to me, ill suited [to the task]. As I hear, he has already resigned the employment, perhaps because he does not understand everything equally well. I send the first sheet. The printing has not yet proceeded beyond four sheets. They intend to make a start with the year 1669, because they have heard that you intend to take up the same task."[89] On April 4, 1671, he refers to the "incompetent translator" as Sterpin and says that, apparently acting on Oldenburg's advice, he has asked the publisher, Pauli, not to continue. He also informs Oldenburg that "the *Transactions* for 1669 are all printed, but I hope he will not go on with the rest." He adds that "a certain other learned German has offered his services to Schulz in turning them into Latin; he will send them to you for approval before they are printed; if the bargain is made."[90] Vogel may not be an accurate informer, but it seems very likely that the "Scotsman" and Sterpin are one, because on April 4, 1671, Vogel is informing him that "I had indeed seen that Sterpin was an incompetent translator. But I had not expected such enormous errors from him. As soon as I received your letter I took care to let Pauli the publisher know your opinion of this translator. And I had already warned him previously to leave off the work, The *Transactions* for 1669 are all printed, but I hope he will not go on with the rest."[91]

Sterpin wrote to Oldenburg from Copenhagen on February 24, 1670–1671, expressing his intention to continue the translation despite the discouraging reception they received. In the letter he identifies himself as "being borne in France, that Contrey, England and Dennemark have almost been equall sharers in it, but oweing most of my education to England."[92] Oldenburg writes to Vogel on April 17, 1671, that he has written Sterpin to forbid him to continue with the translation.[93] Oldenburg indicates in his cautionary notice inserted in *Philosophical Transactions* for September 18, 1671: "An Advertisement necessary to be given to the Readers of the Latin Version, made by Mr. Sterpin at Copenhagen, of the *Philosophical Transactions* for A. 1669; printed at Frankfurt on Main by Dan. Pauli, A. 1671."[94]

Oldenburg was not any happier with the second translator, Christopher Sand, who is described as a "radical theologian."[95] He was born in Königsberg in 1644, lived in Hamburg for some time, and died in Holland in 1689. Sand did write to Oldenburg about the translation, but only when it was well under way. When Oldenburg complained of the ascription of *Philosophical Transactions* to the Royal Society on the title page, Sand disclaimed any responsibility, saying: "I am as it were a bookseller's hack;

what I do I do for him."[96] Sand wrote to Oldenburg several times for advice on the translation (in fact, he seems to have made a nuisance of himself), but it apparently had no effect on the course of his work. As nearly as I can reconstruct the sequence of events from the correspondence and bibliographic citations, the booksellers Henry and Theodor Boom printed Sand's translation of the years 1666–1667 in Amsterdam in 1672, following with the translation of 1665 and 1668 in 1674, when the Sterpin volume that had originally been reprinted by them in 1671 was reprinted again to complete the series. Amsterdam imprints as late as 1681 covering the year 1670 have also been cited,[97] as well as a Frankfurt printing of the 1669 volume in 1671 and an edition in Rouen in 1672. In addition, there was a one-volume quarto edition of the years 1665–1669 in Leipzig in 1675. This, curiously enough, in view of all the clamor for a Latin edition, seemed to mark the end of the efforts to continue to issue *Philosophical Transactions* in Latin.

Oldenburg was also frequently entreated to produce a French version of *Philosophical Transactions.* Denis writes on June 22, 1668, complaining of his difficulties with the English edition:

> I have often wished that instead of the copies in English which you send to France you would send just one in French. I would gladly have had it printed at my expense both for your own reputation and the gratification of an infinite number of the inquisitive, who would be delighted to be able to read and understand them by themselves, instead of which there are only three or four who see them. There would be more reason to pay in France for the translation of your *Transactions* than there is to pay for the translation of the French journal in Italy, Holland and Germany.[98]

Denis writes again on February 28, 1668–1669, expressing his gratification about the forthcoming Latin translation but still hoping for a French one.[99] There were apparently several attempts to introduce French translations. Collins writes to Vernor on April 4, 1671, that "Father Bertit writes that they are translated into French, and that he hath read them, but by reason of his inquiries after a translation, I conceive they are not printed."[100] Harcourt Brown says that the Académie des Sciences had a translation made of *Philosophical Transactions* as they appeared in London,[101] and refers to the help given the Académie de Physique de Caen for the translation of *Transactions Philosophiques* by Jacques Savory, a "poète latin de Caen,"[102] but none of these seem to have appeared in print. The editors of the Oldenburg correspondence refer in their notes to a French edition "prepared in France about this time, which was offered for sale by the Libraire Thomas Scheler of Paris in 1965,[103] but this may also be a reference to a manuscript version. A French translation, first by Frangois de Bremond and continued by Pierre Demours, covering the years 1731 to 1736 finally did appear in Paris between 1738 and 1760,[104] but French readers for the most part had to depend on other means of access to *Philosophical Transactions.*

There was also talk from time to time of an Italian edition. For example, Dodington writes to Oldenburg on April 11, 1672, from Venice about "a worthy inquisitive studious friend of mine who is resolved to translate & print them at Padua incouraged to it by many hundreds who will contribute to ye charge."[105] Weld refers to several volumes of *Philosophical Transactions* that were translated into Italian and published at Naples, but he is probably referring to the Italian translation of the Lowthorp abridgment covering the first fifty volumes of *Philosophical Transactions* issued in Naples in five volumes between 1729 and 1734.[106]

REPRINTS AND ABRIDGMENTS

There were numerous other ways in which access was provided to *Philosophical Transactions* beyond the limited efforts to translate it into other languages. There seems to have been only one attempt to reprint it in English beyond whatever reprints may have been produced for the individual issues and the facsimile and microform editions produced in the twentieth century. It was published in English—strangely enough, in Wittenberg, Germany, by four university professors—to try to compensate for the difficulty and the cost of procuring the original edition through the usual channels.[107] The reprints appeared in quarto with Latin prefaces beginning with volume 47 (1753) printed in 1768 and ending with a seventh volume for 1763 printed in 1774.[108] The "histories" of the Royal Society produced by Sprat and Birch, which to some extent tried to fill the role provided by the periodic *Histoires* of some of the Continental societies, filled the gap somewhat by printing accounts of papers that had not appeared in *Philosophical Transactions.* Thomson's "history" is essentially not a history but a classified arrangement of the articles in *Philosophical Transactions* to the end of 1800, with references to the original edition, and therefore serves to some extent as an index.

These efforts could do little to increase access to *Philosophical Transactions.* That burden was taken up by the large number of abridgments in several languages, and in the anthologies of selected articles listed in the checklist that follows. It also provides a measure of the impact made by *Philosophical Transactions* and the degree to which its contents were disseminated in the eighteenth century. "How few will read or purchase forty-four large volumes of the *Transactions* of the Royal Society, which, in abridgment, are generally read to the improvement of philosophy," Samuel Johnson wrote in 1739, referring to another addition to the Lowthorp abridgment.[109] In addition to the general abridgments in English and other languages, there were abridgments devoted to special subjects. This was particularly true in the field of medicine because *Philosophical Transactions* devoted a considerable amount of space to medical subjects.

The checklist does not begin to measure the full extent of the penetration of *Philosophical Transactions* because it does not reveal many of abstracts, reviews, and reprints of articles that appeared in a large number of periodicals both in Great Britain and on the Continent.[110] *Philosophical Transactions* was reviewed regularly by the leading periodicals of the times, including *Journal des Sçavans, Acta Eruditorum,* and *Giornale de' Litterati* of Nazari, which used it as its primary source of news about science. Nazari's eagerness to procure *Philosophical Transactions* is reflected in his letters to Oldenburg, outlining his elaborate arrangements to have them sent to him and expressing his dismay when he did not receive them.[111] When he was unable to get them, he took the articles second hand from *Journal des Sçavans.* A further indication of the extent to which *Philosophical Transactions* was abstracted and reviewed is an index to the *Journal des Sçavans* published in 1753 that lists 18 columns of entries covering the journal from 1665 to 1735.[112] Another indicator is Callisen's compendious bibliography, which in volume 25 carries many pages of references to journals that carried reviews of *Philosophical Transactions,* including *Commercium Litterarium ad Rei Medicae et Scientiae Naturalis* (Nuremberg, 1731–1745), *Commentarii Lipsienses Litterarii* (Leipzig, 1753–1757), *Göttingsche Anzeigen von gelehrten Sachen* (Göttingen, 1753–1801), and *Journal de Médicine* (Paris, 1754–1794).[113]

THE CHECKLIST

General

1. *Philosophical Transactions, Giving Some Accompt of the Present Undertakings, Studies, and Labours of the Ingenious in Many Considerable Parts of the World.* Ed: Henry Oldenburg & others London, 1665–, vol. 1–. Title varies. (The following data are derived from Callisen and other lists in the literature, but many print variants are known to exist.)[114]

Editors

Vols. 1–11: Henry Oldenburg (nos. 1–136), 1665–1667; vol. 12: Henry Oldenburg and Nehemiah Grew (137–142), 1678–1679; vols. 13–14: Robert Plot (143–166), 1682/1683–1684; vol. 15: William Musgrave (167–178), 1684–1685; vol. 16: Edmond Halley (179–191), 1686–1687; vols. 17–18: Richard Waller (192–214), 1691–1694; vols. 19–28: Hans Sloane (215–337), 1695–1713; vols.

29–30: Edmond Halley (338–363), 1714–1719; vols. 31–34: James Jurin (364–398), 1720–1727; vol. 35: William Rutty (399–406), 1727–1728; vols. 36–46: Cromwell Mortimer (407–497), 1729–1750. Beginning with vol. 47 the responsibility for *Transactions* was assigned to a committee.

Printers and Publishers

Vols. 1–13 (1665–1683) John Martyn (with James Allestry in some of the earlier issues); vol. 14 (1684) Moses Pitt and Sam Smith; vol. 15 (1685) Sam Smith and Henry Clements; vol. 16 (1686–1687) Joseph Streater and Sam Smith; vols. 17–24 (1693–1705) Sam Smith and B. Walford; vol. 25 (1706–1707) B. Walford; vol. 26 (1708–1709) Henry Clements; vols. 27–28 (1710–1713) Henry Clements, W. Innys, and D. Brown; vols. 29–30 (1714–1719) W. Innys; vols. 31–36 (1720–1730) W. and I. Innys; vols. 37–38 (1731–1734) W. Innys and R. Manby; vols. 39–42 (1735–1743) T. Woodward and C. Davis; vols. 43–48 (1744–1754) C. Davis; vols. 49–58 (1755–1768) L. Davis and C. Reymers; vols. 59–64 (1769–1774) Lockyer Davis; vols. 65–66 (1775–1776) W. Bowyer, J. Nichols, and Lockyer Davis; vols. 67–69 (1777–1779) K. Nichols and Lockyer Davis; vols. 70–(90) [volume numbers were not assigned after vol. 81 (1791)] (1780–1800) J. Nichols, Lockyer Davis, and P. Elmsley. (Not all the printers are listed because they are not always given in imprints or colophons.) (From Callisen 23:808.)

Supplements

Issues 61 (July 1670) and 62 (August 1670) (Katzen). These appear not to be "supplements," but responses to issues raised in the respective numbers: for example, the "supplement" to issue 61 was not published until 1678, although it is sometimes found bound after that issue. It is a commentary on the issue by a reader who regards his priorities as being usurped.

Two series that are sometimes listed as parts of *Philosophical Transactions* can also be considered separate publications. *Meteorological Journal* began by being paged in as an integral part of *Philosophical Transactions* in vol. 65, part 1 in 1775, but became a separately paged supplement with vol. 80 in 1790. *Croonian Lectures* had a similar history, although only 8 were published in association with the *Philosophical Transactions* of the 40—some recorded, all except the first, which was paged into vol. 41, appeared as separately paged supplements.

2. *Philosophical Collections, Containing an Account Of Such Physical, Anatomical, Chymical, Mechanical, Astronomical, Optical, or Other Mathematical and Philosophical Experiments and Observations as Have Lately Come to the Publishers' Hands, as Also an Account of Some Books of This Kind Lately Published.* Ed.: Robert Hooke, London, John Martyn, 1679–1682. Nos. 1–7. Im-

print varies, no. 2, London, M. Pitt, nos. 3–7, London, R. Chiswell. Published in place of *Philosophical Transactions,* of which no volumes were issued from 1679 to 1682.

3. *Philosophical Transactions Giving Some Account of the Present Undertakings, Reprinted According to the London Edition.* Eds.: Georg A. Langguth, Georg R. Böhmer, Johann D. Titius, Johannes E. Zeiher. Wittenberg, Carl C. Duerr, 1768–1774, 11 vols. A reprint of vols. 47–57 (1751–1757) with Latin prefaces. Two volumes appeared each year.

Indexes

4. *General Index or Alphabetical Table to All the Philosophical Transactions from the Beginning to July 1677. Also a Catalogue of the Books Mentioned and Abbreviated in the Transactions Digested Alphabetically.* London, John Martyn, 1678. Bound with vol. 12. A *General Index* is also cited for vols. 12–17, bound with vol. 17.

5. *General Index to the Philosophical Transactions. From the 1st to the End of the 70th Volume (1665–1780).* Ed.: Paul Henry Maty. London, Davis & Elmsley, 1787. Followed by: *A Continuation of the Alphabetical Index of the Matter Contained in the Philosophical Transactions of the Royal Society of London. From vol. 71 to vol. 110, 1781–1820 inclusive.* London: W. Bulmer and W. Nicol, 1821.

6. *An Index to the Anatomical, Medical, Chirurgical and Physiological Papers, Contained in the Transactions of the R. Society of London, from the Commencement of That Work to the End of the Year 1813, Chronologically Arranged.* Ed.: James Briggs. London, Callow, 1814. Another edition by Briggs covering the period 1665–1817 (1818) has been cited.[115]

7. *Table des Mémoires Imprimés Dans les Transactions Philosophiques de la Société Royale de Londres depuis 1665 Jusques en 1735 Rangée par Ordre Chronologique, Par Ordre des Matières, & Par Noms d'Auteurs.* Ed.: François de Bremond. Paris, Piget, 1737. 2d ed. Paris, 1739.

Abridgments

Dutch

8. *Genees-en Heelkundige Proeven en Aanmerkingen, Nervat in de Philosophical Transactions, van Derzelver Eerste Uitgave, Tot den Tegenwoordigen Lyd.* 1745? Abstracts in translations from vols. 1–9, 1665–1674, of *Philosophical Transactions,* including some of the selections and remarks taken from Mihles's *Medical Essays and Observations* (London, 1745). (Checklist no. 11.)

9. *Natuurkundige Aanmerkungen, Waarneemingen en Ondervindelingen: Getrokken Uit & Philosophical Transactions en Uit Het Engelsch Vertaalt.* Ed.: Pieter Le Clercq. Amsterdam, 1735. 2 vols.

10. *Verhandelingen der Natuur-Genees-, Heel-en Ontleedkundige Berigten en Waarneemungen Getrokken Uit de Philosophical Transactions.* Amsterdam. 4 vols. (No other references to this item were found.)[116]

English

11. *Medical Essays and Observations Relating to the Practice of Physic, and Surgery Abridg'd From the Philosophical Transactions From Their First Publication Down to the Present Time.* Ed.: S. Mihles. London, Birthand, 1745. 2 vols. The articles are arranged in the order in which they appeared: vol. 2 begins with no. 323 and ends with no. 441; each volume has a subject index. Another edition in Lübeck has been cited.[117]

12. *Memoirs of the Royal Society, Being a New Abridgement of the Philosophical Transactions the Whole Carefully Abrdig'd From the Originals, and the Order of Time Regularly Observ'd, With a Translation of the Latin Tracts from 1665 to 1735.* Ed.: Benjamin Baddam. London, G. Smith and T. Cooper, 1738–1741. 10 vols. Title and imprint varies. 2d ed., 1665–1740, 10 vols. London, J. Noursc, 1745.

13. *Miscellanea Curiosa Being (Containing) a Collection of Some of the Principal Phenomena in Nature, Accounted for by the Greatest Philosophers of This Age; Together with Several (Being the Most Valuable) Discourses Read Before (Read and Delivered to) the Royal Society (as Also a Collection of Curious Travels, Voyages, Antiquities, and Natural Histories of Countries, Presented to the Same Society).* Ed.: W. Derham. London, Jeffery Wale and John Senex, 1705–1707, 3 vols. Title varies. 2d ed., London, James & John Knapton, 1708–1727. Adding: R. Mead's "A discourse of the influence of the sun and moon" and Fontenelle's preface to his "Of the usefulness of mathematical learning." 3d ed. of vol. 1 was published in 1726.

14. *Philosophical Transactions and Collections Abridged and Disposed Under General Heads.* London, Mathews and others, 1705–1781. 10 vols. Title, editors, and publishers vary. Other eds. of earlier vols. were published to add to later vols. Each series is in 4 parts: 1. Mathematical papers, 2. Physical papers, 3. Anatomical and medical papers, 4. Philological (historical) papers. Vols. 1–3 (1665–1700) by John Lowthorp, London 1705; vols. 4–5 (1700–1720) by Henry Jones, London, H. Parker, 1721; vol. 6 (1720–1732) by Andrew Reid and John Gray, London, W. Innys, 1733; vol. 7, by Eames and Martyn, London, Innys, 1734; vols. 8–10 by John Martyn, 1747–1756. Many other editions are cited.

15. *Philosophical Transactions of the Royal Society of London, From their Commencement, in 1665 to the Year 1800, Abridged. With Notes and Biographic Illustrations.* Eds.: Charles Hutton, George Shaw, Richard Pearson. London, C. & R. Baldwin, 1809. 18 vols.: 1, 1665–1672; 2, 1672–1683; 3, 1683–1694; 4, 1694–1702; 5, 1703–1712; 6, 1713–1723; 7, 1724–1734; 8, 1735–1743; 9, 1744–1750; 10, 1750–1755; 11, 1755–1763; 12, 1763–1769; 13,

1770–1776; 14, 1776–1780; 15, 1781–1785; 16, 1785–1790; 17, 1791–1796; 18, 1796–1800. General index at the end of vol. 18.

French

16. *Abrégé des Transactions Philosophiques de la Société de Londres, Ouvrage tradult de l'Anglais et Rédégiée par M. Gibelin.* Paris, Buisson, 1787–1791. 14 vols. 1e Partie: Histoire naturelle, 2 vols. 1787; 2e Partie: Botaniques, 2 vols., 1790; 3e Partie; Météorologie, 15 vols., 1787; 4e Partie: Physique expérimentale par M. Reynier, 2 vols., 1790; 5e Partie: Chimie, par M. Pinel, 1 vol., 1791; 6e Partie: Anatomie et physique animale, par M. Pinel, 1 vol., 1790; 7e Partie: Médecine et chirurgie, par M. Pinel, 1 vol., 1791; 8e Partie: Matière médicale et pharmacie, T. 1, par M. Willemet et Bosquillon, 1789, T. 2, par M. Bosquillon et Pinel; 10e Partie: Mélanges, observations, voyages, par M. M. A. L. Millin de Grandmaison, 1 vol., 1790; 11 et 12e Parties: Antiquités et beaux arts, par A. L. Millin de Grandmaison, 2 vols., 1789–1790. 9e Partie, apparently not published.

17. *Mémoires de Physique Pure. Sans Mathématiques, de Toutes les Académies de Science Rassemblés en un Seul Corps.* Lausanne, Antoine Chapius, 1754. Also includes *Saggi degli Accademia del Cimento,* 1665–1667, and *Philosophical Transactions* covered selectively. Subscriptions were solicited for a second volume that does not seem to have appeared. The series was projected to cover twenty subject areas, beginning with agriculture and ending with "physicomathematiques."

18. *Collection Académique; Composée des Mémoires, Actes ou Journaux des Plus Célebres Académies et Sociétés . . . Traduit en Français et Mis en Order par une Société de Gens de Lettres.* Ed.: Jean Berryat. Dijon & Paris, 1755–1780. 17 vols. Divided in two parts: Partie Française, 1754–1784, 7 vols., and Partie Étrangere, 1752–1768. Vols. 1–3 as *Récueil des Mémoires ou Collection des Pièces Académique.* Other editions are cited, e.g., a Partie Étrangeres in 13 vols., Paris & Dijon, 1755–1780.

German

19. *Auserlesene Abhandlungen praktischen und chirurgischen Inhalts aus den Londoner philosophischen Transaktionen und Sammlungen der Jahre 1699 bis 1720.* Ed.: Nathaniel Gottfried Leske. Lübeck & Leipzig, Donatius, 1774–1778, 5 vols. According to Callisen (23:810:3), Leske used the Lowthorp abridgment, particularly the 2nd ed. of vols. 4–5 (1700–1720) by Henry Jones. Bd. 1 includes 1699–1720 (1774), Bd. 2: 1720–1744 (1775), Bd. 3: 1744–1750 (1776), Bd. 4: 1751–1757 (1777), Bd. 5: 1758–1765 (1778).

20. *Neue Anmerkungen über alle Theile der Naturlehre aus den englischen Transactionen zusammengezogen.* Copenhagen, 1753–1756. 3 vols.[118]

Italian

21. *Compendio delle Transazioni Filosofiche della Società Reale di Londra, Opera Compilata, Divisa per Materie, ed Illustrata del Signor Gibelin . . . e Recata in Italiano da una Società di Dotte Persone con Nuovo Illustrazione, e Tavole.* Venice, Antonio F. Stella, 1793–1798, 20 vols.: 1–3. Storia naturale; 4–6: Botanica; 7–9: Fisica sperimentale; 10–11: Notomia e fisica animale; 12–13: Materia medica e farmacia; 14–15: Medicina e chirurgia; 16: Chimica; 17–19: Antichita, belle arti, invenzioni e macchine; 20: Miscellanee. Vols. 5–16 have the imprint of the Tipografia Pepoliana, vols. 17–20 of Antonio Curti. An Italian translation of Gibelin's French abridgment.

22. *Saggio delle Transazioni Filosophiche della Società Regio, Compendiate da Giovanni Lowthorp & Benianinino Mottes, Tradotte dall' Inglese nell' Idioma Toscana dal Tommaso Dereham.* Naples, Moscheni, 1729–1734. 5 vols. A translation of the Lowthorp abridgment covering the first 50 vols. of *Philosophical Transactions.*

Translations

French

23. *Transactions Philosophiques de la Socété Royale de Londres.* Paris, Piget, 1738–1760. 8 vols. Covers 1731–1744 of *Philosophical Transactions,* 1731–1736, translated by François de Bremond and 1737–1744, by Pierre Demours. Several editions of some of the volumes appear to have been issued.[119]

German

24. *Abhandlungen zur Naturgeschichte, Physik und Oekonomie aus der Philosophischen Transaktionen und Sammlungen mit einigen Anmerckungen übersetzt.* Ed.: Nathaniel Gottfried Leske, Leipzig, Weygand, 1779–1780, 1 vol., T. 1 & 2. Covers the first 14 vols. of *Transactions* to 1693, and omits those articles that were extracted from such journals as *Journal des Sçavans, Giornale di Venezia,* etc., and translated into English. The preface to part 2 promised a second volume that does not appear to have been published.

Latin

25. *Acta Philosophica, Societatis Regia in Anglia, Auctore Henrico Oldenburgio Conscripta.* Translators: Johann Sterpin and Christoph Sand. Amsterdam, Hen. & Theo. Boom, 1665–1681, 7 vols. Variants are noted that indicate that there were several printings and editions (Andrade), a Frankfurt edition in octavo in 1671 (4th ed., 1674), and a Leipzig edition

in quarto in 1675. William Osler (Bibliotheca Osleriana, Oxford, Clarendon Press, 1929) shows vol. 2 (March 1664–1665 to February 1665–1666) of the Amsterdam edition, 1674, and vol. 2 (March 1665/66–February 1666), 1672, etc. Vol. 5 (1670) 1680 was translated by Johann Sterpin.

NOTES

1. The full original title was *Philosophical Transactions Giving some Accompt of the Present Undertakings, Studies and Labours of the Ingenious in Many Considerable Parts of the World*, a title that indicates that Oldenburg intended to report more than just the activities of the Royal Society.

2. Henery Oldenburg. *Correspondence*. Edited and translated by A. Rupert Hall and Marie Boas Hall. (Madison: University of Wisconsin Press, vols. 1–9, 1966–1973); (London: Mansell, vols. 10–11, 1975–1977); (London: Taylor and Francis, vols. 12–13, 1986).

3. A. N. da C. Andrade, "The Birth and Early Days of the *Philosophical Transactions*," *Notes and Records, Royal Society of London* 20 (1965): 9–27; P. George, "The Scientific Movement and the Development of Chemistry in England, as Seen in Papers published in the *Philosophical Transactions* from 1664–1665 until 1750," *Annals of Science* 8 (1952) 1: 302–322; A. R. Hall and M. B. Hall, "Some Hitherto Unknown Facts about the Private Career of Oldenburg," *Notes and Records* 18 (1963): 94–103; and "Further Notes on Henry Oldenburg," *Notes and Records* 23 (1968): 33–42; M. B. Hall, "Oldenburg and the Art of Scientific Communication," *British Journal of the History of Science* 2 (1965): 277–290; R. K. Merton, "Science, Technology, and Society in Seventeenth-Century England," *Osiris* 4 (1938): 360–632; G. R. Potter, "The Significance to the History of English Natural History of John Hill's *Review of the Works of the Royal Society*," University of California, Department of English, Essays, and Studies 14 (1943): 157–180; C. A. Rivington, "Early Printers to the Royal Society," *Notes and Records* 39 (1984): 1–21.

4. H. Brown, *Scientific Organizations in Seventeenth-Century France* (Baltimore: Williams and Wilkens, 1934), 185

5. Andrade, "The Birth and Early Days," p. 12; Brown, *Scientific Organizations*, p. 187.

6. H. Brown, "History and the Learned Journal," *Journal of the History of Ideas* 33 (1972): 365–78 (p. 367).

7. Oldenburg, *Correspondence* 11, pp. 319–25.

8. W. Graham, *English Literary Periodicals* (New York: Thomas Nelson, 1930), p. 3.

9. D. Mornet, *Les Sciences de la Nature en France an xviiie siècle* (Paris: Armand Colin, 1911).

10. Brown, *Scientific Organizations*, p. 198.

11. C. Bellanger, *Histoire Générale de la Presse Française*, vol. 1 (Paris: Presses Universitaire de France, 1969), p. 139.

12. C. R. Weld, *A History of the Royal Society*, vol. 2 (London: J. W. Parker, 1848), p. 564.

13. R. Hahn, *The Anatomy of a Scientific Institution: The Paris Académie of Sciences, 1665–1803* (Berkeley: University of California Press, 1971), p. 64.

14. C. B. O'Keefe, *Contemporary Reactions to the Enlightenment (1728–1762)* (Geneva: Libraire Slatkine, 1974), p. 46.

15. *Phil. Trans.* 1: 213–14.

16. Oldenburg, *Correspondence* 4, p. 72.

17. Oldenburg, *Correspondence* 10, p. 570.

18. S. B. Barnes, "The Beginnings of Learned Journalism, 1665–1730." Ph.D. dissertation, Cornell University, 1934, p. 133.

19. Weld, *A History*, p. 325.

20. T. Sprat, *The History of the Royal Society of London for the Improving of Natural Knowledge* (London: Martyn, 1667); T. Birch, *The History of the Royal Society of London for the Improving of Natural Knowledge*, 4 vols. (London: A. Millar, 1756–1757); T. Thomson, *History of the Royal Society from Its Institution to the End of the Eighteenth Century* (London: Baldwin, 1812). (This work is perhaps better characterized as an index to *Transactions*.)

21. Thomson, *History of the Royal Society*, p. 12.

22. *The Records of the Royal Society for the Promotion of Natural Knowledge*, 4th ed. (London: n.p., 1940), p. 93.

23. Advertisement to twelfth issue of *Transactions*.

24. Weld, *A History*, p. 327.

25. R. K. Bluhm, "Henry Oldenburg, F.R.S. (c. 1615–1677)," *Notes and Records* 15 (1960): 183–197, (p. 190).

26. Andrade, "The Birth and Early Days," p. 16.

27. Oldenburg, *Correspondence* 4, p. 59.

28. Oldenburg, *Correspondence* 4, pp. 111, 230.

29. Rivington, "Early Printers," p. 5.

30. Thomson, *History of the Royal Society*, p. 7.

31. Weld, *A History*, p. 326.

32. Barnes, "The Beginnings of Learned Journalism," p. 133.

33. Merton, "Science, Technology, and Society," p. 411.

34. *The Records of the Royal Society*, p. 93.

35. L. Rostenberg, "John Martyn, Printer to the Royal Society," *Papers of the Bibliographical Society of America* 46 (1952): 1–32.

36. A. D. Farr, "The First Human Blood Transfusion," *Medical History* 24 (1980): 143–162 (p. 146).

37. It does not seem to be quite as rare as suggested (Farr, "The First Human Blood," p. 148). In eight copies examined, one had the July 22 issue only (University of Chicago), two had both issues bound in sequence, although the page numbers of the first issue duplicate the second (New York University, Cleveland Medical Library), and only the revised issue in the others (American Museum of Natural History, Engineering Society of New York, University of Michigan, New York Public Library, and the University of Texas at Austin).

38. R. P. Bond, *Studies on the Early English Periodical* (Chapel Hill: University of North Carolina Press, 1957); W. B. Todd, "Early Edition of the *Tatler*," *Studies in Bibliography* 15 (1962): 121–33; and "A Bibliographical Account of the *Gentleman's Magazine*, 1731–1754," *Studies in Bibliography* 18 (1965): 81–109.

39. This may represent a later printing or is more likely a misprint.
40. Most frequently spelled "Martyn."
41. There was the same distribution of variants for issues 19 and 20 in all of the copies examined, except that they were reversed in one case (University of Michigan).
42. Rivington, "Early Printers," p. 4.
43. *National Union Catalog, Pre-1956 Imprints,* vol. 508 (London: Mansell, 1968–1981), p. 131.
44. Oldenburg, *Correspondence* 5, p. xxvi.
45. It has been indicated that the page numbered 852 in issue 41 (November 16, 1668) should be 825 (Oldenburg, *Correspondence* 5, p. 276). It is nevertheless correctly numbered in at least one copy (University of Michigan).
46. A. C. P. Callisen, *Medicinisches Schriftsteller-Lexikon der jetzt lebenden Aerzte, Wundaerzte, Geburtshelfer, Apotheker, und Naturforscher aller gebildeten Völker,* 33 vols. (Copenhagen: B. de Graaf, 1830–1845), vol. 25, no. 80; J. Dryander, *Catalogus Bibliotecae Historico-naturalis Josephi Banks,* vol. 1 (London: Blumer, 1708), pp. 1–4.
47. Oldenburg, *Correspondence* 11, p. 248.
48. Professor William B. Todd of the University of Texas at Austin has pointed out other printing variations in the text of the variant issues he examined with me, which indicate they were resettings of the texts (personal communication).
49. Birch, *The History of the Royal Society,* vol. 2, p. 18.
50. Oldenburg, *Correspondence* 8, p. 32.
51. When the Royal Society took over responsibility for *Philosophical Transactions* in 1750, it began to publish in half volumes, two parts to a volume, at first one part for each year, but from 1762 two half volumes each year; M. F. Katzen, "The Changing Appearance of Research Journals: An Analysis and Case Study," in A. J. Meadows (ed.), *Development of Science Publishing in Europe* (Amsterdam: Elsevier, 1980), p. 186.
52. *Phil. Trans.* 7: 5050.
53. Oldenburg, *Correspondence* 10, p. 493.
54. Oldenburg, *Correspondence* 8, p. 318.
55. D. Stimson, *Scientists and Amateurs, a History of the Royal Society* (New York: Henry Schuman, 1948), p. 100.
56. Thomson, *The History of the Royal Society,* p. 7.
57. Weld, *A History,* p. 521.
58. Spelled "Davis" in the colophons.
59. The total membership when *Philosophical Transactions* was proposed in 1664 was 148 and did not exceed 220 any time before 1700, even including "Persons of other nations" (M. Hunter, *The Royal Society and Its Fellows, 1660–1700* [Chalfont St. Giles: British Society for the History of Science, 1982], p. 112).
60. A. R. Hall and M. B. Hall, "The First Human Blood Transfusion: Priority Disputes," *Medical History* 24 (1980): 461–65.
61. Weld, *A History,* p. 522. The number of fellows was about 300 in 1740 and did not exceed 380 until 1770. See also H. Lyons, *The Royal Society 1660–1940: A History of Its Administration under Its Charters* (Cambridge: Cambridge University Press, 1944), p. 341.
62. Oldenburg, *Correspondence* 3, p. 272.

63. Oldenburg, *Correspondence* 3, p. 333.
64. Oldenburg, *Correspondence* 3, p. 546.
65. Oldenburg, *Correspondence* 5, p. 80.
66. Oldenburg, *Correspondence* 7, p. 29.
67. Oldenburg, *Correspondence* 6, p. 171.
68. Oldenburg, *Correspondence* 8, p. 12.
69. Oldenburg, *Correspondence* 10, p. 295.
70. Oldenburg, *Correspondence* 11, p. 191.
71. Oldenburg, *Correspondence* 11, p. 14.
72. Oldenburg, *Correspondence* 8, p. 12.
73. E. B. Barnes, *The International Exchange of Knowledge in Western Europe 1680–1689* (Chicago: University of Chicago, Graduate Library School, 1947).
74. C. S. Duncan, *The New Science and English Literature in the Classical Period* (Menasha, Wis.: George Banta, 1913), p. 14.
75. Oldenburg, *Correspondence* 2, p. 646.
76. Oldenburg, *Correspondence* 5, p. 470.
77. *Phil. Trans.* 8: 141.
78. Oldenburg, *Correspondence* 5, p. 187.
79. Oldenburg, *Correspondence* 9, p. 403.
80. M. B. Hall, "Oldenburg, the *Philosophical Transactions*, and Technology." In J. G. Burke (ed.), *The Uses of Science in the Age of Newton* (Berkeley: University of California Press, 1983), pp. 21–47.
81. Oldenburg, *Correspondence* 4, p. 6.
82. Oldenburg, *Correspondence* 4, p. 83.
83. Oldenburg, *Correspondence* 5, p. 204.
84. Oldenburg, *Correspondence* 7, p. 264.
85. Oldenburg, *Correspondence* 7, p. 313.
86. Oldenburg, *Correspondence* 8, p. 200.
87. Oldenburg, *Correspondence* 5 p. 141.
88. Oldenburg, *Correspondence* 5, p. 248.
89. Oldenburg, *Correspondence* 7, p. 456.
90. Oldenburg, *Correspondence* 7, p. 547.
91. Oldenburg, *Correspondence* 7, p. 547.
92. Oldenburg, *Correspondence* 7, p. 468.
93. Oldenburg, *Correspondence* 7, p. 573.
94. Oldenburg, *Correspondence* 7, p. 469.
95. Oldenburg, *Correspondence* 8, p. 143.
96. Oldenburg, *Correspondence* 9, p. 57.
97. Oldenburg, *Correspondence* 3, p. 66.
98. Oldenburg, *Correspondence* 4, p. 474.
99. Oldenburg, *Correspondence* 5, p. 412.
100. S. P. Rigaud ed., *Correspondence of Scientific Men of the Seventeenth Century*, vol. 2 (Hildesheim: Georg Olms, 1965), p. 164.
101. H. Brown. *Scientific Organizations*, p. xiv.
102. H. Brown, "*L'Académie de Physique de Caen*," *Mémoires, Académie des Sciences, Arts et Belles Lettrres de Caen* 9 (1938): 117–208 (p. 14).
103. Oldenburg, *Correspondence* 5, p. 415n.

104. Checklist no. 23, this chapter. See J. E. McClellan III, *Science Reorganized: Scientific Societies in the Eighteenth Century* (New York: Columbia University Press, 1985), p. 164–65.

105. Oldenburg, *Correspondence* 9, p. 18.

106. Checklist no. 22, this chapter; Weld, *A History*, p. 523.

107. W. Stieda, "Die Universität Wittenberg und die Londoner *Philosophical Transactions*," *Neues Archiv für Sächsiche Geschichte und Altertumskunde* 40 (1919): 138–53.

108. Checklist no. 3, this chapter.

109. A. D. Atkinson, "Dr. Johnson and the Royal Society," *Notes and Records* 10 (1953): 131–138; checklist no. 14, this chapter. The Lowthorp abridgment was apparently in short supply from time to time and there were numerous editions. The second issue of *Farmer's Magazine* in 1776 includes a long quotation from the Lowthorp abridgment because "that Book is scarce and in very few hands" (vol. 1, p. 61).

110. Oldenburg, *Correspondence* 8, p. xx.

111. Oldenburg, *Correspondence* 8, pp. 154, 229.

112. *Table Général des Matières Contenue dons le Journal des Savants* (Paris: 1753).

113. Callisen, *Lexikon*, vol. 25, p. 808.

114. Callisen, *Lexikon*, vol. 25, p. 808.

115. Callisen, *Lexikon*, vol. 3, p. 1053.

116. S. Scudder, *Catalogue of Scientific Serials of All Countries, Including the Transactions of Learned Societies in the Natural, Physical, and Mathematical Sciences: 1663–1876.* (Cambridge, Mass.: Harvard University Library, 1879; reprinted New York: Kraus, 1965), no. 20.

117. F. H. Garrison, "The Medical and Scientific Periodicals of the 17th and 18th Centuries, with a Revised Catalogue and Checklist," *Bulletin of the History of Medicine* 2 (1934): 477–503, no. 11:4:f.

118. Scudder, *Catalogue of Scientific Serials*, no. 635.

119. Scudder, *Catalogue of Scientific Serials*, no. 1597.

Chapter 11

Medical Publishing Societies in Early Modern Britain

ABSTRACT

This chapter reviews the relationship between medical journals and publishing societies in eighteenth-century Great Britain. The importance of the perception that these journals were being issued under the auspices of societies is revealed by the number of times this kind of sponsorship was invoked by new medical journals. This kind of endorsement was projected even when a society so designated exercised only a nominal responsibility for the journal and, even when, in some instances, no organization can be said to have existed.

Among the papers in the British Library of Sir Hans Sloane (president of the Royal Society of London from 1727 to 1741), there is a draft of a proposal he made in 1735 for establishing what he called a "publishing society" (Clark 1964:54). Sloane's "publishing society" does not appear to have materialized, but the term itself can be applied to some of the medical societies established in Great Britain in the last half of the eighteenth century.

This chapter reviews examples of these societies. Some of them appeared to have existed primarily for the purpose of publishing journals. This was true even when the journal's association with a society was only nominal or even when the society did not exist except in name. The illusion was created that some sort of communal judgment was being exercised over the publication's contents, even though the methods of managing this kind of institutionalized peer review were seldom articulated or, when described, often were not applied.

Published in the *Bulletin of the Medical Library Association* 82 (1994): 277–82.

Sometimes even the suggestion of the involvement of an august society could lend credence to a publication. An example is *Medicina Curiosa, or a Variety of New Communications in Physick, Chirurgery and Anatomy* (this is only a small part of the actual title), which appeared for only two issues in London in 1684. It has been called the first "serious" medical journal published in Great Britain (Johnston-Saint 1939:117). The title can be translated as "medical curiosities," an expression that reflects some of the interests of both editors and the public in that period. Examples of such "curiosities" in medicine and nature can be found in many of the learned journals of the period, including *Philosophical Transactions* and *Journal des Sçavans,* which in the seventeenth and eighteenth centuries were among the major repositories of medical articles. One of the first contributions made by the eminent Robert Boyle to *Transactions* was a report of a "very odd monstrous calf."[1] Thomas Basset, the editor of *Medicina Curiosa,* may have been seeking credibility for his medical curiosities when he pledged in the first issue, "Whatsoever I shall write I shall always submit to the London College of Physicians, whom the world will own for great and meet judges" (Johnston-Saint 1939:118).

SCOTTISH PUBLISHING SOCIETIES

Two well-known and popular series of medical journals published in Edinburgh, Scotland, beginning in 1731 are good examples in which the mantle of a society is assumed, even when the society did not exist or existed in name only. Both of these series had a great deal to do with establishing the prominence in Edinburgh as a medical educational center in the eighteenth century. The first, *Medical Essays and Observations by a Society in Edinburgh,* was edited from 1731 by Alexander Monro (1697–1767). The society under whose auspices the journal was issued was formed by a small group of professors at the Edinburgh Medical School, principally for the purpose of publishing this journal. The members met fairly regularly during the first year to consider papers read at the meetings, but after 1732, the burden fell entirely on Monro's shoulders (Emerson 1979:154). Despite the fact that an elaborate method of conducting peer review had been announced in the first issue (Kronick 1990:1321), Monro wrote more than 20 percent of the attributed papers that appeared while he was editor and is credited with having rewritten many of the contributions he received as personal letters (Shapin 1971:91).

Monro devoted several pages in the preface of the first volume of *Medical Essays* in 1731 to discussing his aspirations for the journal and the methods by which he hoped they could be achieved. His purpose, he said, was to elevate the practice of medicine in Great Britain, to provide a place for medical reports that were unique and characteristic of the region, and to provide a publication addressed specially to physicians. (Articles of medical interest could be found in publications such as *Philosophical Transactions* and *Histoire et Mémoires* of the Académie des Sciences in Paris.) The editors, Monro added, reserved the right to correct and improve copy sent to them and to return the articles to the author for review. Their purpose, he said, was not to reject any essays that did not conform to their ideas but to make them more acceptable to readers. To that end, he added a long list of criteria and instructions to authors. During the first year of the society's existence, submitted papers were read and discussed at the monthly meetings. In the following years, because of lax meeting attendance, Monro took over the complete responsibility of selecting and editing the subsequent volumes (Emerson 1979:158).

In the preface to volume 5 of *Medical Essays* (1742), Monro announced, "A society [was] being formed . . . for the Improvement of Natural Knowledge in which all the branches of medicine are included, and the members of our Society [were] being adopted into the new one." In recognition of this union with the Philosophical Society of Edinburgh, the fifth volume was issued in two sections, one devoted to medical papers and the other to papers derived from the new society. *Medical Essays* was continued after the sixth volume by *Essays and Observations, Physical and Literary* from 1754 to 1765, under the auspices of the Philosophical Society but still under Monro's direction. Over three-quarters of the papers in the first volume were by physicians (Shapin 1974).

Monro's journals were continued into the nineteenth century under various titles by the equally celebrated Andrew Duncan and were frequently reprinted and translated. The medical papers, still ostensibly under the auspices of the Philosophical Society, were continued in the second series under the title *Medical and Philosophical Commentaries by a Society in Edinburgh*, which appeared from 1773 to 1795. It was not until the seventh volume, published in 1780, that Andrew Duncan (1744–1828) put his name on the title page as editor and dropped the statement "by a Society in Edinburgh." In the preface to this volume, Duncan finally confessed, "When, however, I am now to take upon myself what has hitherto been presented to the Public as the work of a Society, it may naturally be presumed that the abilities of an individual will be found unequal to the undertaking. To remove this apprehension, I may observe that the compilation of this work will not hereafter be more dependent on my exertions than it has already been. I shall still be regularly aided by some Gentlemen, whose reputation can receive no addition from drawing up a proper analysis of the work of others."

Duncan, nevertheless, found it necessary to add, "No one who undertakes the office of a Collector can be supposed to vouch for the truth of what is merely communicated to him by others. The credit due to such circumstances, must in every instance, depend upon the veracity and discernment of the observer" (Porter 1992:12). The Philosophical Society, which by this time had received a royal charter, then began to publish its own *Transactions*. A significant number of medical papers were included for the first three years (1788–1790), but the numbers dropped off sharply shortly thereafter (Shapin 1971:299).

THE LONDON PUBLISHING SOCIETIES

The last quarter of the eighteenth century was a fertile period for the development of both medical societies and medical periodicals in London. Several of the periodicals were issued under the auspices of societies that had broader functions than publishing alone; others were formed primarily for educational purposes, to present papers before the membership, and to publish them selectively (Clark 1964:554). These were truly what Sir Hans Sloane called "publishing societies." It was an age of improvement, promotion, and propagation in numerous fields: religious, social, intellectual, scientific, and medical. Many of the societies inaugurated in this period, however, left no published records of any kind.

In some cases, records may not have been necessary. The Pylorus Club, for example, was made up of members of hospital staffs who were waiting for permanent appointments, and the society served solely as a means for intellectual exchange (Strohl 1963). There was also a Society for Propagating Human Nature and Knowledge. The information about this society was derived from a bookseller's catalog that offered a book issued under its auspices entitled *A Philosophical Essay on Fecundation* and provided no additional information about it. There were at least sixteen provincial medical societies in England in the eighteenth century (Averley 1989:1), none of which appear to have published journals or proceedings of any kind, although some, such as the Convivo-Medical Society in Gloucestershire, where William Jenner first shared his ideas on cowpox, clearly expressed the intention to do so (Averley 1989:32).

Of the medical periodicals issued in London in the last half of the eighteenth century, ten identified themselves with societies. The first was *Medical Observations and Inquiries by a Society of Physicians*, published in six volumes from 1757 to 1784. The society had a small but distinguished membership consisting largely of licentiates of the Royal College of Physicians, who, at that time, were prevented from being admitted as fellows. Among the members were William Hunter and John Fothergill. The society met from time to time to discuss medical papers and to select those to be printed. This journal has the distinction of having ten citations to its papers listed in the Garrison-Morton *Medical Bibliography*. This record for papers published in the eighteenth century is exceeded only by *Philosophical Transactions*.

In the first volume of *Medical Observations*, the editor, who is not named (as is frequently the case with publications issued under the auspices of societies in this period), provides some guidance to prospective contributors in the preface: "Hypothetical disquisitions, points of controversy, numerous and needless quotations, in short whatever has rather a tendency to show parts and erudition of the writer, than to advance medical knowledge will be suppressed" (Porter 1992:13). In the preface to the fourth volume (1771), the editor assured readers, "A strict impartiality in selecting papers for publication is maintained." The editor then adds, "Some ingenious essays are at present withheld from the Public, to give the respective author time to make them more complete. This is only mentioned to prevent those, who may not see their paper in this volume, from withdrawing their correspondence." The editor then goes on to state a policy with which modern editors are familiar: "An apology is necessary for one article in the collection, as it has already appeared in another work. Before this was known, that article had already been printed off. The Society therefore hopes, that those gentlemen who intend to publish their works apart, will, for the future, render such apologies unnecessary." A total of six volumes were printed from 1754 to 1784 at intervals of four to eight years. There apparently was no great urgency to publish in this period. Some of the volumes appeared in several editions.

The success of this publication may have been one of the factors that at long last stimulated the Royal College of Physicians to publish a journal of its own in 1768. The college, formed under Henry VIII in 1518, found little occasion to publish except for regulatory purposes, with the exception of *Pharmacopoeia Londinensis*, which first appeared in 1618. George Clark, historian of the college, informs us that the journal "editor [again, not named] was to receive twenty pounds for each volume, the papers were to be read in the Theater after which the College was to choose which were to be published and to decide on the publication of inquiries and other matters" (Clark 1964:579).

The fellows made their decision to publish at the end of December 1766, gathered enough papers by the summer of 1767, and finally published them in 1768. In the first volume, the college declared its intention to solicit contributions from authors other than its own fellows. The members of the Society of Physicians in London were invited to do so, an offer they declined until sometime after 1781 when some of the tensions between the college and the licentiates had lessened. The preface to the first volume of the college's *Medical Transactions* advised potential contributors "things that are commonly known should not be submitted, and negative reports are welcome where appropriate." A further note was added: "The College thinks it proper to declare that they do not as a body mean to vouch for the truth of any relation, or to give authority to any opinion contained in the papers here published."[2] The first volume contained twenty-one papers, primarily those read before the college the previous year. Almost half of the papers were by two of the fellows, William Heberdeen and George Baker. The second volume was published in 1772 (four years later), the third in 1785 (after fourteen years), and the fourth in 1813, after an interval of twenty-eight years. Two additional volumes appeared the first in 1815 and the last in 1826.

OTHER LONDON MEDICAL EDITORS

Samuel Foart Simmons (1750–1813) edited the successful *London Medical Journal* in eleven volumes from 1781 to 1790. Like Monro and Duncan, he claimed on the title page that the journal was issued under the auspices of a society; in this case, "by a Society of Physicians." It is fairly clear, however, that it was an individual endeavor almost from the beginning. In the preface to the first volume, Simmons says that the journal had its origins in a "Society of physicians who met together occasionally to converse on medical subjects," and that in response to their desire to make "their meetings of use to the public," they decided to issue a monthly publication to serve as a repository for original papers. The preface refers to the "editors" but also reveals that "one of their members, who was the first and most active promoter of this business, undertook the office of editor, to arrange the materials, superintend the printing of the work &c. while others engaged to assist in the different branches for what they deemed themselves to be qualified." The preface also stated, "Gentlemen of the Medical Profession in every part of the world [are earnestly entreated to contribute] accounts of useful cases, epidemical diseases, and such other matters as may tend to the improvement of medical science." The first issue consisted almost entirely of abstracts and extracts from medical works, particularly those of foreign origin. The first original contribution appeared in the second issue, but it rapidly attracted many others. The journal was welcomed warmly, particularly by the foreign press. The *Medicinisch-Chirurgische Zeitung* in 1790 referred to it as "this treasured medical work."[3]

In 1791, in the preface to the first volume of *Medical Facts and Observations* (the successor to *London Medical Journal*), Simmons made it clear that he was and had been the sole editor. Because the journal had been received in an "indulgent manner," the editor had been induced to persevere in bringing it out at stated quarterly periods, much longer than suited his other avocation. But that mode of publication having at length been attended with great inconvenience both to his professional engagements and in the deliberate management of the work, he became desirous of conducting his future labours in a way more convenient and satisfactory to himself. In other words, he said, he would publish a volume as often as he had gathered "about fifteen sheets in octavo" (i.e., approximately two hundred and fifty pages).

During the time that he edited *London Medical Journal*, Simmons was also a member of the Society for Promoting Medical Knowledge. In 1784, the society published a volume of thirty-one *Medical Communications* without naming an editor. The preface to the first volume refers to "the editors" and states that the purpose of the society was "collecting and publishing such papers on medical subjects as they think worthy of being preserved." Of the communications in the first volume, fourteen were by fellows of the Royal Society of London; Simmons was responsible for five of them. The preface to the first volume indicates that "the Society think proper to mention that they are already in possession of some materials for, a second volume, and they flatter themselves that when their design shall become generally known, the number of such correspondents will increase." The second volume, nevertheless, did not appear until seven years later, in 1790. The dates the papers were presented to the society are recorded as ranging from February 1785 to March 1790; ten of the papers were presented in 1785 and nine in 1790, with a scattering of papers in the intervening years.

Another "publishing society," also formed in 1784, was the Society for the Improvement of Medical and Chirurgical Knowledge, of which John Hunter was a founding member. Again, no one is identified as having editorial responsibility. It is questionable whether the publications of either the previous society or this one can be called journals or periodicals. The three volumes of *Transactions* of Hunter's society were issued in intervals of eight and twelve years. Hunter, Mathew Bailie, and Hunter's brother-in-law, Everard Home, were responsible for a majority of the papers. Other papers may have originated with the members of the Lyceum Medicum Londinensis, another society formed by Hunter with the collaboration of George Fordyce. That society included, along with medical students, six teachers of various medical subjects in the hospital schools of London. The nonstudent members of the society were required to read, in alphabetical order of their names, papers on a medical subject at the meetings. There was also a competition each year (a gold medal was the prize) for papers submitted on a subject announced by Hunter and Fordyce (Dobson 1969:242).

A glimpse of how editorial affairs may have been handled in this period is revealed by a 1787 letter to Hunter from his former house student William Jenner, a devoted naturalist. Jenner had sent Hunter a paper, which he asked Hunter as a fellow to read before the Royal Society of London. Hunter, in responding, apologized for his inaction, because, he said, "the whole of this winter we have had nothing but dispute in the Society, and giving up of the Secretaryship &c and are not yet settled, but when we are I will give in your paper, but shall take a copy of it, that in case they should not publish it in the *Transactions*, it may probably be published in the Medical Society, who will make more use of it than in the *Transactions*" (Ottley 1835:170).

Hunter did read Jenner's paper at the Royal Society later that year, and it was published in *Philosophical Transactions* the following year, but not before it was sent back to Jenner with suggestions for revision (Le Fanu 1951:10–111). Another communication of Jenner's, which did not fare as well with the Royal Society, was the first paper on his experiments with cowpox. This time, Hunter was no longer available to intervene, and the paper was sent to another fellow of the society, Everhard Home, who recommended it be rejected because it was based on a single subject (the young John Phipps), and, therefore, he said (with some justification), he "lacked faith" in it (Booth 1982).

The mission of the Medical Society of London, which included surgeons and apothecaries as well as physicians, encompassed more than hearings and publishing papers. The society was founded in 1773 to form a united front against the Royal College of Physicians but published no papers until 1787, almost fifteen years later. The preface to the first volume of the society's *Memoirs* states: "Medical papers, which may tend to the advancement of the sciences, will be received by the Society and such as may be deemed worthy of publishing, carefully preserved until sufficient matter for a volume shall be collected." To stimulate contributions, the preface included the announcement that the society had "now resolved to give a Silver Medal annually to the author of the best memoir that shall be communicated within the year, which shall be decided by the Fellows of the Society, who being the judges of its merits cannot be candidates for the Prize." The response to this invitation must have been meager because the five further volumes of *Memoirs* were published at intervals of three to six years.

There is not much to say about the three remaining medical journals in London that were associated nominally with societies. One was *Select Papers on Different Branches of Medicine, by a Society Instituted for the Improvement of Physical Knowledge* in 1767. The title page indicated that it would "be continued occasionally." Because only one volume appeared, it can be assumed that the occasion did not arise for another volume. *Medical Records and Researches Selected from the Papers of a Private Medical Association* was published in 1798. The copy examined did not include a preface, so little can be said about the society. The volume contained the contributions of only three London physicians and, when another volume appeared in 1813 with the same title, *Private Medical Association* no longer was cited as a source of the papers (Le Fanu 1984:34).

The last of the periodicals to appear under the auspices of a "publishing society" in this period differs from most of the others in that the editor is named. It is *London Medical Review and Magazine, by a Society of Physicians and Surgeons*, which appeared in six volumes between 1799 and 1802. The editor was William Blair (1765–1822), a London physician who also authored books on military medicine and venereal diseases. The prospectus bound with the journal says it is the work of a society of physicians and surgeons "who had experienced the necessity of such a work, and who were convinced that much benefit would accrue from it to the profession, as well as the public at large." The members of the society had "agreed to associate together for the purpose of collecting, arranging and printing such materials as could be procured." The preface to the first issue states "that they are determined to resist the progress of empiricism, under every form and pretense." To this is added what seems like a gratuitous policy: "they must also take the liberty of suggesting that no consideration will prevail on them to insert an author's review of his own publication." (Such reviews were not an infrequent practice in this period.)

There remains only one more medical journal in this period, included for the sake of completeness. *Medical Magazine, or General Repository of Practical Physic and Surgery* had an even more tenuous and ephemeral association with a medical society than did the other journals. Only a first number issued in 1774 is listed in any of the national catalogs consulted. Le Fanu, in his checklist of early British medical periodicals, indicates that another number or volume was issued in 1775 (Le Fanu 1984:58x). There is some corroboration for additional issues in the published catalog of an eighteenth-century medical educator and editor, Ernst Gottfried Baldinger, with the numbers 1,190–1,197, indicating that Baldinger had accumulated eight issues under this title. The title also appears in a German bibliographic journal in a list of monthly periodicals appearing in London in 1775.[4]

The title page of *Medical Magazine* indicates that it was issued by a "Society of Gentlemen" who may or may not have all been physicians or, for that matter, gentlemen. The designation of a responsible "Society of Gentlemen" appears on the title pages of many eighteenth-century general magazines, including *Critical Review* of Tobias Smollett, a physician turned journalist.

CONCLUSION

The environment in which medicine is practiced and in which research is pursued today is vastly different than it was in eighteenth-century Great

Britain. Medicine then was still more of an art than a science. To a greater degree than today, medicine then shared many of the characteristics of the humanities in that knowledge was largely noncumulative. For this reason, obsolescence of medical literature was not a great problem. This is indicated by the fact that medical journals of that period were reprinted and translated decades after they were first issued. It can be said with some assurance that deadlines were not much of a problem for eighteenth-century editors. Periodicals were more like today's multiauthored monographs and were reviewed in the same fashion. The incentives to publish obviously were not very great, or there wouldn't have been such long delays in publishing. Peer review of papers submitted to societies was carried out largely through reading and discussing them before the membership. The importance placed on the perception that contributions were reviewed by a society is demonstrated by the number of times this was implied rather than applied.

It is a great leap from the medical journalism of the eighteenth century to that of today. In contrast to the situation in the eighteenth century, biomedical journalism today presents a complex and diffuse picture in which new disciplines and new mission-oriented subjects are constantly created and combined. The evolution of medical journalism between the end of the eighteenth century and the end of the twentieth has not yet been thoroughly examined. However, it is obvious that there were many highly significant changes in the production and management of scientific communications during that period. On the other hand, some things have not changed at all. The scientific and medical community still has a responsibility to monitor the quality of communications. In addition, these communications continue to be distributed, primarily in the form of journals, a format that in many ways has not changed from that of the eighteenth century, and perhaps may not be the most appropriate to the needs of the medical community today.

NOTES

1. *Philosophical Transactions* 1665, 1 (10): 20–1.
2. *Medical Transactions* 1768, 1: ix.
3. *Medicinisch-Chirurgische Zeitung* 1790, 10: 17–21.
4. *Journal für Kunstgeschichte und zur Allgemeines Literatur* 1775, 1: 160.

REFERENCES

Averley, Gwendoline. "English Scientific Societies of the 18th and Early 19th Centuries." Ph D. diss. University of Durham, 1989.

Booth, C. C. "Medical Communications: The Old and New: The Development of Medical Journals in Britain." *British Medical Journal* (Clinical Research Editions) 285 (1982): 105–108.

Clark, G. A. *A History of the Royal College of Physicians*. 2 vols. Oxford: Clarendon Press, 1964–1978.

Conradi, Johann W. *Catalogus Bibliothecae Medico-Physicae Ernest Gottfried Baldinger*. 2 vols. Marburg: Krieger, 1805.

Dobson, Jessie. *John Hunter*. Edinburgh: E & S. Livingston, 1969.

Emerson, Roger L. "The Philosophical Society of Edinburgh, 1737–1747." *British Journal of the History of Science* 12 (1979): 154–91.

Johnston-Saint, P. "The First English Medical Journal." *Medical Press and Circular* 201 (1939): 117–18.

Kronick, David A. "Peer Review in Eighteenth-Century Scientific Journalism." *Journal of the America Medical Association* 263 (1990): 1321–22.

Le Fanu, William R. *A Bio-Bibliography of Edward Jenner, 1749–1823*. London: Harvey and Blythe, 1951.

Le Fanu, William R. *British Periodicals of Medicine, 1649–1899*. Oxford: Welcome Unit for the History of Medicine, 1984.

Ottley, D. "The Life of Hunter." In *The Works of John Hunter*. Edited by J. F. Palmer. Vol. 1. London: Longman, 1835–1837, pp. 170–71.

Porter, Roy. "The Rise of Medical Journalism in Britain to 1800." In *Medical Journals and Medical Knowledge: Historical Essays*. Edited by W. F. Bynum, Stephen Lock, and Roy Porter. London: Routledge, 1992, pp. 6–28.

Shapin, Steven. "The Royal Society of Edinburgh: A Study of the Social Context of Hanoverian Science." Ph.D. Diss. University of Pennsylvania, 1971.

Shapin, Steven. "Property, Patronagae, and the Politics of Science: The Founding of the Royal Society of Edinburgh." *British Journal for the History of Science* 7 (1971): 1–41.

Strohl, E. Lee. "The Fleece Inn Medical Society." *Surgery, Gynecology, and Obstetrics* 117 (1963): 371–74.

Chapter 12

Economic Aspects of Scientific Journalism in the Seventeenth and Eighteenth Centuries

ABSTRACT

The proliferation of scientific journals, starting in the middle of the seventeenth century, marked the beginnings of a new epoch in the dissemination of scientific information. The new medium brought with it the influences of a long heritage going back to scribal cultures, which even before the invention of printing had produced almanacs and newsletters in writing. The economics of periodical publications varied little from those associated with other kinds of printing in this period. They involved the relationship of fixed and variable costs and their effect on edition size in relationship to anticipated markets. Examples of these aspects of scientific publishing are examined in scientific journals published independently and under the auspices of scientific societies in the seventeenth and eighteenth centuries.

The "invention," as John Ziman calls it (1966:105), of the scientific journal has been heralded as one of the greatest contributions of the scientific revolution in the seventeenth century. The period from 1665, when the first scientific periodical appeared, to 1800, with which we are concerned here, is a long span of time. The great social, intellectual, economic, and scientific changes that took place during this period, important as they were, had little impact on the way scientific journals were produced or distributed. Inventions tend to draw on and adapt the existing conventions, social systems, and technologies that precede it. Many of the characteristics and technologies that define the scientific journal were already in place

---

Published in *Publishing Research Quarterly* 11 (1995): 62–79

and some of them had their origins in what Elizabeth Eisenstein has called the period of "scribal culture" (Eisenstein 1979).

Beginning slowly in the seventeenth century, scientific periodicals and periodicals associated with various aspects of technology and commerce proliferated widely in the eighteenth century in Europe and Great Britain. They were initiated and sponsored by individuals working with publishers and printers and, in some cases, by publishers or printers themselves who may have hoped to profit financially from these ventures. Scientific societies and other institutions organized to further goals in commerce, trade, and other special interests, also sought to achieve some of their objectives by issuing periodical publications. In many cases, they were supported by their members or by official bodies, but they were also faced with the same economic problems as the producers of independent periodicals. There seem to be few studies of the economic aspects of publishing these kinds of journals in this period. A brief review suggests that while profitability may, in some cases, have been an objective, it was often not achieved.

Periodicity, which provides the periodical with its essential characteristic, existed in several formats before the scientific journal appeared. Astrological almanacs are inherently periodicals because they are tied to predictions and announcements relating to events occurring at particular times. If almanacs introduced the concept of a time-bound form of publication, it is essentially to the newspaper that we need to look for the introduction of what Eric Allen calls "periodicity in print" (Allen 1930). Newspapers also owe their origins to the scribal period, and in fact news continued to be distributed serially in manuscript to select audiences well after the printed newspaper appeared. The technologies involved in printing changed very little during this period, and the mechanization of paper production did not occur until the end of the eighteenth century. An important factor in publishing periodicals in this period was the high cost and uncertain availability of paper, which encouraged the production of small editions.

PRODUCTION COSTS AND SIZE OF EDITIONS

The economics of scientific journalism in the seventeenth and eighteenth centuries were similar to those of publishing in general in that period. It involved both fixed costs, such as composition and editing—which go down rapidly with the size of the edition and variable costs such as the price of paper, presswork, and stitching—and binding, which varies with the size of the edition. The reluctance of publishers to invest money that might be tied up in unsold copies for a long time kept press runs of jour-

Table 12.1. Relationship of Publishing Costs to Edition Size
***Hamburger Patriot* 1762**

Edition	400 TH*	GR*	1400 TH	GR	5000 TH	GR
Satz	15	4	15	4	15	4
Druck	4	8	15	4	54	4
Druckerei**	14	15	22	18	52	4
Korrektur	5	10	5	10	5	10
Zensur	2	4	2	4	2	4
Total	65	7	142	18	42	120
Paper costs	36%		57%		69%	

*1 Thaler = 24 Groschen
**Druckereigemeinkosten
Source: Kirchner (1970), 180

nals low. In many cases, it led to reprints of earlier editions that had sold well. Joachim Kirchner (1928–1931), an important historian of German periodicals, reported data he found in the Breitkopf archives for the *Hamburger Patriot*[1] published in 1762. They illustrate this phenomenon very well (see table 12.1).[2] The translations of the rubrics in this table provide no difficulties with the exception of the term *Druckereigemeinkosten*. Literally it means "general costs relating to presswork" and therefore can be considered another one of the variable costs as the table indicates. The comparisons are easier when they are converted into unit costs in the following table (see table 12.2). In contrast to the first table, the fixed costs vary considerably with the size of the edition, while the variable costs remain constant.

I have no similar data for twentieth century journals, except for that derived from a study directed by Fritz Machlup entitled *Information through the Printed Word* (Machlup 1978). From one volume of this study devoted

Table 12.2. Unit Publishing Costs Relating to Edition Size

Edition Size	400	1400	1500
Satz	.92 Gr.	.26 Gr.	.07 Gr.
Druck	.26	.26	.26
Druckerei*	.88	.46	.28
Papier	1.42	1.41	1.40
Korrektur	.33	.09	.03
Zensur	12	.09	.03
Total	3.90	2.50	2.00

*Druckereigemeinkosten
Based on Kirchner (1970), 180.

to journals, we can derive the data in table 12.3. It shows the median manufacturing and operating costs for twenty-one scholarly journals published by commercial, society, and university presses in 1974. Paper, printing, and binding (the variable costs) combined make up less than one-third of the manufacturing costs. The cost factors in Kirchner's and Machlup's data do not readily lend themselves to comparison. Distribution costs do not seem to be included in accounts for the earlier journals, except perhaps for occasional references to discounts to booksellers, and indications of differences of costs between local and distant subscribers. The impact of the mechanization of the production of paper, however, is clearly demonstrated.

Two other features that periodicals share with the general book trade are the sale of books-in-parts or "number books," as Wiles calls them (Wiles 1957), and books sold by subscription. Marketing books by subscription was introduced in the early seventeenth century in England and readily adopted by journal publishers soon after. We can find subscription lists printed in many of the journals of this early period. They range from under a hundred subscribers to over a thousand. These numbers may not include single-issue sales by booksellers, and may not represent the total number of copies distributed. Journals were regarded by many subscribers in the same way they regarded books-in-parts, a factor that led to the reprinting of single issues in order to enable late subscribers to complete volumes. The attempt of publishers to satisfy the desire of subscribers to have a complete run of a particular title also led to changes of titles in some continued series. In Germany, for example this was sometimes accomplished by adding the terms *neue*, then *neuere*, *neuerer*, and finally *neuesten* to an existing title. Another indication that the distinction between periodicals and books had not been firmly established is the fact

Table 12.3. Manufacturing and Operating Costs Twenty-One Scholarly Journals 1974 (Percentages)

Manufacturing	
Composition and graphics	33.7
Paper, printing, binding	16.2
Total	49.9
Operating Expenses	
Editorial	29.2
Distribution	20.9
Total	50.1
Grand total	100.0

Source: Machlup (1978), vol. 2, 152

that single journal issues were frequently reviewed in the periodicals of that time in the same way that monographs were reviewed.

SCIENTIFIC SOCIETY AND INDEPENDENT JOURNALS

Science periodicals, and periodicals in general in this era, fall into two primary categories: those initiated by an individual or a publisher, and thus could be considered independently supported, and those issued under the sponsorship of an organized society. The two earliest and most important vehicles for the dissemination of scientific information in the seventeenth century both made their debut in 1665. *Journal des Sçavans* and *Philosophical Transactions* are also similar in being hybrids of the two categories cited above. They were ostensibly independent but also closely associated with societies. The differences between them reflect the differences between the two countries with which they are associated and the differences between the Académie Royale des Sciences in Paris and the Royal Society of London, with which they are linked.

The Académie Royale des Sciences was founded in 1666 and was heavily subsidized by the government, which paid the appointed members of the academy pensions and provided them with laboratories and observatories. Like the Royal Society, established four years earlier, the Académie received the right of imprimatur, which is the right to license books on its own behalf. The right was used frequently to publish books by members of the academy and to have them printed by the royal printing house. The Académie did not publish a periodical in the seventeenth century, with the brief exception of 1692 and 1693, but instead relied on multiauthor works published sporadically until it began to issue annual histories and memoirs in 1702. In the interim, as Gillispie says "the monthly *Journal des Savants* was virtually an academic organ" (Gillispie 1980:98). It was an organ, however, not only for the Académie des Sciences but also for the Académie des Inscriptions et Belles Lettres, whose members frequently formed part of its editorial board.

The Royal Society of London, on the other hand, received very little from Charles II with its charter, except for his blessings. The society was supported largely by the dues and gifts of its fellows. This was a consequence, Hiram Caton says, of the fact that "the Royal Society was conceived less as a research institute than as a forum for gentlemen" (Caton 1988, 69). There were several books printed with the Royal Society's imprimatur, but the society had a direct financial interest in only a few. One example is Francis Willoughby's *History of Fishes*. It was not a very successful venture, as we can see from this account taken from the society's records, and strongly discouraged the society from similar ventures (see table 12.4).

Table 12.4. Royal Society Account for Willoughby's *History of Fishes* (1686) 1685–1722

Receipts	*£*	*s (shillings)*	*d (pence)*	*Expenses*	*£*	*s (shillings)*	*d (pence)*
Gifts	162	17	6	Production	389	7	5
Sales	111	1	5	Miscellaneous costs	36	13	11
Total	273	18	11				
Deficit	132	2	7				
Total	406	1	4	Total	406	1	4

Source: Bluhm (1958), 100, table 6

Though some French historians may disagree, it has generally been recognized that *Philosophical Transactions* was the first scientific journal, despite the fact that the *Journal des Sçavans* preceded it by three months in 1665. In addition to being the vehicle for papers from all the Parisian academies, *Journal des Sçavans* had a broad subject scope that included law, history, theology, and belles lettres in addition to science. A comparison of the subject contents of the *Journal des Sçavans* and the *Mémoires de Trévoux*, its Jesuit counterpart, in the period 1717 through 1750, shows that although both journals contained scientific material, other subject matter dominated their pages (see table 12.5). Oldenburg was almost contemptuous when writing to his patron Robert Boyle on March 6, 1665–1666: "All their Journaux des Sçavans. From the beginning of januar last, till now in most of wch I find, what is Philosophicall to be taken out of our Transactions; ye rest being, generall, Extracts and Abbreviats of theologicall, Historico-Politicall and such like Books" (Oldenburg 3:48).

In his letter to Oldenburg on January 14, earlier that year, his faithful correspondent Justel tended to agree with him: "Those who manage our journal," he said "are rather historians than philosophers, that is why you

Table 12.5. Space Coverage by Subject: *Journal des Sçavans* and *Mémoires de Trévous* 1715 and 1750 (Percentages)

	1715 MT	*1715 JS*	*1750 MT*	*1750 JS*
Belles letters	14	20	16	15
Theology	24	17	10	2
Law	2	9	2	9
History	35	24	33	29
Science	25	30	39	41

MT = *Mémoires de Trévoux*
JS = *Journal des Sçavans*
Adapted from Ehrard and Roger (1965)

see nothing in it dealing with physics" (Oldenburg 3:12). In spite of these comments, *Journal des Sçavans* must be considered an important vehicle for scientific communication all through the seventeenth and well into the eighteenth century. Many of the important European scientists contributed papers to *Journal des Sçavans*, including Huygens, Parent, Marriotte, Clairaut, Cassini and others. In a study published in 1936, Sergescu identified more than three hundred mathematical papers published in the *Journal* by 1791 (Sergescu 1936, 569).

COSTS, DISTRIBUTION, AND PROFITS

Henry Oldenburg, a German immigrant to England and one of the first secretaries of the Royal Society, was the initiator of *Philosophical Transactions*. We are fortunate that for the early history of *Transactions*, we can refer to the published Oldenburg correspondence, which extends from about 1662 to Oldenburg's death in 1677.[3] The correspondence naturally is full of references to the *Transactions*, but I can only summarize them briefly here. Oldenburg insisted continually in his letters and in prefaces to journal issues that *Transactions* had no formal relationship to the Royal Society. However, the society records show that it "was ordered" into existence by the society's council on March 1, 1664–1665 (Birch 1756, 2:18). There is also substantial evidence that the officers and fellows took a close interest in the publication until it was taken over officially by the society in 1753. Oldenburg had hoped to make one hundred and fifty pounds a year out of the publication to boost his meager income (Oldenburg 2:xxv), but he attested frequently that he never earned more than forty pounds a year from that source (Oldenburg 4:49). We discover what Oldenburg's terms were with the original, official society printers, Martin and Allestre, when it became necessary to print *Transactions* in Oxford during the plague in London in 1665. In a letter from Oxford, dated October 11, 1665, from Robert Moray (a member of the society's council from 1662 to 1672), Oldenburg is informed that the Oxford printer, Davies, had agreed to pay him three pounds "for every printed sheet," and that, as was the case in London, he would share the cost of the copperplates by half, and that he had agreed to an edition of 1,000 rather than the usual 1,200 (Oldenburg 2:563). The correspondence does not seem to provide any information about subscription prices or costs of individual issues. One exception can be found in the charges that Oldenburg lists in his letter, April 24, 1671, to Leibniz for some books the society's printer was sending him. The shipment included three volumes of *Transactions*, which are listed at one pound or about seven shillings each, so that each issue (the volumes all contained twelve) might have cost about seven pence. The volumes were sent, Oldenburg says, "by our bookseller Martin and by Schulz at Hamburg to Zunner at Frankfurt," where ostensibly they would be made available to Leibniz (Oldenburg 8:12). Oldenburg's letters frequently include references to issues and volumes that he is sending to various correspondents, but an exchange of funds rarely is mentioned. It is possible that he may have received copies for that purpose as a part of his contract with the printers.

When *Transactions* was resumed in 1682, following a long hiatus after Oldenburg's death in 1677, it became the responsibility of successive secretaries of the society, who presumably managed the publication with various London printers on the same terms as Oldenburg. When the society finally took over fiscal responsibility for *Philosophical Transactions* in 1752, it resolved the financial problem in a way similar to what societies have been doing ever since—by placing the responsibility on the membership. The society decided to raise the admission fee for fellows from three guineas to five, and to give every fellow a free copy. From his survey of the Royal Society's account books from 1752 to 1768, R. K. Bluhm reported that even with the sale of *Transactions* to non-fellows, the society was not able to meet the cost of production. As we see here (in table 12.6) the society ended up with a deficit of 771 pounds, nine shillings, and one pence for the fifteen-year period covered. Bluhm's data is more detailed than what is presented here, but it does not include any information about manufacturing or distribution costs.

Despite its close connections with the French academies, *Journal des Sçavans* was never published under their auspices. The *privilége*, or license to publish the journal, was received directly from the crown, which monitored publishing activities in France very closely. From 1665 to 1746, the *privilége* was issued in turn to seven individuals who were closely associated with the government in one way or another. The grants were from ten to thirty years, though some grants were relinquished well before the end of the designated periods. Two of the grantees held the *privilége* twice:

Table 12.6. *Philosophical Transactions* Accounts, 1753–1768

	£	s	d
Expenditures			
Production costs, 1753–1786*	2917	5	0
Receipts			
Sales, 1754–1768	1717	7	11
Dues**	428	8	0
Total receipts	2145	15	11
Deficit	771	9	1

*Paper, printing, and engraving
**Receipts of 2 GNs (Guinea, equivalent of 21 shillings) for 204 admissions
Source: Bluhm (1958),101, table 7

the Abbé Jean-Paul Bignon, who also served at times as the king's librarian and president of the Académie; and the bookseller Jean Cusson, who also was the publisher of the journal from 1665 to 1714. The holders of the *privilége* (with the exception of Cusson) negotiated a contract with a publisher for an annual sum to be paid to the editorial staff, which was appointed by the authorities.

Jean-Pierre Vittu, who contributed the entry on *Journal des Sçavans* to the *Dictionnaire des Journaux* (Sgard 1991:645–54) from which some of this information is drawn, complained of the lack of editorial archives, lists of subscribers, and similar information regarding the journal (Vittu 1988:167). We do have some data about the honorariums or fees, which were negotiated with the publishers and paid to the editorial staff. From 1723, when Bignon took charge of the journal to the end of the century, it ranged from a low of 1800 livres to 4800 annually. Nevertheless, the contract to publish the journal was much sought after and was considered an addition to the prestige of the publishing house that acquired it. No subscription prices are cited in Vittu's article in *Dictionnaire des Journaux*, but he tells us elsewhere: "The cost of postage which quickly doubled the price of five sous [a fourth of a livre] for each issue made it difficult for those who lacked connections with travelers or merchants to purchase it. This was a factor which encouraged the development of pirated editions" (Vittu 1988:169).

Journal des Sçavans was pirated in several places in Europe almost from the very beginning, and Parisian publishers began in 1724 to issue a duodecimo edition in addition to the quarto in order to save paper and be competitive with the pirated editions (Vittu 1988:169). As one of three journals in France that had official status, including *Mercure* and *Gazette, Journal des Sçavans* held the exclusive right to publish a journal in France related to science and belles lettres. Any other periodical, which might seem to infringe on this territory, was required to pay the publisher of the *Journal* a negotiated indemnity if it was to be distributed in France. *Année Littéraire,* for example, paid 5,000 francs a year and *Journal Politique et de Littérature* paid 22,000 francs, according to Bachaumont's *Mémories Secret* (Trenard 1969:161).

SCIENTIFIC SOCIETY AND INSTITUTIONALLY BASED JOURNALS

Despite their close association with societies, neither *Journal des Sçavans* or *Philosophical Transactions* (at least for the first seventy-five years or so) can be considered as bona fide society publications. Both the Académie des Sciences and the Royal Society attempted to provide accounts for their activities (*Histoire*) and contributions made by their members (*Mémories*) by publishing retrospective reports. For the Académie, Jean-Baptiste DuHamel (the society's first secretary), in his *Historia* published in Latin in Paris in 1698, covered the years 1667 to 1696. Bernard le Bouvier Fontenelle (who served for a long period as permanent secretary) in his *Histoire de l'Académie Royale des Sciences de 1666 a 1699* covered the same period in French.[4] The Académie also published retrospective collections of papers presented at its meetings, such as *Divers Ouvrages de Mathematique et de Physique par Messieurs de l'Académie a Paris*[5] before it began to publish its proceedings regularly in 1702. The Royal Society was served in a similar fashion by Thomas Sprat in his *History of the Royal Society* published in London in 1667, which covered the origins of the society and extolled its virtues and benefits, and also contained contributions made by some of the fellows. Thomas Birch also began with the society's early history and included the minutes of the membership and council meetings up to the 1750s as well as papers that had been presented to the society and not printed in *Philosophical Transactions* in this period.[6]

Society and institutional journals represented quite a different picture than independently produced journals. First, they had a captive audience in their membership, and second, they also had the potential of continued financial support either from their members or an official sponsor, patron, or governmental agency. *Saggi di Naturali Esperienze*, published from the work of the Accademia del Cimento in Florence in 1667, was not offered in the book trade at all, but was used as presentation copies by the Medicis, who sponsored the academy and its publication. By no means did all of the scientific, learned, "economic," and other special societies that proliferated in Europe in the eighteenth century publish proceedings. Some of them that did publish produced periodicals sporadically at long intervals and long after the papers had been publicly presented to the society. The economics of these publications varied considerably and only a few examples can be presented here.

The Academy of Sciences at Bologna established in 1690 did not publish the first volume of its *Commentarii Societatis* until 1731, and then included only the papers presented before 1723. Richard Rosen tells us: "The publisher printed the transactions at his own expense but received all profits from the sale. Of the thousand copies printed, 150 were turned over (free of charge) to the Secretary who then distributed them, at his discretion, but usually to those who had published in the volume" (1971, 87). The Royal Scientific Society at Upsala, on the other hand, found it difficult to print its transactions (published 1720–1739) without liberal contributions from some of its wealthier members (Jackson 1923). Some societies, such as the Patriotic Society of Hamburg, chose not to risk publishing altogether and directed their resources toward other goals. Founded in 1724, the Patriotic Society did not publish a journal until 1790 when the members may have become aware that publication, in addition to disseminating information, would also bring them recognition (Lowood 1991:111). The Dublin Society[7] founded in 1731 finally decided to publish its *Proceedings* on a regular basis in 1765, but the journal remained purely an administrative report until 1800 when the society began to publish its *Transactions*, which also included substantive papers (Brown 1987:15).

Some provincial academies in France published very little and then frequently at long intervals. Founded in 1726, the Académie des Belles Lettres de Marseille published only thirty-one numbers in the first series of its *Recueil* between 1727 and 1767. No issues appeared in 1729, 1731, 1732, and 1733, years in which it was decided that there were not a sufficient number of papers presented to justify publication (Dassy 1887:180). The thirty-first issue in 1767 was published in an edition less than the usual one hundred. The distribution of two copies to the Académie Française and one each to each academician and associate would have accounted for about fifty copies. Public sale accounted for very few (Dassy 1887:181). The second series (no. 1–12, 1768–1786) of *Recueil* appeared after science and arts were added to the society's name and it became the Académie des Sciences, Lettres et Beaux-Arts in 1768. Originally, the printers for the academy published the *Recueil* at their "own risk and peril." The printer chosen in 1740 declined the honor, saying that the enterprise would put him at great expense with the likely expectation of a profit of no more than three or four livres, while his costs would exceed one hundred. The Académie applied to the municipality for help. The city officials glowingly recognized the honor that the Académie brought to the city and agreed that it would be appropriate for the municipality to support the publication if the cost did not exceed what they called "the modest sum of fifty francs [livres]." In the authorization, they specified that fine paper and attractive typefaces should be used. One free copy should be bound in Morocco for the Duke de Villars, the society's patron, sixty copies to the gentlemen of the Académie and its associates, and ten copies for the city government. Since the edition at that time consisted of only one hundred copies, that left only twenty-nine to be sold or otherwise utilized. The municipality was not always consistent in its authorizations. In responding to the Académie's request in 1779, the officials no longer specified fine paper and suggested that future *Recueils* be printed "without luxurious typography" in the usual one hundred copies, of which ninety would be at the disposition of the Académie (Dassy 1887:885–86).

The Académie Royale des Sciences, de Lettres et des Beaux-Arts de Bruxelles published only five volumes of its *Mémories* from 1772 to 1788. Some sense of the arrangements the society made with the publishers can be revealed by the report the secretary made at the February 25, 1788, meeting. He advised the members (1) that the printer no longer wished to supply the academy with eighteen bound copies of the memoirs, (2) that he no longer wished to pay for the engravings or for the special paper on which they were to be printed, and (3) that he would not accept any instructions from the members of the academy concerning the price at which the memoirs were to be sold (Quetelet 1872, 20). The Medical Society of Copenhagen, on the other hand, was negotiating from a different position when it decided to begin a new series of its *Acta* in 1783. At that time, Genner tells us, a local bookseller offered to pay the society two thalers for each sheet printed, and to supply copies free to the society if the publication was printed in Leipzig. If, however, it was to be printed in Copenhagen, he said he could not pay the society any fees, but would still provide them with free copies. Another publisher offered to pay the society in cash for copies to sell in Germany and to reimburse the society for the cost of printing. The society accepted the first publisher's offer on condition that he supply forty free copies and pay for the copper plates (Genner 1972:43).

Other examples relating to the economics of society journal publication can be cited, but it appears clear that it was seldom a profitable enterprise. The experience of the Regensburg Botanical Society may have been more the rule than the exception. In 1792, when the first volume of the *Schriften der Regensburger botanischen Gesellschaft* was published in six hundred copies, the society had only sixty-five members. Wolfgang Ilg, tells the tale in his history of the society published in 1984:

> Until 1794, no thanks to the advertisements, 250 copies were disposed of, that is, half of them were sold at a price of one *Gulden* and forty *Kreuzer* a volume and the other half given as gifts or exchanged. [That still left 350 unsold copies.] After this Kohlhaus [the publisher] relinquished 300 copies to the bookseller Boettingen in Leipzig, who managed to dispose of 100 volumes before he went into bankruptcy. Some time afterwards the President of the society learned of the failure of the company, and made an effort to rescue the remaining volumes for the society. He was able after great difficulty, and with the help of a lawyer, to bring back the remaining 200 copies to Regensburg. In the meantime the Society had received no benefits from the sale of the 100 copies, and moreover had to pay twenty-two Gulden and forty-seven Kreuzer for their return. (Ilg 1984:176)

It is not surprising that only one volume of this series seems to have appeared.

INDEPENDENT JOURNALS: PROFIT AND LOSS

There seems to be little data on the profitability of scientific journals in the literature. We do know about the profitability of some general journals, which had a role to play in disseminating scientific information. Pierre Rousseau built a fortune on the basis of his *Journal Encyclopédique*, which he edited from 1756 to 1793. He established a publishing house at Bouillon, the capital of a small independent principality where he was not exposed to French censorship. The firm also issued a medical journal, *Gazette Salutaire*, and a legal journal, *Journal de Jurisprudence*, and also vigorously pursued a trade in monographs. Rousseau is reported to have netted some 20,000 francs a year, and to have amassed 180,000 francs by the 1780s (Chartier 1952:29). Another literary periodical that included reviews of medical and scientific works was *Bibliothéque Britannique*, which appeared in Geneva from 1796 through 1815. It was enthusiastically received when it first appeared and was able to recoup its cost in the first year (Bickerton 1972:540). Each of the three original editors is said to have averaged 1,200 francs a year during the period in which it was published (Bickerton 1972:540–42).

Ralph Griffiths, the proprietor of *Monthly Review*, which began in 1749 and continued until well into the nineteenth century, became wealthy as a publisher. Some of his wealth (£10,000) is said to have been acquired from profits from publishing John Cleland's *Memoirs of a Woman of Pleasure*, also known as *Fanny Hill*. He is said to have "engaged Oliver Goldsmith as a regular writer on the review, in return for board and lodging in his house and a miserably small salary" (Adburgham 1972:111). Griffiths also made a practice of engaging established scientists and scholars to write for him, and in 1793 is known to have paid such scientists as Thomas Beddoes three pounds for each of their reviews (Stanfield 1984:104). The journal also featured the reviews of the transactions of scientific societies (Baesel 1974:32). *Critical Review*, which followed *Monthly* in 1756, is said to have devoted more than 30 percent of its contents to scientific and medical works, particularly when Tobias Smollett was one of its principal contributors. Smollett devoted three monthly installments of *Review* in the first year to the 1755 volume of *Philosophical Transactions*. It was not until 1762, however, that he was able to speak of any income from *Critical Review* (Basker 1988:49, 67, 118).

Göttingische Zeitungen von gelehrten Sachen, which came under the direction of the great Swiss physiologist and poet Albrecht von Haller in 1747, was similar to some of the other scholarly review journals that began under the auspices of German universities in this period. When Haller asked to be relieved of his duties as editor upon becoming president of the newly inaugurated *Göttingen Scientific Society* in 1752, he assured the members that the journal brought in enough money to support a staff (Joachim 1936:39) A visitor to the university, sent in 1784 by the Prussian government to survey German universities, reported that the journal received enough income to support the society (Forster 1969:318).

Acta Eruditorum, the Latin language German counterpart of *Philosophical Transactions* and *Journal des Sçavans*, was one of the important mediums for scientific and scholarly information for the hundred years from its inception in 1682. It existed largely on the basis of a two hundred thaler subsidy from the Duke of Saxony and the support of its editor, Otto Mencke, who carried much of the financial burden (Hensing 1973:35–36).

Lorenz Crell, a medical professor at the University of Helmstedt, is credited with issuing "the first successful discipline-oriented journal in science" (Hufbauer 1971:465). Beginning with his *Chemisches Journal für die Freunde der Naturlehre, Arzneygelartheit etc.* in 1778 through 1781, he was associated with a total of at least seven additional chemical journals, issued as continuations of this title or in parallel with them (Kronick 1991). Hufbauer, who has analyzed six years of subscriber lists to one of these titles, *Chemische Annalen* (1784–1804), has identified one of the reasons for Crell's success: his ability to maintain a circulation that, although relatively small, was fairly stable (see table 12.7). Hufbauer tells us that Crell, at first, "depended on his publisher and book dealers to reach the market. In late 1783 when he was preparing to shift to a monthly schedule, he reconsidered this arrangement. He decided to publish the *Chemische Annalen* at his own cost and to market it on a subscription basis. By doing so, he could garner a larger share of the revenues from sales" (1982:85).

At an annual subscription price of three Reichsthaler and on the basis of the subscription figures, Hofbauer argues, Crell's "annual receipts from subscriptions remained above 1,100 Reichsthalers throughout the 1780s." Copies sold through the book trade provided him with additional income, so the journal must have provided him with a substantial profit. In a note to these comments, using price data derived from Joachim Kirchner's *Die Grundlagen des Deutschen Zeitschriftenwesens* (Kirchner 1928), Hufbauer calculates that on the basis of "at least 500 copies, Crell's net income (after honoraria, production, and postage) was probably 500 Rheichsthaler [i.e., one Reichsthaler per copy or more]" (Hufbauer 1982:85).

Data on the economics of early scientific journals are scarce, but some information can be derived from comments made by editors in their own journals. The first series of John Houghton's *Collection of Letters for the Im-*

Table 12.7. Number of Subscribers to the *Chemische Annalen* 1785–1789*

	1784	*1785*	*1786*	*1787*	*1788*	*1789*
New members	424	123	72	40	24	27
Renewing members	0	324	341	351	359	361
Total subscribers	424	447	413	391	383	388

*Hufbauer (1982), 82, table 7

provement of Husbandry and Trade appeared from 1681 to 1684 and did not resume publication until 1692. At that time, Houghton explained that the hiatus was a result of "pecuniary difficulties . . . but now having a contribution of a guinea a year from some gentlemen and expecting it from more, [gentlemen], I go on, and shall sell them for a penny each" (Lacy 1923:450). Ernest Baldinger, a noted eighteenth-century medical educator, was responsible for at least eight periodicals beginning with his *Nachrichten von jezt lebenden Aerzten und Naturforschern* in 1749. Many years later, in 1790 at the end of the twenty-third issue of his *Medicinisches Journal,* he complains of the pressure of work this has produced. From the first of April to the eleventh of June, he says, he has had sixty-nine travelers to entertain, and although he writes letters every morning from four to seven, he still owes some three hundred responses. Beside his principal duties as a professor of medicine, he is working, he says, for four book printers. All of this, he insists, he does mostly for the benefit of the public, as his journals cost him more than they bring in.

Friedrich Albrecht Gren, a professor at the University of Halle, indicated in the preface to the first volume of his *Journal der Physik* in 1790 that he was publishing the journal at his own cost. In the preface to the first volume he lists only eighty subscribers, all of them in German cities, except for Paul Usteri in Zurich. The journal continued on with two title changes through the first quarter of the nineteenth century, so we may assume either that it was also made available through the book trade. Later he enjoyed an expanded list of subscribers, or both. Eberhard August Wilhelm Zimmermann in the preface to the first volume of his *Annalen der Geographie und Statistik* in 1790 tells his readers: "The initiation of this undertaking has been far more costly and time consuming than I had imagined, particularly because it has been laborious to procure foreign works, due in part to the difficulties with foreign exchanges." The journal nevertheless continued for a further two volumes in 1791 and 1792 in a period when periodicals tended to have very short durations.

Another example of a scientific journal that started as a private undertaking of the editor is August Friedrich Hecker's *Magazin für die pathologische Anatomie und Physiologie.* In the preface to the first volume published in Altona in 1796, he says that he has, with pleasure, undertaken to meet all the costs of production, including the copperplates. He also says that he would like to pay honoraria for contributions that are made to the journal, but makes no commitment to do so. It depends, he says, on the success of the journal. He also does not bind himself to any frequency of publication. Actually, only one volume seems to have appeared. He had better luck with a journal of broader scope and perhaps more popular interest. His *Journal der Erfindungen, Theorien und*

Widerspruche in der Natur-und Arzneiwissenschaft that began in 1792 lasted through the first decade of the nineteenth century.

An editor who made his financial problems more explicit to his readers was Arthur Young, one of the most prolific writers and advocates of progressive agriculture in the eighteenth century. In the preface to the first volume of his *Annals of Agriculture* in 1784, he reports that in the first five issues of the volume his income was £78.16.3 against an expenditure of £90.14.3—despite the fact that he had reduced the press run from 1,000 to 500 after the second issue (Kronick 1965:32). Young returns to the issue at the end of volume 10, where he says he is providing another financial report to disabuse his readers of the notion that he is getting rich from the proceeds of the *Annals*:

> After five years attention to the printing of the *Annals*, the present state of the sale makes it necessary for me to venture some remonstrance to the public; I have heard, that it has been reported, I make five hundred pounds a year by it; as I know not how far such ideas spread nor by whom they may be received, it is but justice to myself, to wish that the truth may be known; I cannot do this better than by stating the account of the last volume viz. the ninth.

His account (see table 12.8) shows a deficit of £71.8.6 for manufacturing and distributing that volume. He continues:

> It is true, the expenses on this volume ran higher than common: those of vol. viii were only £151.1.10 and the sale of it, to this time, amounts to £126.19.3 consequently [*sic*] the balance against that volume is now about £14. Notwithstanding this unfavorable aspect, the sale of the early volumes would have more than answered these deficiencies, if it had not been for reprinting small impressions, of many numbers merely for supply sets, a business yet continuing as much as ever.

Young apparently paid no honorariums to contributors (he wrote much of it himself) because he follows by saying that he has devised the idea of:

Table 12.8. *Annals of Agriculture*, Expenditures and Receipts, vol. 8, 1788

Expenditures	*£*	*s*	*d*	*Receipts*	*£*	*s*	*d*
Printer	84	1	4	Sales	184	1	2
Stationer	70	8	0	Balance	71	6	6
Plates	15	4	0				
Mr. Nicholl	8	13	9				
Advertisements	26	13	11				
Postage and carriage	5	6	2				
Totals	210	7	8		210	7	8

Source: Annals of Agriculture (1788) vol. 10, 589

> offering honorary rewards to the best papers, in each volume, to have been decided by the correspondents themselves, the expense of which, including all charges, would have been about £100. This scheme I mentioned to some of the most distinguished correspondents, but they one and all condemned the plan, assuring me, that should I persist in it, they could not think of continuing to send me any papers, as they corresponded merely through a desire to serve the public, and wished to be free of the idea of any other motive; under the influence of this intimation I laid aside the intention. If it had not thus dropped, I should by this time have found such an expense considerably beyond what the publication would have paid, and particularly as I have found even the receipt of the sale not altogether free from precariousness I had no expectation of meeting with.

After again emphasizing the importance of keeping open such a channel for agricultural information as provided by *Annals* and the difficulties he has had in maintaining it, he states: "The continuation of it on such terms, is inconsistent with my scanty situation; I can run no great hazards; and therefore I have inserted this account of the publication, that if its friends will not exert themselves to extend the sale, it must soon drop." In a footnote, he suggests some methods by which help could be given. Landlords could buy copies for the tenants of their estate, claiming that "there are in this kingdom 117 counties, the present sale does not amount to four a county" (*Annals* 1788, 10:589–94). The appeal must have been successful because the *Annals* continued to appear for a further thirty-six years, until 1815.

CONCLUSIONS

This is a short chapter about a very large subject. Obviously, there is a great deal more to be said about the economics of seventeenth- and eighteenth-century scientific journalism. I have not mentioned the occurrences of reprints of journal issues when some journals may have been produced in what turned out to be short press runs. The compilers of a recent checklist with the title *British Newspapers and Periodicals of 1641–1700* (Nelson and Seccombe 1987) have indicated that *Philosophical Transactions* was one of the three most frequently reprinted serials in the seventeenth century (Nelson and Secombe 1986:72). Despite the difficulties in collating sets, the editors nevertheless identified ten issues that they regard as having been reprinted in the first twenty-seven numbers (Nelson and Seccombe 1987:460–68). Although the gross income from larger press runs may have been larger, the margin of profit for each issue, as we have seen, would be lower.

A sale of less than the full press run would have delayed or may even have reduced the return on the publishers' investment, of which a larger part was for the purchase of paper. Although general periodicals such as *Monthly Review* may have provided large profits, publishing journals of scientific or technical information was often, in general, a risky and costly undertaking.

NOTES

1. It was published for *Patriotische Gesellschaft* in Hamburg, which was one of many "patriotic societies" founded in Germany during this period that concerned themselves with the economic health of their communities.
2. The table also shows the impact of the cost of paper on production costs as a percentage of the manufacturing costs with the increase of edition size.
3. It has been most ably edited in thirteen volumes by A. Rupert Hall and Marie Boas Hall. Quotations from letters written by or to Oldenburg in any one of a number of foreign languages are translations generously supplied by the editors (Oldenburg 1965–86).
4. Paris, 1729–1734, 11 vols.
5. Paris, 1693.
6. London: Millar, 1756–1757, 2 vols.
7. Later *Royal Dublin Society*.

REFERENCES

Adburgham, Alison. *Women in Print: Writing Women and Women's Magazines from the Restoration to the Accession of Victoria*. London: Allen and Unwin, 1972.

Allen, Eric W. "International Origins of Newspapers: The Establishment of Periodicity in Print." *Journalism Quarterly* 7 (1930): 307–19.

Baesel, Don R. "Natural History in the British Periodicals of the Eighteenth Century." Ph.D. diss. Ohio State University, 1974.

Basker, James C. *Tobias Smollett, Critic and Journalist*. Newark, N.J.: University of Delaware Press, 1988.

Bickerton, David N. "A Scientific and Literary Periodical: The *Bibliothèque Britannique* (1796–1815); Its Foundation and Early Development." *Révue de Litérature Comparée* 4 (1972): 527–47.

Birch, Thomas. *The History of the Royal Society of London*. 2 vols. London: Millar, 1756–1757.

Bluhm, L. K. "Remarks on the Royal Society's Finances, 1660–1768." *Notes and Records of the Royal Society* 13 (1958): 82–103.

Brown, Barbara T. "Three Centuries of Journals in Ireland: The Library of the Royal Dublin Society, Grafton Street." In *Three Hundred Years of Irish Periodicals*. Edited by Barbara Hayley and Edna McKay. Dublin: Association of Irish Learned Journals, 1987, pp. 1–28.

Caton, Hiram. *The Politics of Progress: The Origins and Development of the Commercial Republic 1600–1835*. Gainesville: University of Florida Press, 1988.

Chartier, Gustav, and Roland Mortier. *Le Journal Encyclopédique (1750–1793): Une Suite de l'Encyclopédie.* Paris: Nizet, 1952.

Dassy, Louis T. *L'Académie de Marseilles, Ses Origines, Ses Publications, Ses Archives, Ses Membres.* Marseilles: Barlatier-Feissant, 1887.

Ehrard, Jean, and Jacques Roger. "Deux Périodiques Français du 18e Sciècle: Le *Journal des Savants* et les *Mémoires de Trevoux.* Essais d'Une Étude Quantative." In *Livre et Société dans la France du XVII Siècle.* Edited by C. Bolleme, et. al. Vol. 1. Paris: Mouton, 1965, pp. 33–59.

Eisenstein, Elizabeth. *The Printing Press as an Agent of Change.* London: Cambridge University Press, 1979.

Forster, Robert, and Elborg Forster. *European Society in the Eighteenth Century.* New York: Walker, 1969.

Genner, J. *The Medical Society of Copenhagen, 1772–1972.* Odense: Odense University Press, 1972.

Gillispie, Charles C. *Science and Politics in France at the End of the Old Regime.* Princeton, N.J.: Princeton University Press, 1980.

Hensing, Ulrich. "Acta Eruditorum (1682–1782)." In *Deutsche Zeitschriften des 17 bis 20 Jahrhundets.* Edited by Heinz Dietrich Fischer. Pullach bei Munich, 1973, pp. 29–47.

Hufbauer, Karl. "Crell." In *Dictionary of Scientific Biography.* Edited by Charles C. Gillilspie. Vol. 3. New York: Scribners, 1971, pp. 464–66.

Hufbauer, Karl. *The Formation of the German Chemical Community (1720–1795).* Berkeley; University of California Press, 1982.

Ilg, Wolfgang. *Die Regensburgische botanische Gesellschaft: Ihre Entstehung, Entwicklung und Bedeutung.* Regensburg: 1984. *(Denkschriften der Regensburgischen botanischen Gesellschaft Bd. 42)*

Jackson, Benjamin D. *Linnaeus.* London: Witherby, 1923.

Joachim, Johannes. *Die Anfänge de Königlichen Societät der Wissenschaften zu Göttingen.* Berlin: Weidemann, 1936.

Kirchner, Joachim. *Die Grundlagen des Deutschen Zeitschriftenwesens.* 2 vol. Leipzig: Hiersemann, 1928–1931.

Kirchner, Joachim. "*Der Hamburger Patriot;* eine Untersuchung zur Frugeschichte des Zeitschrift." In *Ausgewählte Schriften.* Stuttgart: 1970, pp. 173–94.

Kronick, David A. "Scientific Journal Publication in the Eighteenth Century." *Pa pers of the Bibliographical Society of America* 59 (1965): 28–54.

Kronick, David A. *Scientific and Technical Periodicals of the Seventeenth and Eighteenth Centuries: A Guide.* Metuchen, N.J.: Scarecrow Press, 1991.

Lacy, Mary G. "An Early Agricultural Periodical." *Annual Report of the American Historical Association* 1 (1923): 445–54.

Lowood, Henry K. *Patriotism, Profit, and the Promotion of Science in the German Enlightenment.* New York: Garland, 1991.

Machlup, Fritz, et al. *Information through the Printed Word: The Dissemination of Scholarly,. Scientific, and Intellectual Knowledge.* 3 vol. New York: Praeger, 1978.

Nelson, Carolyn, and Matthew Secombe. *Periodical Publications 1641–1708: A Survey with Illustrations.* London: The Bibliographical Society, 1986. (Occasional Paper No. 2.)

Nelson, Carolyn, and Matthew Secombe. *British Newspapers and Periodicals of 1641–1700: A Short-Title Catalogue of Serials Printed in England, Scotland, Ireland*

and British America. New York: Modern Language Association of America, 1987.

Oldenburg, Henry. *Correspondence*. Edited and translated by A. Rupert Hall and Marie Boas Hall. Madison: University of Wisconsin Press, vols. 1–9, 1965–1973; London: Mansell, vols. 10–11, 1975–1977; London: Taylor and Francis vols. 12–13, 1986.

Quetelet. A. "Premier Siècle de *l'Académie Royale de Belgique*." In *Centième Anniversitaire de Fondation (1722–1872)*. Vol. 1. Bruxelles: Hayez, 1872, pp. 1–120.

Rosen, Richard L. "The Academy of the Institute of Bologna, 1690–1804." Ph.D. Diss. Case-Western Reserve University, 1971.

Sergescu, R. "Les Mathematiques dans le *Journal de Sçavans*: Premiere Période 1660–1701." *Osiris* 1 (1936): 568–83.

Sgard, P., ed. *Dictionnaire des Journaux. 1600–1789*. Paris: Universitaire, 1991.

Stanfield, Dorothy A. *Thomas Beddoes MD 1760–1808: Chemist, Physician, Democrat*. Dordrecht: Reidel, 1984.

Trenard, Louis. "La Press Franqaise des Origens a 1788." In *Histoire Généralee de la Presse Française*. Edited by Claude Bellanger, et al. Vol. 1. Paris: Presse Universitaire de France, 1969, pp. 27–404.

Vittu, Jean-Pierre. "Diffusion et Reception du *Journal des Savants*." In *Diffusion et la Lecture des Journaux Française Sous l'Ancien Régime*. Actes de Colloque Internationale Nimègue, 3–5 June 1987. Amsterdam: APA Holland University Press, 1988, pp. 167–75.

Wiles, R. M. *Serial Publications in England before 1750*. Cambridge, Mass.: Cambridge University Press, 1957.

Ziman, John M. *Public Knowledge*. Cambridge, Mass.: Cambridge University Press, 1966.

Chapter 13

Antecedents to and Influences on the Early Modern Scientific Journal

The scientific journal was born under the auspices of the recently founded Royal Society of London with the first issue of *Philosophical Transactions,* which appeared in March 1665. It has been called "perhaps the most important of all the innovations associated with the Society in its formative years" (Hunter 1990:24). John Ziman referred to it as an "invention" that was one of the greatest contributions of the scientific revolution in the seventeenth century (Ziman 1966:105). This innovation occurred over two hundred years after the invention of printing and was preceded by many other forms of communication methods, many of which are still prevalent today and helped influence and form the journal. Long before 1665, scientific and technical information had been disseminated for thousands of years in many forms.

Science is a cumulative form of knowledge. It grows at the rate that new knowledge is submitted to the judgment of peers and when some form of consensus can be achieved. We can define science as the acquisition of knowledge about nature through observation and rationalization, and technology as the process of using this knowledge to meet human needs. In that event, the dissemination of news about science and technology has existed since certain regularities were noticed in nature and communicated to others to examine, accept, or reject. Without communication, the distinctive cumulative character of science is lost. Science, therefore, cannot exist without communication in one form or another. Control of fire, which perhaps is one of the first great discoveries of man, was so important that Greek mythology attributed it to the god Prometheus who, it is told, stole it from the gods for the benefit of mankind. The ability to control fire is credited with the beginnings of civilization along with all the good and evil that accompanies it. Knowledge of initiating and controlling fire, as far as we know, was transmitted without any organized system of information dissemination. Nevertheless, it was so widely dispersed that there is no primitive people we know about that had knowledge of it, nor do we know the processes by which it was disseminated. George Sarton says in the first volume of his *History of Science*: "Our documentation is often very imperfect. For example, primitive men applied a large amount of knowledge before they were conscious of having any; if they were not conscious of it, how could we be?" (Sarton 1959:ix).

The process of manufacturing glass is said to go back to about 2500 BC, and the production of the first glass vessels occurred about a thousand years later: "neither the process nor the tools underwent significant modification for nineteen centuries" (Harden 1954:337). It was not until the last two decades of the eighteenth century, however, with Priestly, Cavendish, and Lavoisier, that we began to acquire durable scientific theories about the nature of fire, and the first book describing the methods of manufacturing glass was not printed in Europe until the early part of the seventeenth century (Charleston 1957:217).

ORAL TRANSMISSION

Before the invention of writing, the primary ways of transmitting scientific and technical knowledge was verbally or by means of witnessing; two modes of communication that still are very actively in use today. Evidence is bountiful about highly gifted craftsmen who worked with great technical skills and knowledge and handed this down verbally from generation to generation. We marvel at the cathedrals and feats of engineering of the distant past, which included aqueducts and other large projects. Great advances also occurred in agriculture and husbandry and in examples of ancient handiwork, which are exhibited in our museums. Egyptians developed great skills in quarrying huge granite blocks for their monuments (Bromehead 1954:569–70). Very little of the technological developments in the millennia preceding the scientific revolution are documented in the literature of its time. In fact, there were a great many developments that occurred before there were any writing systems at all, and we know about them largely through artifacts and conjecture (Neuberger [1930] 1969).

Many industrial processes such as tanning, paper making, processing woven cloth, grinding corn, and others were in wide use in Europe and

Great Britain beginning with the period of the Roman Empire about which there are few records (Gimpel 1976:14). In his biography of Lavoisier, McKie tells us: "Some kind of practical chemical knowledge was common to all the early peoples. The art of cooking, baking, brewing, pottery and metalworking are all really chemical in nature. In Babylon and in Egypt these arts and crafts, including the making and coloring of glass, were so highly developed that they excite our admiration today in the specimens that have survived to our time" (McKie 1952:34).

The evolution of technical skills and science is in some ways analogous to the evolution of biological species; productive and proven changes are passed on from one generation to another. Scientific knowledge grows by small increments that are added to an ideological structure that has survived and has been modified over the ages. The most effective and valid increments survive and are modified on the basis of new experiences (experiments). Perhaps technology, like nature, does not proceed by leaps, but it is also a result of a process of incremental changes. A well-known axiom states that *natura non facit saltum.* ("Nature does not proceed by leaps.") The source of this axiom is uncertain, but is attributed by some to Leibniz and Linnaeus, who used it as a principle in his biological classifications (Gould 1992:10). Basalla rejects the analogy of the biological process of evolution, "the world of things born," with the evolution of "things made," which are only modifications of antecedent artifacts (Basalla 1988:2). A corollary to this axiom is that each age carries with it some of the baggage of the proceeding age, an axiom that applies to communications media as well as other technical changes. Oral tradition remains in memory and is passed on from generation to generation when it is repeated. Vansina tells us: "The mind through memory carries culture from generation to generation. . . . Rumors that are not contradicted survive and become part, first of the store of oral history, later also of oral tradition" (Vansina 1985:xi, 6).

It appears that the technicians who developed the industrial processes in chemistry, engineering, and other practical arts had little exchange with the scholars in classical society. Therefore, most of the knowledge of the practical arts was stored in memory. Before the time of Homer, all the records of Greek culture are said to have been stored in oral memory; that is, they were verbally transmitted from one generation to another (Havelock 1963: vii). Many individuals in earlier periods are known to have had prodigious memories. Yates tells us that memory training was an important part of the teaching of rhetoric in classical Greece, where a mnemonic system involving the construction of a remembered structure of places and things could be used to fix a sequence of words, ideas, or other kinds of associations that could be recalled at will. Extraordinary feats of memory could be accomplished by use of this device. The elder Seneca "could repeat two thousand names in the order they had been given" (Yates 1966:16). The system was widely known and practiced during the Middle Ages and threads run through the Renaissance and into the early modern period, where it found its place among the discussion of "methods" (Yates 1966:370).

This formed the basis of programs for the improvement of memory that appeared in the succeeding centuries, such as the one popularized by M. Feinagle in the beginning of the nineteenth century (Feinagle 1813). It was based on memorizing the objects in a house and associating desired memories with each of the objects. The troubadours in medieval societies played a role similar to that of the modern journalist, reciting their accounts from memory, except as Dawson informs us, "he was apparently permitted more editorializing" (Dawson 1921:5). The oral tradition prevailed in the early university curricula through lectures and disputations. We cannot be sure what effect oral transmission had on information in the distant past, but we have some knowledge through recent studies of the kind of distortions that can occur in the dissemination of rumors (Rosenbaum 1963). Oral tradition incorporates a vast repertoire from religious history, to narratives in poetry and prose, to folk remedies and customs that form the basis for our cultures.

Oral transmission of information played a large part in the history of science and technology long after the invention of writing and printing, and still continues to do so today. University teaching in the Middle Ages was largely based on oral lectures. The four basic faculties in theology, canon law, medicine, and arts required little in the way of demonstration, and manuscript textbooks were scarce and expensive. The study of medicine was pursued through disputations and other academic exercises, and the practical study of medicine took place under established practitioners. Many literary, religious, and other texts have come down to us as a result of oral transmission. They make up some of the great epics and traditions of our times, in spite of the problems of authorship and variations in texts. Oral transmission of information continues to play an important part in our society in education and the arts.

SCRIBAL CULTURE

The invention of writing made it possible to create a more or less durable record that could be preserved, conveyed to someone at a distance, and passed on to future generations. All kinds of surfaces and materials on which a mark could be made or inscribed were used: stone, metal, clay, papyrus, and, later, paper (Winger 1955:13). Many of these materials are still in use for their original purpose today, even if only for commemorative plaques. The earliest medical documents were discovered in Babylonia early in the twentieth century when an archaeologist found what is thought to have been a part of the library of King Sargon of Akkad who ruled about 3800 BC. The records included clay tablets with reports of physicians, including recipes for various diseases (Ball 1925). Among the earliest Egyptian records is the Edwin Smith Papyrus, a copy made in 1600 BC of a collection of surgical cases originally written in 2600 BC (Breasted 1930). The cases were presented in a formal pattern that included the examination of the patient, a diagnosis and, in some cases, where the physician decided to venture it, a treatment. The great library at Alexandria, which dated from about 300 BC, grew to hold around 490,000 papyri rolls before it was burned in 642 AD (Thiem 1979).

The Greeks had an alphabet by the latter part of the eighth century BC, although the tradition of an oral literature is said to have persisted until well into the sixth century BC. We are told: "It is not until the middle of the fifth century or a little later that a book trade can be said to have existed in Greece" (Reynolds 1974:2). This is not difficult to understand when we learn that the scroll that contained the text of Plato's Symposium was about twenty feet in length, whereas it would have taken up little more than seventy pages in book form (Reynolds 1974). Nevertheless, the book trade flourished in ancient Rome where a publisher, employing large numbers of slaves who wrote from dictation, could produce a book in a very short time (Thornton 1954). Manuscript books continued to be produced in response to demand all through the Middle Ages. Lucien Febvre and Henri-Jean Martin, in their *The Coming of the Book*, cite a Flemish manuscript that contains an order to a scriptorium in 1437 from one "who was almost certainly a wholesale bookseller" for two hundred copies of each of two texts and four hundred copies of a small prayer book (Febvre 1976:28). "Slave labor, with all its drawbacks, made it possible for ancient Rome to manage remarkably well without the printing press. With his trained staff of readers and transcribers, a publisher could turn out an edition of any work at very cheap rates and almost at a moment's notice. There was no initial expense of typesetting before a single copy could be produced; no ruinous extras in the shape or printer's corrections. . . . [T]he complete edition could be ready, if necessary, within twenty-four hours." It has been estimated that the size of the edition could be as much as five hundred or even a thousand (Mumby 1974:23).

The advent and spread of writing systems in most places in the inhabited world vastly expanded the ability to preserve and disseminate scientific and technical knowledge. Goody relates the growth of knowledge to the proposition that "writing, and more especially alphabetic literacy, made possible the kind of scrutiny of discourse (by giving oral communication a semi-permanent form) which made possible the increase in scope of critical activity (and hence of rationality, skepticism, and logic)—to resurrect memories of those questionable dichotomies" (Goody 1972:27). Only a small part of the manuscript documents that survived the onslaught of time, conquest, and other human misfortunes deal with either science or technology. This is probably because those manuscripts that we possess survive primarily in the form of copies and frequently in the form of copies of copies. Most of the scribes were employed to copy texts dealing with subjects other than such mundane ones as agriculture and commerce. Books produced in this way were very expensive and beyond the means of most students. Learning, therefore, well into early modern times was based on the oral tradition.

One of the important products of writing systems was correspondence, which not only enabled ideas to be exchanged at great distances, but in fortunate cases also left records that provide important historical data. Before the invention of printing and of the scientific journal, correspondence (the so-called erudite letter) was one of the few ways in which scholars could establish priority in new ideas and discoveries. Another kind of writing that provided a format that may have had some influence on the form of the scientific paper was the consultations that physicians wrote on request for patients in distant places, called variously *consulti* or *consilia*. They are unitary communications, usually discussing a single case. Several collections of these written by famous physicians of the past still exist. One such collection is that of Ugo Benzi (1376–1439), a celebrated physician and philosopher (Lockwood 1951:47). Saul Jarcho has recently translated and published collections of *consilia* of Giambattista Morgagni (1682–1771) (Morgagni 1984) and Francisco Albertini (1662–1738) (Jarcho 1989). Marcello Malpighi (1623–1694) left a large number of *consulti*, none of which were published during his lifetime (Adelmann 1966). The *consilia* sometimes represented a form of diagnosis at a distance, which also involved the practice of urinoscopy, in which a urine sample was sent to a physician for a judgment. The French physician Theophraste Renaudot, who played an important role in the history of general journalism, published a pamphlet in 1643 in which he described a method of treating patients by correspondence, which he titled *La Presence des Absents*, and undoubtedly further enraged his arch enemy, the Dean of the Parisian Faculty of Medicine, Guy Patin (Packard 1924:188).

Lecture notes is another form of written record that has significance in historical studies. In some cases, lectures notes provide the only record of the teaching of some individuals and, in some instances, have even been used to establish priorities. In the dispute with the second Alexander Monro (1733–1817) over the description of the absorbent function of the lymphatic system, William Hunter (1718–1783) cited the evidence of notes taken by his students. It was confirmed by a set of lecture notes of one of

Hunter's students that were not found until 1959, in Australia, of all places, and were dated 1752, three years before Monro published his description (Eales 1974:285). Joseph Black showed a similar reluctance to publish his work. When it came time to publish his lectures, the editor was required to have recourse to handwritten notes made by his students (Golinski 1992:45).

Elizabeth Eisenstein uses the term "scribal culture" to describe the methods of producing books, messages, and other forms of information before the introduction of printing (Eisenstein 1979:3, 9). The deficiencies of writing systems as a means of disseminating and storing information became vividly apparent once printing was invented in the 1450s and spread rapidly throughout Western Europe. Manuscript books were very expensive and difficult to procure. It required about ten months for a good scribe to produce a copy of the Bible that might then sell for as much as 600 crowns (Caton 1988:112). Manuscripts were more vulnerable to loss and destruction because they tend to survive only in single copies that are frequently fragmentary. When they survive in multiple copies, the copies may vary considerably one from the other. Sigerist points out the difficulties in understanding and evaluating written texts where "dozens of manuscripts of the same book were different from one another and from the original book" (Sigerist 1951:21). This is particularly true of ancient texts such as Galen's major works on anatomy and dissection, which survive only in the form of fragments from which scholars have attempted to create a complete text (French 1978:235). It is estimated that little more than 10 percent of all Greek writings were preserved as they were transcribed from one copy to another. Over two thousand manuscript copies of Aristotle's works dating from the thirteenth and fourteenth centuries have been found, and it has been estimated that probably fifty times as many may have been originally produced because his works were at first written (Fabre 1963:48). It should be pointed out that the problem of determining "true text," that is, a text actually of a particular author, was not entirely eliminated when printing was introduced and fidelity to the author's text depended upon the competence and diligence of the compositor as well as the proofreader.

There is another deficiency that applies to all forms of communication—verbal, written, or printed. There are apparent limits to what can be transmitted by verbal description (Holmes 1992:126). Some information can only be conveyed by personal contact or actual witnessing an experiment or examining a piece of equipment. An example is Robert Boyle's important experiments with the air pump in the 1660s. In their *Leviathan and the Air-Pump*, Shapin and Schaffer show how "despite Boyle's reporting practices, no one was able to build a pump and make it operate without seeing the original" (Shapin and Schaffer 1985:20). Joseph Priestly, another great pioneer in pneumatic science, echoed this view in a letter that he wrote to Caleb Rotherham in 1774:

> I do not expect that the account of my experiments on air will give you so much satisfaction as you expect, owing to the great difference between seeing and reading. I have not yet found any person, though ever so good a philosopher, and who has read my paper ever so carefully, but is surprised to see me make the experiment. But the instructions I have given are sufficient to enable any person to do everything after me. (Schaffer 1984:151)

This is particularly true in technology, as Basalla tells us: "Because *all* of technology can never be translated into words, pictures, or mathematical equations; the practitioner with a hands on knowledge, be it of eighteenth century textile machinery or twentieth century computers, will always have a role to play in the dissemination of technical innovations" (Basalla 1988:83).

This difference between book learning and practice is still highly recognized today in the establishment of apprenticeships, internships, and residencies, and the prevalence of hands-on laboratories in scientific and technical training. In his essay "The Mind's Eye," Ferguson explains "Many features and qualities of the object that a technologist thinks about cannot be reduced to unambiguous verbal descriptions; they are dealt with in his mind by a visual, nonverbal process (Ferguson 1977:827).

ADVENT OF PRINTING

The great ideological and intellectual changes that took place in the centuries following the invention of printing in the fifteenth century have been well established in many appraisals. To Eisenstein (1979), the invention of printing provided a watershed between the ancient and the modern world, while it undoubtedly supplied a new and important technology for communicating information. The new information that was disseminated was also in response to a new philosophy of the purposes and pursuit of knowledge. Mellottée cites Didot, who wrote in 1851: "The discovery of printing separated the ancient world from the modern world," to which Mellottée adds: "There is perhaps no other invention which by its effects so changed the face of things as the invention of printing" (Mellottée 1905:11).

Roger Chartier has characterized Eisenstein's book as "without question one of the most imposing monuments erected in homage to Gutenberg's invention." Its essential theses, he says, is that "printing and the constantly growing recourse to printed works represents a fundamental discontinuity of which the effects are felt in all the domains of social and intellectual life" (Chartier 1981:191–92). These changes were not precipitous, nor were they all a consequence of the introduction of this novel method of disseminating information by a mechanical process. Under the following headings, Kernan summarizes Eisenstein's conclusions about some of the changes introduced by printing: (1) books were made more easily available, (2) the production of accurate (similar) texts was more possible, (3) the standardization of formats was encouraged, and (4) the product it provided was more durable, that is, it promoted the survival of texts by means of multiplying the number of copies (Kernan 1987:53).

One notable and highly visible improvement introduced by printing was the quality and reliability of illustrations, especially in botany and anatomy. Nevertheless, manuscripts survived for a long time as a method of information dissemination, even as they do today under restrictive governments. Chartier identifies three principal areas in which manuscripts were found: in clandestine ideology, in the maintenance of administrative secrets for state purposes, and as a result of religious persecutions (Chartier 1977:499). Scribes continued to be useful in preparing good copy for the printer from the sometimes crabbed manuscripts of university professors (Reeds 1983:265). The spread of printing was almost instantaneous as presses continued to be set up all over Europe. The growth continued throughout the seventeenth and eighteenth centuries. The total number of presses in Paris alone increased to 195 in 1701, and the number of shops with four presses doubled from 8 to 16 between 1644 and 1701 (Martin 1969:699). The output of the fifteenth-century press was impressive. It has been estimated that at least 30,000 to 35,000 different editions were printed between 1450 and 1500, representing 10,000 to 15,000 different texts. If an average printing of each of these editions was no more than 500, the total number of books printed before 1500 would be about twenty million (Febvre 1976:248).

Lest we completely accept the axiom that the medium consecrates the message, we should remember George Sarton's study of the books on science that were printed in the period up to 1501. He concluded:

> The early presses consecrated the hegemony of Aristotelianism as harmonized with Christian theology. . . . The majority of the incunabula authors were contemporaries, and most of these were mediocre. . . . However, if the printed literature of any time were examined in the same way that we were able to examine the total printed literature of the fifteenth century, mediocre publication would swamp the others in the same manner, for the simple reason that mediocre authors are always more numerous than distinguished ones, and each of them on the average are more prolific. (Sarton 1938:82–84)

The data confirmed an earlier judgment Sarton made that "printers were business men (they were so from the very beginning) and it is probable that, then as now, they tried their best to meet the needs of the public"

(Sarton 1938:55). Nevertheless, elsewhere Sarton also says:

> The discovery of printing was one of the great turning-points in the history of mankind, and it was of special importance for the history of science. It changed the very warp and woof of history, for it replaced precarious forms of tradition (oral and manuscript) by one that was stable, secure, and lasting; it is as if mankind had suddenly obtained a trustworthy memory instead of one that was fickle and deceitful. (Sarton 1957:3)

In addition to scholarly works on the classical authors, the press poured out almanacs, popular books on medicine and hygiene, and manuals on various crafts derived from the manuscript "book of secrets" tradition that was welcomed by scholars who were beginning to be more and more involved with practical matters (Eamon 1990:113). To disregard the importance of the invention of printing to Stillman Drake is "as absurd as if social historians in the future were to regard the replacement of the horse by the automobile as having introduced no fundamental change" (Drake 1970:45).

Important as it was, the advent of printing was not the sole, or even the most important, influence on the development of the scientific journal. Nor did printing render earlier forms of communications technology obsolete. There is ample evidence all around us that few methods or technologies used in the transmission of information ever go out of use and are in some cases even enhanced by newer technologies. The axiom holds true for incising messages on stone and writing or printing on paper, which the information specialists call asynchronous, as there is an interval between the time they are created and the time they are received. It is less true for some of the new electronic technologies, in which messages can be both asynchronous and synchronous, some of which rapidly become obsolete (Paisley 1993:222).

It took two full centuries before the conditions arose that made the scientific journal not only possible but also necessary and distinctive. The revolutionary changes in information dissemination that took place in the seventeenth century were a result not so much of technical factors as of philosophical, ideological, social, and economic factors. Among these are the increase in the seventeenth century in political and economic stability, the growth of literacy in all classes, and an increase in secular thinking (McLean 1972:49). Merton has called attention to other factors that gave impetus and direction to those pursuing the new science. They included the growth of capitalistic enterprise, the expansion of foreign trade, and the increase in marine traffic (Merton 1970:71). Surplus wealth, which had always existed among the ruling classes, the kings, princes, and dukes, now became available to more and more commoners such as merchants, manufacturers, and large landowners. The most important factor, however, which led to the introduction of the scientific journal, was an ideological one, the scientific revolution; it would perhaps be fair to say that it was preceded by a philosophical revolution.

THE PHILOSOPHICAL REVOLUTION

There is a voluminous literature on the philosophical revolution and the scientific revolution that was one of its primary manifestations. Ornstein refers to the period as "a revolution in the established habits of thought and inquiry, compared to which most revolutions registered in history seem insignificant" (Ornstein 1929:21). Events are usually regarded as revolutionary when they occur in a brief period of time. The scientific revolution of the seventeenth century, however, had deep roots going back to classical times, to the classical authors who nourished it, and to the Middle Ages when these authors held sway. The Reformation challenged the authority of the church and, by extension, the authority of the classical authors. The humanists turned their vision away from the world to come to the world that was present. The Renaissance prepared the way for the development of a new outlook on nature and methods of examining it.

Some of the ancient ideas still persisted for a long time. Johann Theodor Eller (1689–1760), in a memoir he submitted as a member of the Berlin Academy of Sciences and Letters, was still asserting in 1746 that the two basic principles in nature were fire and water (Gough 1968:53). "Speculative philosophy," which, in Maupertius' words, "studies the objects that do not have the quality of being material," still played a significant role in the Berlin Academy (Aarsloff 1989:198). Recent students of medieval philosophy and science have made it clear that the Middle Ages was not devoted entirely to a barren cultivation of an authoritative body of knowledge consecrated by famous classical authors, but was a part of the "continuous intellectual tradition emanating from the medieval universities" (Lindberg 1978:viii).

Some of the characteristics of medieval intellectual life were antagonistic to the emerging philosophical views of the seventeenth century. Originality and competition for recognition "were largely foreign to the medieval mind which commonly sought to cloak the truly original under the tradition of earlier periods" (Merton 1970:169). Science and medicine were largely based on the written word, which acquired an almost sacerdotal quality, particularly when associated with the names of those recognized as great sages. It is no coincidence that the scientific journal, which was dedicated to the dissemination of new knowledge, had to wait for the origins of modern science. The predominant intellectual goal in the Middle Ages was the development of self-consistent and integrated philosophical systems that took the form of monographs and dealt with generalities rather than particulars. Medieval and Renaissance science, says Butterfield, was pretty much a matter of catching up with the science of the ancients. "It is not always realized to what a degree the sciences in the middle ages were a matter for what we today would describe as literary transmission and came into European history as a heritage from ancient Greece and imperial Rome" (Butterfield [1957]1965: 89).

When we say that science in the Middle Ages was a "handmaiden to theology," and that scientists had to take care in their ideas and discoveries not to conflict with prevailing religious ideas, we must remember that some of the scientists of the seventeenth and eighteenth centuries and even later also had to deal with this restraint. We must also remember that there were many members of the clergy who participated in the development of science in all periods.

Twentieth-century students of medieval science have made it clear that the Middle Ages was not completely devoted to a barren cultivation of an authoritative and unchallengeable body of knowledge consecrated by the classical authors and the church (Duhem 1914–1959; Thorndike 1923–1958; Crombie 1953). It has been made clear that many of the "new ideas . . . of the scientific revolution had their roots in medieval thought" (Lindberg 1978:viii). The contribution in technology also has been acknowledged: "The literature of the fourteenth and fifteenth centuries is extraordinarily rich in treasures of a technical nature, which at times were real manuals, and at times disconnected reflection on their own work or procedures employed in the various arts" (Rossi 1970:15).

Instruments such as the quadrant and the astrolabe were among the products of medieval technology that served both the astronomers and the seamen of the early modern period. The medieval alchemists, despite their aura of mystery and secrecy, contributed basic laboratory techniques and tools that were used by the chemical experimenters following the new philosophy. The classical scientific writings continued to command attention. In the seventeenth century, the august College of Physicians of London could still be called a "citadel of Galenic orthodoxy" (Dewhurst 1963:20). Upon assuming his position as Savilian professor at Oxford in 1703, Edmond Halley, despite his active interest in the new science and technology, found it necessary to take up the study of Greek geometry, a subject he had been interested in for some time (Ronan 1970:186). The Middle Ages left rich scientific heritage transmitted from Greek and Arabic science, particularly in mathematics, astronomy, and optics, which provided a solid basis for the development of these sciences in the seventeenth century.

The terms "Renaissance" and "humanism" have no fixed chronological limits and broad ideological significance and differences to various scholars. They represent contemporary ideologies that both converge and diverge from each other. To Sarton, the Renaissance roughly covers a period from 1450 to 1600; it begins, he says, approximately with the introduction of printing and ends approximately with the scientific revolution, when the seeds of both the Renaissance and humanism were beginning to bear significant scientific fruit (1957:vi). The Renaissance was characterized by a rebirth of classical humanism, with its emphasis on human values and classical aesthetic sensitivities, and an accompanied liberation of the spirit and the freedom to think new thoughts and share them with others. The rebirth of interest in the art, philosophy, and science of the Greeks and Romans had threads reaching into the Middle Ages. It was manifested by the rediscovery and translation of classical texts. "It was then that ancient Greek mathematical and scientific works were rediscovered that had been unknown to, or utterly neglected by, medieval writers. Correct texts were made, and appropriate commentaries were written also for other ancient works that had been known only in part or only in corrupt versions" (Drake 1970:44).

In reviving and expanding the thoughts and interests of the classical authors, the humanists turned their attention to the world around them instead of contemplating the hereafter. In the past, humanism and science have been regarded as antithetical aspects of the Renaissance. Humanists were regarded as being preoccupied with literary style (Crombie 1952:268) and admiring the ancients as much for their eloquence as for their wisdom. Drake characterizes Galileo's opinion of the humanists when he quotes him as saying that the oratory presented at the humanistic societies were "lovely flowers of rhetoric" that are followed by no fruit at all (Drake 1966:1195). Both Sarton and Thorndike have been cited as promulgating the idea that the humanists "turned the minds of prospective scientists from fact to form, from substance to grammar, from empirical investigations to adoration of ancient authorities" (Cochrane 1976:1039–40).

While these comments may with some justice be applied to some of the humanists of the Renaissance who were primarily interested in the philological aspects of the classics, they are by no means characteristic of all of them. Sarton also pointed out the contributions of the humanists in rejecting some of the fantastic notions of the ancient writers, discrediting the appeal to authority to answer questions, and being more receptive to developments in technology than their predecessors (Cochrane 1976:1051). It may not be too broad a generalization to say that all of the Renaissance scientists were humanists, but not all of the humanists were scientists. The examples are numerous. Kepler was absorbed in the great classical texts and drew inspiration from them (Grafton 1992:72). Agricola began his career as a popular teacher of Greek. Among his contributions were glossaries of Latin and Greek terms in German that he used in his monographs on fossils and metals. He also "created single-handedly a comprehensive set of science of the underworld and wrote one of the great masterpieces of early modern technology, the *De re metallica* for which he is now chiefly remembered" (Hannaway 1992:555–59). The Minim priest Marin Mersenne (1588–1648) is another example. He wrote works on logic, biblical exegesis and theory, before engaging himself in the study of mechanics, optics, and the theoretical aspects of music (Dear 1988). There are numerous other examples of humanists who were also scientists, and the tradition continued through the early modern period through the eighteenth century and into the nineteenth. Albrecht von Haller (1708–1777), one of the most celebrated scientists of the eighteenth century, sought an appointment in literature in Bern after getting his medical degree at Leiden, and wrote poetry and published novels before taking a post in physiology at the University of Göttingen. Georg Christopher Lichtenberg (1742–1799), an outstanding physician, is considered one of the outstanding critics of German literature in his period as well as "one of the more notable scientists of the *Aufklärung*" (Dougherty 1988:78). While he was professor of mathematics and physics at the University of Göttingen, he wrote on a wide variety of subjects in the arts, belles letters, and science in *Göttinger Taschenkalendar*, of which he was the editor and principal author for twenty years (Mautner 1952). And, of course, there is also Goethe (1749–1822), who wrote some of the greatest poetry of his age as well as works on botany, comparative anatomy, physics, and geology (Magnes 1949). The concept of science and the humanities as separate and opposed forms of scholarship is said to have developed only in the nineteenth century as a product of the structure of new university disciplines and curriculae (Blair 1992).

Particularly in the field of medicine, scholars began to question the teachings of such venerable figures as Hippocrates and Galen and trust only the evidence seen from their own eyes. Paracelsus burned the books on Galen in public and suffered the consequences for it. Vesalius created an entirely new human anatomy and, with the help of skilled artists, created one of the greatest anatomical books of the Renaissance. But even he found it difficult to deny the presence of a nonexistent intraventricular septum in the heart by which Galen supported his theory of the movement of the blood. The interchange of skills between the artists and the scholars is also reflected in the collaboration of scholars with craftsmen in many fields (Hall 1959; Zilsel 1942). In this period, says Rossi, "a new historical reality, made possible that collaboration between scientists and technicians and that penetration of technology into science and vice versa which was at the root of the great scientific revolution of the seventeenth century" (1970:31).

Humanists who were confronted with technical problems encountered in the classical texts sought the help of workers in the crafts. The craftsmen in their turn benefited by being exposed to ideas derived from classical texts. The works of Vitruvius, which came to light only in the Renaissance, are an indication of the high level of technology that existed in classical times (Neuberger 1969). Renaissance scholars were turned increasingly toward practical problems. Simultaneously, artisans began to acquire intellectual interests.

There was also no line drawn between the crafts and the "fine arts." Artists such as Leonardo and Brunelleschi were skilled in many crafts including architecture, sculpture, and hydraulics. Brunelleschi, Donatelli, Ghiberti, and Ghirlandario all began their careers as goldsmiths (Rossi 1970:16, 22). Leonardo wrote a letter to the Duke of Milan around 1482 applying for an appointment to the court. He began the list of his qualifications with a description of his ability to fabricate mobile bridges as well as "plans for burning and destroying those of the enemy," make military weapons, and construct public and private buildings. Only at the very end does he mention his abilities as a sculptor and painter (MacCurdy 1955:1152–3). The close union between the arts and the crafts persisted into the eighteenth century, as can be seen in the history of the Royal Society of Arts, which in 1754 began its existence as the Society for the Encouragement of Art, Manufactures, and Commerce. Among the early premiums the society awarded were prizes for drawing, producing cobalt, and raising madder (Hudson 1954:9). The new natural philosophers needed the skills of the craftsmen to fabricate the necessary instruments and apparatus to carry out the experiments that were the focus of the scientific revolution.

Despite some of the criticism of his philosophy, Francis Bacon (1561–1626) is still regarded as the harbinger and the prophet of the scientific revolution. Thorndike, nevertheless, has called him "a crooked chancellor in a moral sense and a crooked naturalist in an intellectual and scientific sense" (Thorndike 1923, 7:88). Ravitz, however, reminds us that, although none of Bacon's experiments added much to science, his ideas about the function of science and how it should be pursued were influential not only in England, but everywhere his works were read (Ravitz 1989:116ff). There were many efforts to throw off the shackles of scholasticism and challenge some of the thinking of the classical philosophers. The contributions of Francis Bacon are regarded by many as watersheds in this endeavor and in developing modern science. Richard Foster Jones echoes Ornstein when he says, "Sir Francis Bacon bears the same relationship to the movement under discussion [the scientific revolution] as, for example Karl Marx bears to the development of Communism, but, of course, to much better purpose" (Jones [1961]1982:viii).

There are numerous examples of Bacon's influence during the seventeenth and eighteenth centuries. His ideas are said to have been a factor in the founding of the Academia Naturae Curiosorum founded in Germany in 1651, although *New Atlantis* was not available in German until the Frankfurt edition of Bacon's collected works in 1665 (Winau 1977). Although Pelsener accuses Bacon of naiveté and calls to our attention the fact that Bacon only mentioned Galileo twice in his works, and then only in his writings from Italy, and to the fact that he completely ignored the discoveries of Kepler, he still reminds us that Voltaire called Bacon the "father of experimental philosophy"(Pelsener 1932:173). When the faculty of the University at Göttingen was forming its academy, Albrecht von Haller (1708–1777), one of its founders, looked to Bacon's *New Atlantis* for the concept of the functions of a scientific society (Sonntag 1975:380). The *Discours préliminaire* to the 1755 edition of *Collection Académique*, a publication in thirteen volumes that sought to bring together the scientific literature of the principal academies in Europe, begins with a review of Bacon's philosophy, stating with him that "observation is the first step of philosophy, and the facts which observations accumulate must be regarded as the basic substance of our general ideas and our science (*Collection* 1755, 1:vi). And finally, to cite only a few of the accolades for Bacon in this period, the first article of *Journal Encyclopédique*, launched on January 1756, was dedicated to a panegyric on Bacon who is heralded as one of the precursors of the encyclopedists.

Bacon's program for the new science called for the accumulation of factual data from which generalizations could be drawn cautiously. Knowledge of this kind, he thought, should be pursued in a cooperative manner with scientists working collaboratively or sharing the results they had acquired individually (Jones 1961[1982]:xii; Zilsel 1945:325). He also recommended that the knowledge that currently existed in all areas of science and technology should be systematically collected. He described these efforts as an "Inventory, or an enumeration and view of inventions already discovered and in use" (Anderson 1948:33). He called these inventories "histories." The Royal Society of London began to compile a "History of the Trades" (Houghton 1941), but it remained for the Académie Royale des Science of Paris to complete a magnificently produced *Description des Arts et Métiers* in the middle of the following century, between 1761 and 1788.

Many of Bacon's ideas about how science should be pursued and managed are incorporated in the description of an institute devoted to the new experimental science, which he described in his *New Atlantis*, a utopian novel published after his death (Webster 1967:394). The institute to which the visitors to the island are conducted, the guide tells them, is "an order, or society which we call 'Salomon's House.'" The account (it was not complete when Bacon died) ends with a kind of table of organizations for the institute, outlining the functions of the staff members. Most of these functions are related to the creation and dissemination of news of science and technology.

There are twelve members who travel abroad to gather information relevant to the functions of the institute, called "Merchants [perhaps more aptly Collectors] of Light." There are three that collect experiments from the literature, oddly called "Depredators." An additional three "collect the experiments of all mechanical arts, and also of liberal sciences," a function that seems to overlap with those of the "Merchants of Light" and "Depredators." These three are called "Mystery-men," another ambiguous designation that may refer back to the "secrets" of medieval craftsmen, such as alchemists. In addition, there are three that "try new experiments," called "Pioneers or Miners"; three that "draw the experiments of the former four into titles and tables"; three that are involved with what might be called development, who spend their time "looking into the experiments of their fellows and cast about how to draw out of them things of use and practice," called "Dowry-men or Benefactors." All of these activities are followed by a conference among the members of the entire institute, after which three, called "Lamps," are assigned to "direct new experiments," and three to "execute the experiments," called "Inoculators." Finally, there are three that "raise the former discoveries by experiments into greater observations, axioms and aphorisms," called "Interpreters of Nature."

The staff also includes "novices and apprentices" to assist the research and development members as well as "a great number of servants and attendants, men and women." Feminists may wish to note that women are referred to only among the "servants and attendants." The scientific enterprise needs all these functions and activities fulfilled in order to function optimally. Although there are echoes here of the humanistic societies of the Renaissance, this represented to Bacon an ideal institute in an ideal society. There do not seem to have been any research institutes or scientific societies organized on this elaborate model, but the staff designations clearly suggest many of the functions significant in the development and progress of science and technology. The most significant element and the most important one for the development of the scientific journal is the emphasis on experiment and discovery. It reflected the view that, although we must be aware of and reexamine the knowledge of the past, we must also seek out new experiences in nature (Bacon 1942:299–300).

NEWS OF SCIENCE AND THE EXPERIMENT

Several recent writers have commented on the widespread use of the word "new" in the literature of the seventeenth century. This sense of newness was stimulated by the geographic discoveries of the fifteenth and sixteenth centuries that opened minds to new kinds of discoveries in nature. Thorndike commented on the frequency in which the word "new" appeared in titles of seventeenth-century works, such as Bacon's *Novum Organum* (1620) and Galileo's *Dialogues Concerning Two New Sciences* (1638) (Thorndike 1951), and Lindberg speaks of "a rhetoric of 'newness'" in this period (Lindberg 1990:xvii). In *The Discovery of a New World*, John Wilkins was open to considering the "plurality of worlds" and the possibility of inhabitants on the moon (Shapin 1996:5). The excitement and curiosity created by the "new philosophy" helped to enhance a receptivity to wonders including those published by the philosophers, particularly when they included news of curious and wondrous things. The sense of excitement in this new age is reflected by the physician Henry Power (1623–1668), who in 1664 wrote in his *Experimental Philosophy*: "This is the Age wherein (methinks) Philosophy comes in with a Spring-tide. . . . These are days that must lay a new Foundation of a more magnificent Philosophy, never to be overthrown that will Empirically and Sensibly canvas the Phenomenon of nature" (Hall 1991:55).

A similar exuberance is exhibited by Edward Browne (1644–1708), the eldest son of Thomas Browne and also a physician, who in a travel report published in 1673 described the period he was living in as "an age so curious and inquisitive and withall so industrious, that every day addeth new Information and Accounts both to our county and foreign parts" (Spillane 1984:27).

There is general agreement that the establishment of the "experiment" as a primary means of acquiring new scientific knowledge was one of the most important contributions to the scientific revolution. "Seventeenth-century England witnessed the rise and institutionalization of a program devoted to systematic experimentation by a literature explicitly describing and defending practical aspects of that program" (Shapin 1988:373)

The idea began to be established that scientific knowledge grew by small increments that could lead to broader concepts and could eventually be incorporated into acceptable generalizations or laws. Nature, said Fontenelle, must be pursued in "small parcels" before we can attempt to construct grand syntheses (Delorme 1949:124). The scientific revolution is characterized by the change from attempting to create full blown and all-encompassing systems to the task of solving separate discrete problems about the way nature works. The goal of preserving the products of these inquiries was emphasized in the statements made by the editors of new journals in their prefaces. The experimental method molded the new scientific literature and the way it was recorded and disseminated. In the preface of the first issue of *Patriotisches Archiv für die Schweiz*, published in Basle in 1789, the editor indicates that, above all, he is interested in "fakta, fakta."

Experiments that produced new knowledge had to be shared in order to be validated by a consensus of peers. In his *Genesis and Development of a Scientific Fact*, Ludwik Fleck says that scientific facts are "socially constructed" (Shapin 1980)—that is, they must be communicated in such a way that they that they can be verified. In his *Structure of Scientific Revolutions*, Kuhn defines the "paradigm" as a "concept based on a form of consensus or agreement, which therefore requires the development of an adequate means of communication" (Kuhn 1962:x). The early scientific societies, such as Accademis del Cimento in Florence and Académie des Sciences in Paris, tried to accomplish this goal by promoting collaborative research and anonymity of authorship. It has been suggested that the insistence on anonymity may have been one of the reasons for the breakup of the Accademia del Cimento (Hanlon 1959:2), and it soon became evident that it was not acceptable to the members of Académie des Sciences as well (Hahn 1971:19–27). By the end of the eighteenth century, the position of the experimental approach to research was firmly established. Article 3 of the Regulations of Société Philomathique stipulates: "The Society will maintain a register of experiments to replicate. These experiments will be chosen according to the preferences of those who support the basic theory involved. The Society will name the commissioners who will be charged with repeating the experiments which they have chosen from an examination of the register. There will always be experiments in process of being repeated" (*Rapports* 1792, 1:ii).

This attitude is also reflected in Thomas Beddoes' contempt in his review of Lamarck's *Mémoires de Physique et d'Histoire Naturelle* in Griffith's *Monthly Review* in 1793, where he comments that the author "did not find himself under the necessity of making a single experiment" (Stanfield 1984:107). I. B. Cohen indicates "the prevalent methodological emphasis upon experimentation, an emphasis that made the 'experimental history' a typical scientific product of the day" (Cohen 1985:37). This is particularly significant for the scientific journal because an experiment represents a single unitary experience that is most appropriate for distribution in that format.

This is not to insist on the fact that experimentation in science originated in the seventeenth century. In his *Budget of Paradoxes*, DeMorgan states: "No knowledge of nature without experiment and observation: so said Aristotle, so said Bacon, so acted Copernicus, Tycho Brahé, Gilbert, Kepler, Galileo, Harvey, etc., before Bacon wrote. No derived knowledge until experiment and observation are concluded: so said Bacon, and no one else" (DeMorgan 1915:1:76). Undoubtedly, there were also experiments before the early modern period. We can perhaps even go back as far as Archimedes (287–212 BC), who said "Eureka" when he found the principle of specific gravity by discovering that a body displaced an amount of water equal to its weight (Sarton 1959, 2:77).

Scholars in the thirteenth century such as Robert Grossteste conducted an experiment when he explained the shape of the rainbow by using a spherical lens (Crombie 1953:74). William Gilbert's (1546–1603) experiments on the properties of the magnet are considered "the first major original contribution to science published in England (Singer 1959:221). Santorio (1561–1636) designed experiments to study metabolism by gathering physiological data on the relationship between ingestion of food and its effect on body weight, and Otto von Guericke (1602–1686), the mayor of Magdeburg, was one of the early investigators of electricity and invented the air-pump, which Boyle used to good advantage in his experiments (Heathcote 1948–1950). Francesco Redi (1621–1697) who was a member of Accademia del Cimento and conducted most of the experiments for the education and entertainment of the Medici court, said: "I do not put much faith in matters not made clear to me by experiment" (Findlen 1993:42) And we must not forget William Harvey (1578–1657), whose experiments on the circulation of the blood revolutionized the study of physiology.

Before the invention of the scientific journal, new information of this kind had limited means of dissemination or storage. A scholar could write a letter to a colleague, who might share the information with his friends and correspondents, or the scholar could collect his experiments and observations in a monograph. These modes of dissemination continued to be used throughout the scientific revolution and indeed through today. The scientific journal provided an important alternative that made it possible for scientists to communicate the results of their experiments without waiting to compile a collection to merit the publication of a book. The journals by no means consisted entirely of reports of experiments; they also included observations and experiences that in some instances can be classified as random and uncontrolled experiments if a similar kind of scientific rigor is applied to them.

Bazerman studied the evolution of the "experimental report" (reports on experiments) by looking at every fifth volume of the first scientific journal, *Philosophical Transactions* of the Royal Society of London, from volume 1 to volume 40, and every tenth after that to volume 90 (1665–1800). He selected all those that used "the word experiment either in the title or running text." After eliminating all those in which "the author was only referring to prior experiments," he found that only one hundred of about one thousand articles qualified as reporting "new experiments" (Bazerman 1988:63). The vast majority of the articles in the early scientific journals were not reports of experiments, especially in the taxonomic sciences, which are based primarily on observation. In fact, a large part of the contents of the early scientific journals consist of the same kind of letters that scientists of that period sent to their friends and colleagues. It was this commerce in letters that the scientific journal was designed to facilitate. This helps to support the idea that any revolutionary idea or practice takes time to establish itself and to flourish as it does today.

REFERENCES

Aarsloff, Hans. "The Berlin Academy under Frederick the Great." *History of the Human Sciences* 2 (1989): 198–206.

Adelmann, Howard B., ed. *The Correspondence of Marcello Malpighi.* 5 vols. Ithaca, N.Y.: Cornell University Press, 1985.

Anderson, Fulton H. *The Philosophy of Francis Bacon.* Chicago: University of Chicago Press, 1948.

Bacon, Francis. *Essays and New Atlantis.* New York: Walter J. Black, 1942.

Ball, Otto F. "Early Medical Journalism." *Journal of the American Medical Editors' Association* 5 (1925): 56–58.

Basalla, George. *The Evolution of Technology.* Cambridge, Mass.: Cambridge University Press, 1988.

Bazerman, Charles. *Shaping Written Knowledge: The Genre and Activity of the Experimental Article in Science.* Madison: University of Wisconsin Press, 1988.

Blair, Ann, and Anthony Grafton. "Reassessing Humanism and Science." *Journal of the History of Ideas* 53 (1992): 535–40.

Breasted, James Henry, trans. *Edwin Smith Papyrus.* Chicago: University of Chicago Press, 1930.

Bromehead, C. N. "Mining and Quarrying." In *A History of Technology,* vol. 1. Edited by Charles Singer, E. J. Holmyard, and A. R. Hall. New York: Oxford University Press, 1954, pp. 559–71

Butterfield, Herbert. *The Origins of Modern Science, 1300–1800.* New York: Free Press, [1957] 1965.

Caton, Hiram. *The Politics of Progress: The Origins and Development of the Commercial Republic, 1600–1835.* Gainesville: University of Florida Press, 1988.

Charleston, R. T., and L. M. Angus Butterworth. "Glass." In *A History of Technology,* vol. 3. Edited by Charles Singer, F. J. Holmyard, A. R. Hall, and Trevor J. Williams. New York: Oxford University Press, 1957, pp. 206–48.

Chartier, Roger, and Daniel Roche. "L'Histoire Quantitative du Livre." *Revue Française d'Histoire du Livre* 16 (1977): 478–501.

Chartier, Roger. "L'Ancien Régime Typographique: Réflexions sur Quelques Travaux Recents." *Annales* 36 (1981): 191–209.

Cochrane, Eric. "Science and Humanism in the Italian Renaissance." *American Historical Review* 81 (1976): 1039–57.

Cohen, H. Floris. *The Scientific Revolution: A Histographical Inquiry.* Chicago: University of Chicago Press, 1994.

Cohen, I. Bernard. *Revolution in Science.* Cambridge, Mass.: Harvard University Press, 1985.

Collection Académique, Composée de Mémoires, Actes ou Journaux des Plus Célebrès Académies et Sociétés Litteraires Étrangèrs. 13 vols. Paris et Dijon: 1755–1780.

Crombie, A. C. *Augustine to Galileo; The History of Science* AD *400–1650.* London: Heinemann, 1952.

Crombie, A. C. *Robert Grossteste and the Origins of Experimental Science, 1100–1700.*

Oxford: Clarendon Press, 1953.
Dawson, John. *Toulouse in the Renaissance: The Floral Games, University and Student Life*. New York: Columbia University Press, 1921.
Dear, Peter. *Mersenne and the Learning of the Schools*. Ithaca, N.Y.: Cornell University Press, 1988.
Dear, Peter. "Narratives, Anecdotes, and Experiments: Turning Experience into Science in the Seventeenth Century." In *The Literary Structure of Scientific Argument: Historical Studies*. Edited by Peter Dear. Philadelphia: University of Pennsylvania Press, 1991, pp. 135–63.
Delorme, Suzanne. "Académies et Salons." *Révue de Synthèse* (1949–1950): 115–31.
DeMorgan, Augustus. *A Budget of Paradoxes*. 2d ed. Chicago: Open Court Publishing Company, [1872] 1915.
Dewhurst, Kenneth. *John Locke (1632–1744), Physician and Philosopher: A Medical Biography*. London: Wellcome Historical Medical Society, 1963.
Dougherty, F. W. P. "Correspondence of Lichtenberg." *History of Science* 26 (1988): 77–79.
Drake, Stillman. "The Accademia dei Lincei." *Science* 151 (1966): 1194–1201.
Drake, Stillman. "Early Science and the Printed Book: The Spread of Science beyond the University." *Renaissance and Reformation* 6 (1970): 43–52.
Duhem, Pierre. *Le Système du Monde: Histoire des Doctrines Cosmologiques de Platon à Copernic*. 10 vols. Paris: A. Hermann et Fils, 1914–1959.
Eales, Nellie B. "The History of the Lymphatic System, with Special Reference to the Hunter-Monroe Controversy." *Journal of Medical History* 29 (1974): 280–94.
Eamon, William. "From the Secrets of Nature to Public Knowledge." In *Reappraisals of the Scientific Revolution*. Edited by David C. Lindberg and Robert S. Westman. Cambridge: Cambridge University Press, 1990, pp. 333–65.
Eisenstein, Elizabeth. *The Printing Press as an Agent of Change*. London: Cambridge University Press, 1979.
Fabre, Maurice. *A History of Communication*. Translated from the French by Peter Chartin. New York: Hawthorn Books, 1963.
Febvre, Lucien, and Henri-Jean Martin. *The Coming of the Book: The Impact of Printing 1450–1800*. Translated by David Gerard. London: Humanities Press, 1976.
Feinagle, M. Gregor von. *The New Art of Memory*. 2d ed. London: 1813.
Feingold, Mordechai. "Newton, Leibniz, and Barrow Too." *Isis* 84 (1993): 310–38.
Ferguson, Eugene S. "The Mind's Eye: Nonverbal Thought in Technology." *Science* 197 (1977): 827–36.
Findlen, Paula. "Controlling the Experiment: Rhetoric, Court Patronage, and the Experimental Methods of Francisco Redi." *History of Science* 31 (1993): 35–64.
Fleck, Ludwik. *Genesis and Development of a Scientific Fact*. Translated from the German by Fred Bradley, et al. Chicago: University of Chicago Press, 1979.
French, Roger K., and Geoffrey E. R. Lloyd. "Greek Fragments of the Lost Books of Galen's Anatomical Procedures." *Sudhoff's Archiv* 62 (1978): 234–49.
Gimpel, Jean. *The Medieval Machine: The Industrial Revolution of the Middle Ages*. New York: Rinehart and Winston, 1976.
Golinski, Jan. *Science as Public Culture: Chemistry and Enlightenment in Britain, 1760–1820*. Cambridge: Cambridge University Press, 1992.

Goody, Jack. "Literacy, Criticism, and the Growth of Knowledge." In *Culture and the Creators: Essays in Honor of Edward Shils*. Edited by Joseph Ben-David and Terry N. Clark. Chicago: University of Chicago Press, 1972, pp. 226–43.

Gough, J. B. "Lavoisier's Early Career in Science: An Examination of Some New Evidence." *British Journal for the History of Science* 4 (1968): 52–57.

Gould, Stephen J. "Life in a Punctuation." *Natural History* 1992 (101): 1021.

Grafton, Anthony. "Kepler as a Reader." *Journal of the History of Ideas* 53 (1992): 561–72.

Hahn, Roger. *The Anatomy of a Scientific Institution: The Paris Academy of Sciences, 1666–1803*. Berkeley: University of California Press, 1971.

Hall, A. Rupert. *The Scientific Revolution, 1500–1800*. London: Longmans Green & Co., 1954.

Hall, A. Rupert. "The Scholar and the Craftsman in the Scientific Revolution." In *Critical Problems in the History of Science*. Edited by Marshall Clagett. Madison: University of Wisconsin Press, 1959, pp. 3–21.

Hall, Marie Boas. *Promoting Experimental Learning: Experiment and the Royal Society 1662–1727*. Cambridge: Cambridge University Press, 1991.

Hanlon, C. Rollins. "The Decline and Fall of Scientific Societies." *Surgery* 46 (1959): 1–8.

Hannaway, Owen. "Georgius Agricola as Humanist." *Journal of the History of Ideas* 53 (1992): 533–60.

Harden, D. B. "Glass and Glazes." In *A History of Technology*, vol. 2. Edited by Charles Singer, E. J. Holmyard, A. R. Hall, and Trevor I. Williams. New York: Oxford University Press, 1954, pp. 310–46.

Havelock, Eric A. *A Preface to Plato*. Cambridge: Belknap Press, 1963.

Heathcote N. H. de V. "Guericke's Sulphur Globe." *Annals of Science* 6 (1948–1950): 293–305.

Holmes, Federic L. "Do We Understand Historically How Experimental Knowledge Is Acquired." *History of Science* 30 (1992): 119–36.

Houghton, Walter E. "The History of Trades." *Journal of the History of Ideas* 2 (1941): 35–60.

Hudson, Derek, and Kenneth W. Lockhurst. *The Royal Society of Arts 1754–1954*. London: John Murray, 1954.

Hunter, Michael. "First Steps in Institutionalization: The Role of the Royal Society of London." In *Solomon's House Revisited: The Organization and Institutionalization of Science*. Edited by Tore Frängsmyr. Canton, Mass.: Science History Publications, 1990, pp. 13–30.

Jacob, James R. *The Scientific Revolution: Aspirations and Achievements, 1500–1700*. Atlantic Highlands, N.J.: Humanities Press, 1998.

Jarcho, Saul. *Clinical Consultations and Letters of Ippolitto Francesco Albertini, Francesco Torti and Other Physicians*. Boston, Mass.: Francis Countway Library of Medicine, 1989.

Jardine, Lisa. *Ingenious Pursuits: Building the Scientific Revolution*. New York: Talese, 1999.

Jones, Richard Foster. *Ancients and Moderns: A Study of the Rise of the Scientific Movement in Seventeenth-Century England*. Reprint. New York: Dover Publications, [1961], 1982.

Kernan, Alvin. *Printing Technology, Letters, and Samuel Johnson*. Princeton, N.J.: Princeton University Press, 1987.
Kuhn, Thomas S. *The Structure of Scientific Revolutions*. Chicago: University of Chicago Press, 1962.
Lindberg David C., ed. *Science in the Middle Ages*. Chicago: University of Chicago Press, 1978.
Lindberg, David C. "Introduction." In *Reappraisals of the Scientific Revolution*. Edited by David C. Lindberg and Robert S. Westman. Cambridge: Cambridge University Press, 1990, pp. xvii–xxvi.
Lockwood, Dean Putnam. *Ugo Benzi: Medieval Philosopher and Physician 1376–1439*. Chicago: University of Chicago Press, 1951.
MacCurdy, Edward. *The Notebooks of Leonardo da Vinci*. New York: George Braziller, 1955.
McKie, Douglas. *Antoine Lavoisier: Scientist, Economist, and Social Reformer*. New York: Henry Schuman, 1952.
McLean, Antonio. *Humanism and the Rise of Science in Tudor England*. New York: Neale Weston Academic Publications, 1972.
Magnes, Rudolf. *Goethe as a Scientist*. Translated by Heinz Norden. New York: Henry Schuman, 1949.
Martin, Henri-Jean. Livre, *Pouvoir et Société au xviie Siècle (1598–1701)*. Geneva: Droz, 1969.
Mautner, Franz H., and Franklin Miller, Jr. "Remarks on G. C. Lichtenberg, Humanist, Scientist." *Isis* 43 (1952): 223–31.
Mellottée, Paul. *Histoire Économique de l'Imprimerie. Vol 1: L'Imprimerie Sous l'Ancien Régime, 1439–1789*. Paris: Hachette, 1905.
Merton, Robert K. *Science, Technology, and Science in Seventeenth-Century England*. New York: Howard Fertig, 1970.
Morgagni, Giambattista. *The Clinical Consultations. The Edition of Enrico Benassi (1935)*. Translated and revised by Saul Jarcho. Boston: Francis A. Countway Library of Medicine, 1984.
Mumby, Frank A. *Publishing and Bookselling*. 5th ed. London: Jonathan Cape, 1974.
Neuberger, Albert. *The Technical Arts and Sciences of the Ancients*. Translated by Henry L. Brose. Reprint. New York: Barnes & Noble, [1930], 1969.
Nicolaisen. W. F. H., ed. *Oral Tradition in the Middle Ages*. Tempe, Az.: Medieval & Renaissance Texts and Studies, 1995.
Ornstein, Martha. *The Role of Scientific Societies in the Seventeenth Century*. Chicago; University of Chicago Press, 1929.
Osler, Margaret J., ed. *Rethinking the Scientific Revolution*. Cambridge: Cambridge University Press, 2000.
Packard, Francis L. *Guy Patin and the Medical Profession in Paris in the Seventeenth Century*. New York: Hoeber, 1924.
Paisley, William. "Knowledge Utilization: The Role of New Communication Technologies." *Journal of the American Society for Information Science* 44 (1993): 222–34.
Pelsener, J. "Gilbert, Bacon, Galilee, Kepler, Harvey et Descartes, Leur Rélations." *Isis* 17 (1932): 171–208.
Porter, Roy, and Mikulas Teich, eds. *The Scientific Revolution in National Context*. Cambridge: Cambridge University Press, 1992.

Rapports Généraux des Travaux de la Société Philomathique de Paris. Vol. 1. Paris: 1792.

Ravitz, J. R. *The Merger of Knowledge with Power: Essays in Critical Science*. London: Mansell, 1989.

Reeds, Karen. "Publishing Scholarly Books in the Seventeenth Century." *Scholarly Publishing* 14 (1983): 259–74.

Reynolds, Leighton D., and Nigel G. Wilson. *Scribes and Scholars: A Guide to the Translation of Greek and Latin Texts*. 2d ed. New York; Oxford University Press, 1974.

Ronan, Colin A. *Edmond Halley: Genius in Eclipse*. London: Macdonald, 1970.

Rosenbaum, Jean B., and Meyer Subrin. "The Psychology of Gossip." *Journal of the American Psychoanalytic Association* 11 (1963): 817–31.

Rossi, Paolo. *Philosophy, Technology, and Arts in the Early Modern Era*. Translated by Salvator Attanasio. New York: Harper & Row, 1970.

Sarton, George. "The Scientific Literature Transmitted through the Incunabula." *Osiris* 5 (1938): 22–45.

Sarton, George. *Six Wings: Men of Science in the Renaissance*. Bloomington: Indiana University Press, 1957.

Sarton, George. *A History of Science*. 2 vols. Cambridge: Harvard University Press, 1959.

Schaffer, Simon. "Priestley's Question in Histographic Survey." *History of Science* 32 (1984): 151–80.

Shapin, Steven. "A View of Scientific Thought." *Science* 207 (1980): 1065–66.

Shapin, Steven. "The House of Experiment in Seventeenth-Century England." *Isis* 79 (1988): 373–404.

Shapin, Steven. *The Scientific Revolution*. Chicago: University of Chicago Press, 1996.

Shapin, Steven, and Simon Schaffer. *Leviathan and the Air-Pump: Hobbes, Boyle, and the Experimental Life*. Princeton, N.J.: Princeton University Press, 1985.

Sigerist, Henry E. *A History of Medicine*. Vol 1. Primitive and Archaic Medicine. New York: Oxford University Press, 1951.

Singer, Charles. *A Short History of Scientific Ideas to 1900*. New York: Oxford University Press, 1959.

Sonntag, Otto. "Albrecht von Haller on Academies and the Advancement of Science, the Case of Göttingen." *Annals of Science* 32 (1975): 379–91.

Spillane, John D. *Medical Travelers: Narratives from the Seventeenth, Eighteenth, and Nineteenth Centuries*. Oxford: Oxford University Press, 1984.

Stanfield, Dorothy A. *Thomas Beddoes, M.D., 1760–1808: Chemist, Physician, and Democrat*. Dordrecht: D. Reidel, 1984.

Thiem, Jon. "The Great Library of Alexandria Burnt: Toward the History of a Symbol." *Journal of the History of Ideas* 40 (1979): 507–26.

Thorndike, Lynn. *A History of Magic and Experimental Science*. 8 vols. New York: Macmillan, 1923–1958.

Thorndike, Lynn. "Newness and the Craving for Novelty in Seventeenth-Century Science and Medicine." *Journal of the History of Ideas* 12 (1951): 584–98.

Thornton, John L., and R. I. J. Tully. *Scientific Books, Libraries, and Collectors*. London: Library Association, 1954.

Vansina, Jan. *Oral Tradition as History*. Madison: University of Wisconsin Press, 1985.

Webster, Charles. "The College of Physicians, 'Solomon's House in Commonwealth England." *Bulletin of the History of Medicine* 41 (1967): 393–412.
Webster, Charles. *The Great Instauration: Science, Medicine, and Reform, 1626–1660.* London: Duckworth, 1975.
Winau, Rolf. "Zur Frühgeschichte der Academiae Curiosorum." In *Der Akademiegedanke in 17. und 18. Jahrhundert.* Edited by Fritz Hartmann and Rudolf Vierhaus. Bremen: Jacobi Verlag, 1977, pp. 117–37.
Winger, Howard. "Historical Perspective on the Role of the Book in Society." In *The Future of the Book.* Edited by Lester Asheim. Chicago: University of Chicago Press, 1955, pp. 13–25.
Yates, Frances A. *The Art of Memory.* Chicago: University of Chicago Press, 1966.
Zilsel, Edgar. "The Sociological Roots of Science." *American Journal of Sociology* 47 (1942): 544–62.
Zilsel, Edgar. "The Genesis of the Concept of Scientific Progress." *Journal of the History of Ideas* 6 (1945): 325–49.
Ziman, John M. *Public Knowledge.* Cambridge: Cambridge University Press, 1966.

Chapter 14

Establishing Periodicity in Early Modern Scientific Publications

Almanacs, Newspapers, and Journals

INTRODUCTION

Enthusiasts for the new natural philosophy in the seventeenth century placed their emphasis on observing natural phenomena carefully and putting theories about them to the test of controlled experiments. These observations and experiments could be encapsulated in unitary reports and shared with other natural philosophers through the medium of correspondence. This process created a rich commerce of letters, which helped to define the audience for these kinds of communications. What was missing was a method of disseminating scientific information on a continuous basis and to a wider audience. It was not until 1665, however, that this commerce was organized and institutionalized with the invention of the first scientific journal, *Philosophical Transactions* of the Royal Society of London. Like many other inventions that are influenced by the current technology and practices, the new medium drew from other forms of disseminating news that were in existence at the time: the almanacs, newspapers, and journals that had appeared much earlier all over Europe and Britain. The characteristics that the new communication media had in common with the older media was periodicity, a commitment to a designated or natural interval in which to be published. The continuity of these periodicals was accompanied by an expectation that they would be issued in the future under a unifying title. In reality, many of the publishing ventures did not succeed in establishing all or even any of these characteristics. They did, however, contribute to the emerging journals of the seventeenth and eighteenth centuries the processes for managing their production and distribution. By the use of a unifying title, these periodicals identified and kept their particular clienteles. Periodical publications created a new medium to which authors could submit their messages to a broader audience than they could through correspondence. They also provided an incentive for editors to collect information from authors to meet scheduled datelines. Almanacs and newspapers played some of the roles that were taken over by the scientific journals when they appeared and continued to flourish throughout the seventeenth and eighteenth centuries.

Though these media served different functions and somewhat different audiences, they shared a similar format in the early modern period. In a chapter called "The Press and the Enlightenment" in his book about the enlightenment, Thomas Munck claims: "Printing technology was efficient and adaptable, and the capital overheads required to swap between different types of printing (say from book production to news-sheets) was low—particularly so since eighteenth-century newspapers were done in small formats, mostly printed in continuous prose without headlines, and virtually indistinguishable in appearance from unbound books" (Munck 2000:106).

ALMANACS

Almanacs were among the first books issued from the printing presses in Europe. One of Gutenberg's earliest printed works that appeared in 1448 was an almanac (Capp 1979:26). Printers were quick to take advantage of the market for these kinds of publications, and they proliferated throughout Europe and Britain. Grand-Carteret's bibliography of French almanacs for the years 1600 to 1895 contains 3,633 numbered items, of which 1,311 were published 1799 and earlier (Grand-Carteret 1896). They were distributed as widely in England as well, where in the 1660s it is estimated that four hundred thousand were sold in a single year (Capp 1979:23). One of the first almanacs that was distributed widely was the *Grant Calenrier Compost des Bergers*, which was first published by Guy Marchant in 1493, and later issued under other titles and translated into other languages. It was translated by Robert Copland into English ca. 1518 as *Kalendar & Compost of Shepherds* (Heseltine 1930). It provided a model for the popular almanacs that appeared in the succeeding centuries. The English translation is divided into five sections with the following headings:

The tree of vices and the paine of Hell.
The calendar of saints days; phases of the moon.

Salutary science and the pains of Hell.
Of physic and governance of health.
Astrology and physiognomy.

It includes a calendar, prayers, medical and cooking information, and statements and precepts on morals and astrological treatises (Heseltine 1930).

Almanacs of these periods were very diverse in the kinds of information they included. Despite some of the formal requirements, the contents were still at the discretion of the editors, or compilers, who sought to make their almanacs distinct and different from the competing almanacs published at the same time. Calendar information was, naturally, a sine qua non, but other similar kinds of information were to be found in many of the titles, particularly items relevant to certain dates of the year: feast days, signs of the Zodiac, phases of the moon, the weather conditions prevalent in every month, dates of the fairs. The contents could also be extended to include information of particular interest to a special clientele, such as names of municipal officials, interest tables, and a host of other kinds of information (Johnson 1985:140; Blagden 1969:24; Dodge 1997). In 1762, an almanac devoted to centenarians was cited in the medical journal *Gazette d'Epidaure* (vol. 1:11–14) with the title *Almanach de la Veillesse, ou Notice de Tous Ceux qui Ont Vecu Cent Ans et Plus*. The most popular almanac in England is said to have sold 82,000 in 1761, and *The Ladies' Diary*, another popular almanac, sold around 30,000 about the same time (Capp 1979:239, 246).

Publishing almanacs was a very competitive business, as we can see from the numbers of different titles published in the seventeenth and eighteenth centuries. One estimate is that there were as many as 2,000 different editions of English almanacs in the seventeenth century alone (Bennett 1970:165). Grand-Carteret's bibliography of French almanacs (1896) lists 88 titles for the seventeenth century and over 1,200 in the eighteenth century. Almanacs were a staple of the printing trade in every European country. It was an activity printers could easily schedule in conjunction with their other business. In France, where the press was under the tight scrutiny of the Royal censors, even almanacs were closely controlled. In England, King James I bestowed monopoly for publishing almanacs on the Stationer's Company, the printers guild that it held through the eighteenth-century. The profitability of the enterprise is demonstrated by the balance sheets of the company for 1748. The receipts shown were £1,000 against expenditures of £415 (Blagden 1969:41).

Almanacs during the early modern period continued to include the typical material of the standard almanacs. Despite the formal requirements of including saint's days, phases of the moon, and so on, the contents of any particular title were at the discretion of the publisher, who might choose to address other subjects or direct his publication to a specific audience or craft. In eighteenth-century France, publishers began to include commercial and administrative topics, such as information on the law courts and the names of municipal officers (Johnson 1985). Almanacs have been said to have played an important part in the development of meteorological reporting, in which weather predictions were based on the appearance on the rise and fall of heavenly bodies (Hellmann 1893, no. 12:8).

Medical and health information was a staple of almanac publishing. "A considerable proportion of Tudor and Stuart physicians were themselves astrologers" (Capp 1979:208). Almanacs provided a primary source of health information for a large part of the early modern population. They sometimes included bizarre accounts of monsters and medical abnormalities. *Schreibkalender Schaffausen* in 1781 described a bearded lady whose ears hung down to her shoulders. *Hinkende Merkurius* in 1783 reported a case of a woman whose body was divided half-black and half-white. In addition to stories of two-headed calves and sea monsters, almanacs also contained notices of epidemic diseases and gave health advice and suggested home remedies (Lombard 1925).

One can't write about early modern almanacs without considering the presence and persistence of astrology in their production. Astrology is a complex system of charting relationships between the appearance of celestial bodies and other variables to make predictions about events in human life and in nature. It is based on astronomical observation and knowledge with which it was closely allied in the premodern period. It has been defined as "a system of magic embedded in a scientific and mathematical matrix where it played a role in the intellectual and scientific community throughout most of the sixteenth and seventeenth centuries" (Jerome 1977:124). It was based on a belief that the stars and planets in the firmament had a significant influence on events in personal life and in nature. In the sixteenth century, astrology was regarded as an applied science of astronomy. Astrology had two different aspects: one was natural astrology, which related to natural phenomena such as in agriculture, medicine, and meteorology; and judicial astrology, in which an individual's destiny could be determined at birth based on the position of the stars and could be used to predict events in the future. Astrology is said to have flourished among the early Greeks but was banned by the church in the Middle Ages and reestablished in the Renaissance along with the increased interest in Greek philosophy and science (Jerome 1977:21; Harris 1980:87; Curry 1987:2).

Certain ambivalence existed among even the leading early modern scientists about the validity of astrology (Capp 1979:187). Picodella Mirandola (1463–1494), who wrote a treatise attacking the belief, provides an early example of criticism. He checked weather predictions in astrological almanacs over a period of a hundred and thirty days and found they were correct only seven times (Jerome 1977:48). Boyle was not convinced about the truth of astrology, but he suggested experiments to prove it. Curry cites an entry from Robert Hooke's diary entry about the death of John Wilkins, dated November 11, 1672: "Dyed about 9 in the morning after a suppression of Urine . . . a conjunction of Saturn and Mars," but suggests that this may merely have the use of a meaningless convention (Curry 1987:140). Evidence of the rejection of astrology in the seventeenth century appears in an unsigned letter to Henry Oldenburg, dated July 24, 1675, which has been attributed to John Beale (1603–1683), one of his faithful and early correspondents: "We see with what confidence some men pretend to Astrology even in trifles, yet how lazy as unwilling the whole Tribe is to make proof, either of that force or kind of Subterranean Streams, or of Celestial or Ethereal Influence; no not of the Moon, or any Planet or Constellation, to any considerable accurate" (Oldenburg 1975, vol. 11:417).

Halley and Newton are said to have shared a belief "in comets as instruments of divine intervention" (Capp 1979:258). In fact, Thorndike suggested that Newton might have drawn some of his theory of gravitation from the idea of action at a distance in the literature of astrology (cited by Curry 1987:140). Nevertheless, by this time some astrologers had begun to use data from the Copernican system rather than the heliocentric system. Capp states: "The almanac was a valuable educational medium whatever the reader's level of astronomical knowledge or ignorance" (1979:198). Some contrary evidence is provided by Hetherington, who searched a sample of ninety-three almanacs published from 1607 to 1700 and found only one that supported the Copernican system, and not a single one who mentioned Newton (1975:279). Many years ago, Marjorie Nicholson made an extensive review of English almanacs and reached conclusions that are still valid today: "The majority of almanacs before 1640 pretended to be more than they were—handbooks of general reference. . . . Many of the later almanacs cease to be almanacs in the conventional sense, and play their part in the development of periodical literature. . . . After 1640, however, and particularly after 1660, many of the almanacs are mines of valuable information to both the political and intellectual historian" (Nicholson 1939:4).

There are other ways in which the almanac contributed to the development of newspapers and journals. They helped provide experience in the production of large print runs that would rarely be exceeded by the other two media. They also provided a model of systems of distribution that would be used by newspapers and journals (Chartier 1993:122–3). The services of the post office, which developed in the seventeenth and eighteenth centuries in England and the Continent, supplemented direct sales from printers, bookshops, and book peddlers who worked in urban and country areas. Almanacs were also involved with science and scientific institutions. In some cases, where societies did not publish proceedings or other publications, information about their activities is sometimes to be found in almanacs. Claude Hartmann quotes from a letter, dated April 25, 1784, of the Abbé François Rozier (1734–1793), editor of *Observations sur la physique*, one of the most important scientific journals the last quarter of the eighteenth century: "I am currently a member of a number of academies of which man are silent and known only through their almanacs" (Hartmann 1996:11). Rozier does not identify the society, but it could easily have been Société d'Agriculture, Sciences, Belles-Lettres et Arts d'Orleons, whose records were destroyed in 1940 (Roche 1978, vol. 2:312). In his study of provincial societies in France, Daniel Roche frequently finds it necessary to cite almanacs in his bibliography on each of the many societies that appeared in the provinces in the eighteenth century and earlier (Roche 1978, vol. 2:208–345).

Almanacs also served early scientific societies by providing them economic support through monopolies on producing them. This was not true in England because Royal charter had issued a monopoly on their production in 1557 to Stationers' Company, which had been established as a guild as early as 1403 (Howe 1981). Several other societies, however, received considerable support in this way. Académie Royale de Sciences et Belles-Lettres in Berlin was supported largely by the income from a royal calendar for which they held a monopoly (Armytage 1965:33). Académie des Sciences, established in St. Petersburg in 1724, drew its income largely from customs fees and calendars (Heilbron 1979:120, n.21). In Sweden, the scientific society derived its income from a monopoly granted in 1749. In 1752 they were printing 177,000 copies. When they sought a loan to build an observatory and increased the price of the almanac by 50 percent, they were still able to print 169,000 copies (Sinnerstad 1989:48). Societät der Wisssenschaften, founded in 1752 in Göttingen, was aided in meeting some of its financial problems by assuming a monopoly for almanacs (McClellan 1985:116).

NEWSPAPERS

The range of periodicity for the newspaper over the almanac was both extended and encumbered. It was extended because the printer was not limited to an annual issue, but could select any interval of time to issue his newspaper. It was encumbered in meeting established deadlines because he was at the same time competing with other printers for the attention of readers, and could lose readers by failing to produce scheduled issues at the appointed time. For this reason, the newspaper provided a more congruent model for the journals that followed them in time, although journals did not have the same problems as newspapers about periodicity, because they were not responsible to report current events. Nevertheless, journal publishers had to make some kind of a commitment about the future issues of their journal in order to gain and keep subscribers.

There was no clear distinction between the early modern newspapers and the journals that followed them more than a century later. In the nineteenth century, the Germans made a distinction between *Zeitung*, a newspaper, which is primarily concerned with contemporary news and *Zeitschrift*, which could cover a wide variety of subjects and published at various intervals. However, the term *Zeitschrift* was rarely used the previous centuries and many German journals carried the title *Zeitung* in their titles, even in the eighteenth century. The cognates for these terms in French are *Gazette* and *Journal*. Newspapers and journals share other characteristics beside periodicity. One such characteristic is what the Germans call *Kollectivität*, the act of bringing together disparate items under a single title, but *Aktualität* also implies topicality or relevance to the present and is more particularly the province of the newspaper (Kirchner 1976). In periodical bibliographies such as that of Kirchner (1976) and Sgard (1991), little distinction is made between newspapers and journals. The titles in these bibliographies occupy a spectrum, from those that deal entirely with contemporary events to those that consist entirely of original articles. The problem is compounded by the fact that early modern newspapers resembled the formats of journals. This lack of distinction is reflected in the way the terms were defined in the past. Hatin quotes Académie Français as saying that the term "journal" is a "relation day by day of that which occurs or has occurred in other countries and in other places," while he insists that from the beginning the appropriate term for this function was *gazette* (Hatin [1866] 1965:lxxxii).

REDUNDANCY AND OBSOLESCENCE

There are two characteristics to which the newspaper and the journal have somewhat disparate relationships: redundancy and obsolescence. Redundancy is inherent in the newspaper because much of the same news is published in many different sources. However, it is also true that if the reader has not acquired the news from any other sources, it is, at that moment, not redundant to him. A similar situation can occur in the journal literature but in rather a different way. An example, perhaps, is a paper that Lavoisier had submitted to Académie Royale des

Sciences in 1769 on his famous experiment on the constitution of water. He published a summary of the paper in Rozier's newly founded *Observations sur la Physique* in 1771, whereas it did not appear in Académie's *Histoire et Memoires* until 1773 (Guerlac 1975:70). This puts a rather different meaning on the question of redundancy. Lavoisier was anxious to make knowledge of his experiment known to the scientific community as soon as he could. Publication by the Académie may have been redundant, but it also gave the paper its *primature*, which added to its credibility.

Early modern newspapers borrowed freely from each other, often without any attribution. Redundancy appeared often in the learned and scientific journals that came later. Journal editors justified it in various ways. The editor of *Botanisches Taschenbuch*, David Heinrich Hoppe, in the preface to the first volume published in 1790 promised to include the work of famous writers on botany who were not well known to the pharmacists who make up an important part of his audience. Once the translations, compilations, summaries, and reviews are sifted out of the journal literature of this period, the residue may represent a small component of the original literary contributions.

Rates of obsolescence of the information are quite a different issue in the early modern newspapers and journals. News in newspapers generally becomes obsolete as soon as it is disseminated, while new contributions in the journal literature take quite a different course depending on the progress in any particular discipline. In the preface to the first volume of his *Chemisches Archiv* published in 1783, Lorenz Crell, the editor, pledges to include all the new discoveries in chemistry for the benefit of his readers. However, he says he will attempt to include only those new discoveries that were made known after 1770, a point apparently at which he thinks chemical literature may be obsolete. On the other hand, Wolfgang von Steinwehr, in translating the chemical, anatomical, chemical, and botanical papers of Akademie de Wissenschaften in Berlin in 1749, included those published as far back as fifty years. In medicine, obsolescence was less of a problem when medical journals consisted primarily of discussions and descriptions of cases encountered in practice, and physicians still depended to some extent on ancient learning. We encounter this attitude in reading Johann Jacob Römer's preface to *Albrecht von Haller's Tagebuch der medicinischen Literatur*. "No ordinary physician," he says, "can keep up with the volume of medical literature being issued, because he lacks the time to devote to this task. This problem is alleviated by good newspapers, journals, systematic collections and libraries." He nevertheless adds that the physician must also take pains to familiarize himself with the older literature (Römer [1789], 1974:12).

A rather different attitude is expressed by the editors in the first volume (1773) of *Memoirs of the Medical Society of London*, in which they state that: "The principal part of our knowledge must be derived from comparing our own observations with those of others. . . . Deceased authors cannot solve our difficulties, nor will the observations made in other ages and climates, hold always true in our own." A point of view that seems to combine both of these views of obsolescence can be found in the first volume of *Bibliothèque Phyico-Economique* (1782), in which the editors say: "Everyone owes it to the present and future societies to instruct them about that which is discovered or known to be useful and interesting, in gratitude for the advantage they receive daily about the discoveries, whether by their contemporaries or from those who have lived before them."

NOUVELLES A LA MAIN—MANUSCRIPT NEWSLETTERS

The dissemination of news on a scheduled, even if irregular, basis was well-established long before the invention of printing. The French called them *gazettes manuscrites.* They were developed for rulers, politicians, bankers, and merchants who needed to know what was going on all over the world. Later they evolved into a medium for a larger public (Morgan 1929:3). They also continued to be distributed a long time after the invention of printing. They had the advantage that they were not subjected to external overseers, such as censors. At the same time, the copyists that were employed in producing the copies had to be screened to make sure they were not police spies (Kölving 1988:177). Manuscript newsletters have been classified into three types: those written by correspondents to serve governments or other groups of subscribers; reports of political or economic interest; and erudite or scholarly letters (Carroll 1966:49). The last category functioned in rather a different manner than the composite letters sent to the members of the first two categories cited. Scholars, in sharing letters in a network of correspondents, can be said to constitute in their totality a sort of composite newsletter.

Manuscript newsletters existed in quantities in the seventeenth century and well into the eighteenth century. Wiles, discussing the situation in England, says: "Although the history of newsletters in manuscript has not yet been written, it is well known that during the seventeenth century there were several series of hand-written bulletins of news prepared in comparatively small quantities by men living in London, where great affairs took place or were discussed, and sent to regular subscribers in the country for a suitable fee" (Wiles 1965:9). Henry Oldenburg, despite heavy burdens as the editor of *Philosophical Transactions* and a secretary of the Royal Society of London, evidently thought at one time of starting a newsletter himself. Writing to Robert Boyle, his patron, he says:

> Sir, give me leave to entreat you, that in case you should meet with any curious persons, that would be willing to receive weekly intelligence, both of the state and literary news, you would do me the favour of engaging them to me for it. The expense cannot be considerable to persons that have but a mediocrity: £10 year will be the most that will be expected, £8 or £6 will also do the business. (Bluhn 1960:185)

Oldenburg was well fixed to receive news from abroad. He had a vast correspondence with foreign scholars. The Royal Society had arranged for him to receive his foreign mail at the office of Sir Joseph Williamson, the undersecretary of state, to pay the postal charges, which had to be paid by the recipient. In return, Oldenburg stripped off all the nonscientific material from the letters and reported it to the undersecretary. In any event, as far as I know, the newsletter never materialized, but his efforts may have had something to do with his spending some time in the Tower of London, possibly under suspicion of spying, and failing to edit an issue of *Transactions*.

A newsletter service was established by the Fuggers, one of the leading banking families in Europe based in the ancient town of Augsburg, near Munich. It is probably one of the best-known series of early mercantile newsletters that received reports from all parts of the world. Most of the news relates to political, military, and diplomatic events, but occasionally we read of famines in France and Constantinople, plagues in Rome, Prague, and Venice. There is a report of a miraculous healer who is described as a quack, and notice is taken of a famous alchemist (Klarwill 1924). It is possible that an early German newspaper or newsbook, *Messrelationen*, may have been stimulated by the Fugger correspondence. Michél von Aitzing, or Eyzinger, married a member of the Fugger family and produced the publication twice a year from 1688 to 1698 for the Frankfurt Book Fair that occurred twice a year. *Messrelationen* contained accounts of the events that had taken place since the last fair. Aitzing had such success with the title that he continued to issue it until his death in 1698 (Weil 1934:19–20).

Gazettes manuscrites (manuscript newsletters) were permitted to circulate in France despite the close supervision of the state, as long as two copies of each issue were deposited with the police. They contained news of the activities of the Royal Court and nonpolitical events in the community (Weil 1982:93–95). Hatin says that they became scandal sheets rather than political news sources, and were being constantly sought out by the police (Hatin [1859] 1967, 1:56). The two best known of the French manuscript newsletters constitute a social, literary, and cultural history of the eighteenth century. The newsletter of Louis Petit Bachaumont originated in the famous salon of Madame Doublet, where social leaders and artists gathered and "everything in Paris that commanded attention" was discussed. The newsletter began in 1740 and continued until Bachaumont's death in 1771 (Hatin [1866] 1965:67). *Correspondance Littéraire* of Friedrich Melchoir Grimm is of equal historical importance. Grimm came to Paris from Germany and quickly became intimate to the circle of *philosophes*. When he arrived in Paris, he met the Abbé Rayal, who had been circulating manuscript newsletters to a group of subscribers at one of the German courts. Grimm embarked on a similar venture by first sending out first a biweekly report to a small group of subscribers that included Frederick the Great and King Stansislaus of Poland (Crocker 1966:187). It has justly been called an "upscale manuscript journal" because all his subscribers could pay handsomely for his services (Munck 2000:108). Denis Diderot very early became a collaborator, writing criticism on the theater, literature, and the arts. Knowing his interests in mathematics and in science, one is not surprised to find a piece on probability theory, which he wrote for *Correspondance* in 1781 (Wilson 1972:431).

There does not seem to be a manuscript newsletter in the early modern period devoted to science and technology. Perhaps devotees to these subjects did not have the same need for privacy or exclusivity as Bachaumont and Grimm's subscribers. The price the subscribers had to pay for this service may have been out of reach to most readers except for heads of state and highly placed officials. But there is at least one occasion when reports relating to science and technology were circulated in manuscript. The first monthly issues of *Bulletin des Sciences* of Société Philomathique from July 1791 to October 1792 were sent to members in manuscript (Bertholet 1888:vi).

Manuscript newsletters in England were at first called diurnals because they originally were used to issue daily reports of the activities of Parliament. Many such newsletters were circulated from early in the seventeenth century and continued to be distributed even after the printed newsbooks appeared, even though the manuscript newsletters were more costly (Nelson 1986:22) and are said to have survived into the second half of the eighteenth century (Handover 1960:126) The Duke of Newcastle in 1723 spent £48 to circulate news to a select audience; more than half of this sum was for manuscript newsletters (Cranfield 1962:178). Some of the early printed newspapers were set in a special script type to emulate the manuscript newsletters, perhaps because it represented a more familiar and intimate medium than print.

PRINTED PRECURSORS OF THE NEWSPAPER

Transitional forms from the manuscript newsletter to the printed newspaper encompass a large number of intermediate formats. They began to appear shortly after the invention of printing. "As far back as 1505 single-sheet *Zeitungen* (tidings) had been sporadically printed in Germany; by 1699 there were 877 such publications, all irregular. The oldest regular newssheet known to history was the weekly *Avisa Relation oder Zeitung* founded at Augsburg and composed from reports of agents posted throughout Europe by merchants and financiers" (Durant 1961:57).

This passage seems to refer us back to the Fugger newsletters discussed above. The precursor forms were also known by a variety of other names: news-pamphlets, corantos, broadsheets, and *feuilles volantes*. There is no single characteristic that describes them all. Some appeared sporadically, in single or multiple sheets, but others were printed with some frequency in the form of books or pamphlets.

Among the first news publications issued after the invention of printing was "the *occasionnel*, a newssheet about current events," but some of the other designations may serve as well. Martin says they were produced in large numbers in the fifteenth and sixteenth centuries, although few seem to have survived. They dealt, he says "with astonishing events, miraculous and diabolical manifestations, natural catastrophes and the passage of comets, or described 'monsters'" (Martin 1994:295). Broadsheets were a primary source of news in London from 1590 to 1690. A compilation of broadsheets lists 270 that are described, and calls attention to 170 that were registered with Stationer's Company for the same years that no longer seem to exist. The broadsheet for August 26, 1609, reports: "Strange newes out of Kent, of a monstrous and misshapen child, borne in Old Sandwich, upon the 30 of Julie last, the like (for strangenesse) hath never been seene" (Collins 1943).

Hatin cites an English *Mercurie* that appeared in 1588, of which only three issues are in the British Library, numbers 50, 51, and 54, so they were obviously preceded by 49 other issues. One issue relates the departure of the English Armada and another provides an account of Francis Drake's engagement with the Spanish fleet (Hatin [1866] 1965:lviii). *Nieuwe Tindinghe*, which appeared weekly in Antwerp in 1620 in Latin, has been described as the "first periodical newsheet" (Martin 1994:300). It is difficult, therefore, to give primacy to any particular title in any of these genres.

Growth in interest in the news has been associated with the Thirty Years War (1618–1648), which affected a great many people conscious of the impact of political and military activity outside their own areas. A collection in Stockholm of thousands of copies of weekly corrantos from the Low Countries covering the first half of the seventeenth century reflect these interests (Frearson 1993:2). Newsbooks generally appeared weekly, but sometimes special issues reflect other interests. An annotated bibliography of English corrantos issued in numbered series from 1620 to 1642 describes another interesting collection. An entry in a corranto dated March

6, 1624, informs the reader of "A wonderful accident in Transylvania, with swarmes of flyes of incredible bignesse, with foure wings a peece (Dahl 1952:49). By 1712, twenty-seven single leaf newssheets were regularly issued in London each week (Harris 1987:19). News in the various formats discussed above continued to appear into the eighteenth century. We have some evidence of broadsides being issued as late as the middle of the nineteenth century. An antiquarian book catalog issued in 1999 lists five broadsides dated between 1815 and 1870 (Quaritch Occasional List VI: nos. 21–26, 1999).

There does not seem to be a great deal of interest in the history of science in the material covered in these media. There is one exception, however, which has attracted the interest of science historians. It is a twelve-page newsbook that measures three by five inches published in The Hague in 1608. It describes a visit of the Ambassador of the King of Siam to Prince Maurice of Nassau. On page nine it speaks of a "certaine lunette" produced by a spectacle maker from "Mildelbourg" capable of "discovering and seeing distinctly distant objects." The newsbook of which only two copies are known to exist was reproduced and included with an essay by Stillman Drake, the noted Galelio scholar in a limited edition published by Jake Zeitlin, the distinguished antiquarian bookseller in Los Angeles (Drake 1976). It was thought at one time that Galelio was inspired by the newsbook to grind his own lenses and to go on to manufacture telescopes that made it possible for him to make his important astronomical observation. Stillman Drake says that it is more likely that he may have learned of it in a number of other sources, and adds that "news of it was already widespread throughout Europe" (Drake 1978:138).

ORIGINS AND DEVELOPMENT OF THE NEWSPAPER

It is reported that the first printing of a newspaper took place in Anvers (Belgium), followed by one in Germany in 1615, England in 1622, and in France by 1631 (Hatin [1866] 1965:ix–xxi). The fact that it occurred so late in France was primarily an outcome of the rigid censorship that prevailed there. The *Gazette* that appeared at that time nevertheless became a kind of model for newspapers, which followed it all over Europe and Great Britain. It was founded by that extraordinary seventeenth-century physician Theophraste Renaudot. If Renaudot is not known for any other reason he should be remembered for his aphorism on press freedom, although ironically it was sorely lacking in the *Ancien Régime*. "The Journal," he said, "takes on the nature of a raging flood which increases in the degree that it is resisted" (Laboulbene 1880:1059; Snorrason 1958:209; Sudhoff 1903:455). Renaudot is noted also for the innovative social welfare he introduced to Paris. Hatin's encomium seems justified: "Here is a man with modern ideas, one of the lively spirits for whom progress is a necessity, who in their impatience sometimes take false routes, but in the end, always turn at the end to the profit of society" (Hatin [1859] 1967:63).

Born in Loudon in 1584, he studied at the Medical Faculty in Paris before returning to Montpellier to take a degree in medicine. This put him at a disadvantage with the Parisian Faculty, which was at odds with the Montpellier Faculty, particularly when Renaudot began to introduce his medical and social practices in Paris where he had the support of the powerful Cardinal Richelieu. He also had a large inheritance at his disposal that made his task easier. "At the beginning of the seventeenth century," says Bigourdan in describing Renaudot's contribution, "Paris did not yet have any of the public institutions without which social life would be impossible today: newspapers, auction houses, employment offices, pawnshops; all of these were created in a short time" (Bigourdan 1917:939). Renaudot was struck with the terrible poverty in Paris, and proposed hiring the poor to clean the streets of the city. He organized a Bureau d'Adresse where in his own words: "Everyone can give or receive advice about all of the commodities and necessities for life and human society" (Dekeyser 1949). He also opened a public clinic that was attended by fifteen physicians to provide free medical treatment and drugs to the indigent.

All of Renaudot's activities were based on the Bureau d'Adresse, which was located at his house, *Maison du Grand Coq*, including his medical clinic. He published a monthly list under the title *Feuille du Bureau d'Adresse*, in which he publicized the activities and services of the Bureau. It was also issued as *Cahier du Commoditez Présentés du Bureau d'Adresse*. Few copies of these publications survive (Feyel 1991a:418–22). Harcourt Brown says, "Some scholars are inclined to see in the weekly *Feuilles du Bureau d'Adresse* the first scientific periodical (Brown 1984:21). However, they are most likely to find a precedent in the reports of the weekly conferences that were held under the auspices of the *Bureau* in which Renaudot tried to establish a public forum for the discussion of scientific and philosophical issues of the day. It was open to the public and attracted some of the outstanding scientists of the period. Although all of the contributors were anonymous, the mathematician Jean-Baptiste Morin and the astronomer Ismael Bouillau have been identified as participants (Mazauric 1994:162).

The conferences were held weekly, except for vacation periods, from 1633, and in French rather than in Latin, as was true with the universities. The subjects included political and social economy, philology, military arts, physics, and mathematics, but little emphasis was given to mathematics or mechanics except for perpetual motion. Astronomy and meteorology were also discussed, as was medicine (Bigourdan 1916:941–42). Reports of the conferences were published every week from 1634 to 1641, which may qualify them as the proceedings of a learned society. They were assembled in five collections of *Centuries* of nearly one thousand pages each from 1634 to 1655 (Richardson 1933:313). They were reprinted in French several times and translated into English in two folio volumes under the title *A General Collection of the Discourses of the Virtuosi of France* (London: Dring and Starkey, 1664).

The account of the origins of the *Gazette,* as related by Howard Solomon, tends to paint a picture of Renaudot that is less benign than we may have been lead to think. However, it fits neatly into a concept of the man as a person of aggressive energy and an ability to exploit his connections with authority. Renaudot published the *Gazette* in May 1631, four months after the *Nouvelles Ordinaires des Divers Endroits* was issued by two Parisian printers. He had the government withdraw their *privelege,* an award granted by the government that gave him the exclusive rights in perpetuity to publish a newspaper in France. The original publishers continued to issue the *Nouvelles* for a short time until they were ordered to cease by the government, at which point Renaudot took over publishing the *Novelles* in addition to the *Gazette* (Solomon 1972:112–7). Renaudot established a veritable journalistic empire based on his own presses, which included publishing *Mercure Français,* an irregular annual that had been published since 1613 (Jager 1991:867–69). The *Gazette* was issued weekly in four pages, which increased to eight in 1642 and became a biweekly in 1762. From 1752 it was published in two editions, a regular edition for Paris, and an edition in small characters for distribution in the provinces. It was also reprinted with permission in a few provincial cites, such as Lyon and Rouen from the beginning. Between 1631 and 1752, thirty-eight provincial cities had acquired the *privelege* (Feyel 1991b:446).

After Renaudot's death in 1653, the *privelege* for the *Gazette* remained in the hands of members of the Renaudot family until it was sold to a group of French investors in 1749 (Feyel 1991b:447). The new owners of the *Gazette* prohibited the reprints of the newspaper in the provinces, which previously had been authorized and began to issue a publication called *Affiches de Province*—but it failed to circulate as widely as they had hoped. They then decided to sell thirty-year *priveleges* to thirty-five cities in France to establish a Bureau d'Adresse and the right to publish a weekly *Affiches* of their own (Feyel 1982:8). They appeared under a number of titles in addition to *Affiches. Dictionnaire des Journaux* covers periodical titles published in French from 1600–1789 and includes a total of 1,239 titles. Of these, eight are classified as newspapers (*Gazettes*) and eighty-seven as *Affiches*—that is, titles covered under the *privelege* of *Affiches des Province.* The rest fall into a number of other categories, including general and specialized journals (Sgard 1991:1133–4).

It is difficult to generalize about the subject content of the provincial newspapers. Usually they included the kind of information supplied by *Feuille du Bureau d'Adresse,* dealing with all the commodities and necessities of life and human society. Beyond that, the editors of the provincial papers were free to go wherever their ambitions and imaginations led them while keeping an eye on the government sensors. Analysis of the subjects by Michel Marion covered in *Affiches de Paris* under the editor the Abbé Gueron from 1762 until his death in 1780 provides some insights. He found that 41.40 percent of the content was devoted to science and the arts, 22.22 percent to history, 14.65 percent to belles-lettres, 12.18 percent to theology, and 8.94 percent to law. These figures may not reflect the subject matter of most of the provincial newspapers, and therefore may not be very meaningful, unless perhaps they are compared with the holdings of a group of selected Parisian libraries, in which Marion said sciences and art fell at the very bottom of the list at 9.7 percent (Marion 1980:32).

In his analysis of the early modern provincial press in France, Feyel found that among the principal subjects included were agriculture, meteorology, and medicine. They included subjects such as how to avoid the *vapeurs* and how to save a drowning person. He quotes the editor of *Affiches de Poitou,* who suggested that there should be an exchange of information among the editors of provincial newspapers so as to form "a kind of academy extending throughout the entire realm" (Feyel 1984:40). A perhaps atypical example of the provincial newspapers may be the physician Pierre Dorian who contributed a weekly article to *Affiches de Poitou* from 1773 to his death in 1777. They were generally on medical subjects, but he occasionally wrote on natural history as well (Badouin 1912). Feyel regarded that newspaper as "a journal of economics and an agricultural gazette" (Feyel 1991b:80).

There does not seem to have been a pivotal figure like Renaudot in the history of the English newspaper. However, they developed much earlier and had greater opportunity to expand. The earliest new media in England was in the form of broadsides and news pamphlets in the last quarter of the sixteenth century, but they did not begin to be published serially until 1594 when *Mercurius Gallo-Belgicus* appeared in Cologne, where it was issued semiannually until 1635 (Raymond 1996:8). It was illegal in England to publish domestic news of political and military events, but reporting events that happened abroad was permitted. The first officially licensed newspaper, *Weekly Newes from Italy, Germany, and Bohemia. The Palatinate, France and the Low Countries* appeared in 1622 (Hutt 1973:10). The government up to the period of the Commonwealth had rigidly imposed censorship when it was somewhat relaxed. The new Printing Act under Charles II in 1663 again severely restricted the printing of news. It was placed under a Court-appointed licensure, who established a monopoly by publishing *Oxford Gazette*. It was later moved to London and renamed *London Gazette* and became the only newspaper published with any regularity in the second half of the seventeenth century (Chappell 1970:130–31). The lapse of the Licensing Act in 1695 marked the end of censorship by the government and brought about an increase in the number of newspapers (Harris 1975:139). The number increased tenfold between 1700 and 1799 (Winton 1977:22). All of these newspapers, of course, may not have been in existence at any one time because many of them were of short duration.

The first English newspapers were issued weekly, but even *Nieuwe Tindinghe*, first published as a weekly in 1605, began to appear twice a week after 1620. The first truly daily paper was *Daily Courant*, published in London in 1702 (Martin 1994:300, 414). The first daily journal published in France, *Journal de Paris*, by comparison did not appear until 1777 (Trenard 1969:241). There were four dailies in London in 1760, nine in 1783, and fourteen in 1790 (Black 1991:34). While advertisements were rare in English newspapers in the seventeenth century, they became an important part of the income for many English newspapers in the eighteenth century. By the end of the eighteenth century, some newspapers devoted as much as 74 percent of their space to advertisements (Black 1987:52). This emphasis is reflected in the titles of some of the newspapers of the period such as *Evening Advertiser* and *General Advertiser*. Advertisements for patent and proprietary medicines made up a staple of these papers and provide insight into the health concerns of the period (Brown 1975). In the early eighteenth century, press runs ranged from four hundred for *Flying Post* (issued Tuesday, Thursday, and Saturday) to six thousand for *London Gazette* (issued Monday and Thursday). The press runs for *London Gazette* increased to 8,500 in 1710. Of these copies 2,031 were left unsold, 1,087 were distributed, and 5,382 were sold (Sutherland 1934:114). To the best of my knowledge, they did not follow the good business practices of Renaudot, who collated all the unsold copies for a single year and sold them as annual volumes under the title *Receuil des Gazettes de l'Année* (Solomon 1972).

Newspapers were made available in many ways. Newsagents bought copies from the printer at a discount and sold them. They were offered by subscription outside of the city of London through a network that included provincial printers and through the postal system (Harris 1987:39–45). To order newspapers through the ordinary post was costly, unless you could receive it through the auspices of someone who had franking privileges, like Members of Parliament, who could have them sent to subscribers at no cost to the recipients (Cranfield (1962:179). An important mode of distribution was through the coffeehouses of London. It is estimated that there were three thousand coffeehouses in London at the beginning of the eighteenth century (Cranfield 1962:179). After the lapse of the licensing act in 1695, proprietors of coffeehouses could take one or more newspapers for the use of his customers (Cook 1994:161).

A newspaper is a medium for carrying accounts of events and purveying history in particular space in time. Henry Fielding, in his novel *The History of Tom Jones*, had this to say about newspapers: "[They] consist of just the same number of words whether there be any news or not. They may likewise be compared to a stagecoach, which performs constantly the same course, empty as well as full" (Fielding 1964: 33). However, a newspaper is also a commodity, which must take consumers into consideration if they are to prosper. The standard fare for most English newspapers was information about prices and new books (especially since newspaper printers also sold books), bankruptcies, births, deaths, entertainment, and at times special features such as mathematical problems, essays, or poetry (Wiles 1965:109–110). It is said that Samuel Taylor Coleridge, William Hazlitt, and Charles Lamb wrote for the *Morning Chronicle*, and that Robert Southey and William Wordsworth wrote for the *Morning Post* (Chappell 1970:163). The *Western Flying Post* in one issue carried a column and a half on its front page covering the "Life and Death of the celebrated Dr. Boerhaave" (Wiles 1965:304). The emphasis in the eighteenth-century English newspaper was primarily on foreign and domestic economic and political news (Harris 1987:168). The lively political life of eighteenth-century England was reflected in the newspapers. Some newspapers, as always, served to reflect the political and economic ideologies of their patrons and subscribers (Werkmeister 1967:19ff). Newspapers also served to publish news and reports of learned and scientific societies when no other medium was available or rapid dissemination was desired. The Dublin Society for Improving Husbandry, Manufactures, and other Arts and Sciences announced in the February 24, 1733, issue that it intended from time to time to publish instructions on husbandry and other useful subjects in *Dublin News Letter* (Berry 1915:23).

In 1695, before the lapse of the licensing act of 1662, printing in the provinces was illegal except for the two university towns, Cambridge and Oxford. The first provincial newspaper, however, did not appear until 1701 (Ferdinand 1997:6, 10) After that they appeared all over England, modeled primarily on the London papers. They depended on advertisements for much of their income, providing space among others for quack doctors who advertised regularly in their pages (Cranfield 1962: 22, 42).

In the American colonies on the other side of the Atlantic, newspapers were even more important in disseminating medical and scientific news because it took a long time to be published in the English papers. The first colonial newspaper appeared in 1690 with the title *Public Occurences Both Domestic and Foreign*. On the front page of its first issue, it carried a story stating that "Epidemical Fevers and Agues grow very common in some parts of the Country." This was followed by a report on a small pox epidemic in Boston. It did not seem to have lasted beyond the first issue, and it was not until fourteen years later that another colonial newspaper appeared. It was modeled on *London Gazette* and was titled *Boston News-Letter* (Hutt 1973:21, 23). The existence of a medical society in Boston, said to have been founded in 1735, is known to still be in existence in 1741 because the November 13, 1741, issue printed a case report of a surgery for removal of a bladder stone by a member of that society (Burrage 1923:62). In 1768, *Pennsylvania Chronicle* acted as the official outlet for papers read before the American Philosophical Society (*Early Transactions* 1969:126). Benjamin Franklin's famous electrical experiments were printed in *Pennsylvania Gazette* in 1737 (Cohen 1941:129). This is just a small sample of the reporting of scientific and technical reports appearing in colonial newspapers.

The history of German language newspapers and those of other languages other than French and English are not as easy to encapsulate as the latter two. Outside of France, Europe was divided into numerous principalities and other political units without focal cities such as London and Paris. Nevertheless, German newspaper history has been extensively studied and voluminously documented (Baasch 1930; d'Ester 1942; Groth 1960; Solomon 1972; Schöne 1939). German newspapers had their origins in the *Meszrelationen*, which were printed twice a year for the book fairs at Frankfurt and Leipzig and were distributed widely. They ranged from eighty to one hundred and twenty pages, and in addition to book notices included discussions of anomalous births, plagues, and miraculous cures (Deneke 1969:61). By 1630, there were about two dozen newspapers in various parts of Germany (Stieda 1897:66). The first dailies appeared in Germany in 1660, more than forty years before they appeared in England and more than a hundred years before they appeared in France (Stieda 1897:73). In a voluminous study of the medical content of German newspapers in the seventeenth and eighteenth centuries, Deneke found a wealth of information. The newspapers were full of descriptions of surgical operations, anatomical dissections, unusual diseases, and similar material that he says consumed as much as a third of the space (Deneke 1969:84–85). Knowledge for inoculation against smallpox was spread largely through the medium of the German newspapers in the eighteenth century (Deneke 1969:108–14). Buchner also described the medical emphasis in German newspapers of this period. He cites a report in *Hamburgisher Correspondent* in 1726 that relates that surgeons are so advanced that they are able to cure cancerous growths of the breast, and describes an operation on the left breast that took only six minutes (Buchner 1922:29).

JOURNALS

Publishers of journals had greater freedom in establishing periodicity than the publishers of almanacs or newspapers that were tied either to an annual cycle or needed to be responsive to current events. Journals in the seventeenth and eighteenth centuries could be published weekly, monthly, bimonthly, biannually, or any other interval that the publisher chose. They were bound only by the commitment they had made to their subscribers, and could set the publication intervals at their own convenience, or even find excuses when they found it necessary to circumvent them. Irregularities were caused by war, illness, or any of the other complexities of life. In some cases, the publishers chose not to make any commitments about regularities and promised their readers only that a fixed number of issues would constitute a volume, which would fulfill their obligations. Numerous examples of many of these variations are found in the prefaces of the early journals of the period.

Problems with maintaining a prescribed periodicity occurred with both of the first two major learned journals that appeared in 1665. *Journal des Sçavans*, which was one of the most important depositories of scientific advances in the early modern period, was published as a weekly. However, only twelve weekly issues appeared from January 5, 1665, when it began to the end of March when it was discontinued by the Royal censors, despite the fact that it was a "commercial success" (Brown 1976:81). The Abbé Gallois, who replaced de Sallo as editor, managed to publish fifty-two issues in 1666, but the regularity began to decline the following year when only sixteen were published. Regularity declined even more after that until only four issues were published in 1669. In the year Gallois was elected to the Académie Français, 1673, no issues appeared at all. Two issues appeared under Gallois in 1674, but the journal did not resume regular weekly issues until 1675 when the Abbé de la Roque took over as editor (Birn 1965:22–24).

Philosophical Transactions of the Royal Society of London (a centerpiece of early modern science) had a rather different history of irregularity. Henry Oldenburg, the founder of *Transactions*, and secretary of the Royal Society, began to issue the journal as a monthly in 1665. They continued to appear monthly with few exceptions until his death in 1677, despite the London Plague of 1665 and the Great Fire in London in 1666. In July 1666, Oldenburg published two issues to accommodate Robert Boyle's description of his method of cooling water. At this time, he explained that he had "never meant so to confine himself to a set time, as not to retain the Liberty of taking any other where there is occasion" (Atkinson 1999:113). The monthly issues continued, after a lapse of time after Oldenburg's death, until 1694 when *Transactions* became bimonthly. By the time the Royal Society took over official responsibly for *Philosophical Transaction* in 1750, they had become a quarterly. At that time, the society proudly announced in the preface to the first issue of that year that "in no case have papers been read before the Society, been delayed more than a year. . . . In this respect the Society is far in advance of the French Institute, where papers frequently remain unpublished for two, three and even four years" (Weld 1848:1521, n. 46).

Statements about periodicity or lack of periodicity are frequently found in the prefaces of eighteenth-century learned journals. What follows are only a few examples. In the preface to the second volume of *Physikalische Bibliothek*, published in Rostock in 1776, the editor indicates that he would not be bound to any particular schedule, but would attempt to avoid missing an issue for the book fairs. In the preface to the first volume of *Hamburgische Magazine* in 1747, the editor says the journal would appear every month, except when that becomes too restrictive, in which case it would be issued every two months. In any case, he said, four issues would constitute a volume to satisfy his promise to his subscribers. Johann Heinrich Gottlob von Justi, editor of *Fortgesetzte Bemuhungen zum Vortheil der Naturkunde*, in the preface to the volume for 1759, says: "I will not commit myself to any strict schedule in which it will appear, and promise only that the shortest interval between issues will be six weeks or at most three months." Johann Beckmann, the inventor of the term "technologie," edited *Beiträge zur Oekonomie, Polizey und Cameralwissenschaft*, which first appeared in 1779. The first issue was followed by a second in the same year and a third in the following year. In the third issue, he says he would not attempt to predict how frequently future issues will appear. It depends, he said, on how much time is left over after he has attended to his other responsibilities and on the help he receives from friends and well wishers. In her review of the German press before 1815, Margot Lindemann quotes Carl Phillipp Moritz, editor of *Gnothe Sauton, Magazin der Erfahrungsseelenkunde*, issued between 1783 and 1793, which with some justice can be called the first psychiatric journal. Moritz said: "A definite time for the journal to appear cannot be predetermined; rather it will come out as materials and papers are accumulated" (Lindemann 1969, vol. 1:186). In the preface to the first volume, the editor of *Leipziger Magazin zur Naturkunde und Ökonomie* discussed the publishing practices in which four issues would make up an annual volume. "The gap between the publication of the second and third issue," he explained, "was not our fault. However, this is no place to discuss the reasons for the delay." He promised that it would not occur again.

The problems with establishing periodicity in independent journals were also to be found in the publications of societies, which were faced with the additional responsibility of placing their imprimatur on their publications. *Bath & West & Southern Counties Society for the Encouragement of Agriculture* made this clear in the preface to the first volume (1792) of their *Letters and Papers on Agriculture, Planting etc.* "Though this Society can never wish to withhold those articles of practical communication which is its plan to encourage, yet the importance of publishing will appear too great to be hazarded by any arbitrary adherence to stated periods." This sentiment is also reflected in the preface to the first volume (1763) of *Deliberations et Mémoires de la Société Royale d'Agriculture de la Généralité de Rouen*, where the following statement occurs: "The Society does not obligate itself to deliver a volume each year, but proposes to occupy itself continually with everything that appears advantageous to agriculture." It promises only that "as soon as there are enough memoirs to fill a volume it will be made public."

CONCLUSIONS

The scientific journal did not make its appearance until 1665 when the need for such a medium became apparent. All the elements were then in place to make it possible, including the formats, management techniques, and distribution systems. The technology did not change very much in the early modern period, although there was a growth in the number of printing presses in response to the growth of the middle class and of literacy. We cannot make great claims for the contributions made by almanacs and newspapers in this period. However, they provide a fertile field for studying the contemporary attitudes and their effects on society. A measure of their influence can be found in the degree that terms like almanacs and newspapers are found in the titles of seventeenth- and eighteenth-century journals, such as *Almanach für Ärzte und Nichtärzte*, a popular medical journal, edited by Christian Gottfried Gruner from 1782 to 1796, and *Gazette Salutaire* which was edited by Emmanuel Grünwald from Bouillon from 1761–1793.

REFERENCES

Armytage, W. R. G. *The Rise of the Technocrats: A Social History*. London: Routledge and Kegan Paul, 1965.

Atkinson, Dwight. *Scientific Discourse in Socio-Historical Context: The Philosophical Transactions of the Royal Society of London, 1665–1975*. Mahwah, N.J.: Erbaum Association, 1999.

Baasch, Ernst. *Geschichte des Hamburgishe Zeitungenwesens von den Anfangen bis 1914*. Hamburg, 1930.

Badouin, M. "Une Journaliste Médicale de Province Avant la Révolution." *La France Médicale* 54 (1912): 161–63, 181–84, 206–208, 221–23, 241–43.

Bennett, H. S. *English Books and Readers (1603–1640)*. Cambridge: Cambridge University Press, 1970.

Berry, Henry F. *A History of the Royal Dublin Society*. London: Longmans, Green & Co. 1915.

Bertholet, M. *Notice sur les origines et sur l'histoire de la Société Philomathiques*. Gauthier-Villars, 1888.

Bigourdan, Guillauime. "Les Premières Sociétés Scientifique de Paris au 17e Siècle." *Comptes Rendus de l'Académie des Sciences* 163 (1916): 937–43; 164 (1917): 129–34.

Birn, Raymond. "Le *Journal des Savants* sous l'Ancien Régime." *Journal des Sçavans* 1965: 15–35.

Black, Jeremy. *The English Press in the Eighteenth Century*. London: Croom Helm, 1987.

Black, Jeremy. "Journalism and Its Problems in Late Eighteenth Century." *Journal of Newspaper and Periodical History* 7 (1991): 31–38.

Blagden, Cyprian. "Thomas Carnon and the Almanac Monopoly." *Studies in Bibliography* 14 (1969): 223–43.

Bluhm, R. K. "Henry Oldenburg, F.R.S. (c1615–1677)." *Notes and Records of the Royal Society of London* 15 (1960): 183–97.

Bollème, Geneviève. *Les Almanacs Populaires aux xvii et xviii Siècles: Essai d'Histoire Sociale*. Paris: Mouton, 1969.

Brown, Harcourt. *Science and the Human Comedy: Natural Philosophy in French Literature from Rabelais to Maupertius*. Downsview, Ontario: University of Toronto Press, 1976.

Brown, Harcourt. *Scientific Organizations in Seventeenth-Century France, 1620–1680*. Baltimore: Williams and Wilkins, 1984

Brown, P. S. "The Vendors of Medicines Advertised in Eighteenth-Century Bath Newspapers." *Medical History* 3 (1975): 352–69.

Buchner, Eberhard. *Aerzte und Kuhrfurschers: Interessante Dokumente aus Allen deutschen Zeitungen (17. und 18. Jahrhundert)*. München: Albert Langen, 1922.

Burrage, Walter S. *A History of the Massachusetts Medical Society, 1781–1922*. Norwood, Mass., 1923.

Capp, Bernard. *English Almanacs, 1500–1800: Astrology and the Popular Press*. Ithaca, N.Y.: Cornell University Press, 1979.

Carroll, Dewey Eugene. "Newspaper and Periodical Publication in Countries of Europe, 1600–1959." Ph.D. dissertation. University of Illinois, 1966.

Chappell, Warren. *A Short History of the Printed Word*. New York: Knopf, 1970.

Chartier, Roger. "Book Markets and Reading at the End of the Old Regime." In *Publishing and Readership in Revolutionary France and America*. Edited by Carol Armruster. Westport, Conn.: Greenwood Press, 1993, pp. 117–36.

Cohen, I. Bernard. *Benjamin Franklin's Experiments*. Cambridge, Mass.: Harvard University Press, 1941.

Coleman, William. "The People's Health: Medical Themes in 18th-Century French Popular Literature." *Bulletin of the History of Medicine* 51 (1977): 56–74.

Collins, D. C. *A Handlist of News Pamphlets, 1590–1690*. London: South-West Essex Technical College, 1943.
Cook, Harold J. *Trials of an Ordinary Doctor: James Groenevelt in Seventeenth-Century London*. Baltimore, Md.: Johns Hopkins University Press, 1994.
Cranfield, Geoffrey Alan. *The Development of the Provincial Newspaper, 1700–1760*. Oxford: Clarendon Press, 1962.
Crocker, Lester G. *Diderot: The Embattled Philosopher*. New York: Free Press, 1966.
Curry, Patrick. "Saving Astrology in Restoration England: 'Whig' and 'Torry' Reforms." In *Astrology, Science, and Society*. Edited by Patrick Curry. Woodbridge, Suffolk: Boydell Press 1987, pp. 245–59.
Dahl, Folke. *A Bibliography of English Corantos and Periodical Newsbooks, 1620–1642*. London: Bibliographical Society, 1952.
Dekeyser, L. "Les Origines de la Press Médicale." *Bruxelles-Médicale* 29 (1949): 1753–65.
Deneke, J. F. Volrad. *Arzt und Medizin in der Tagespublizistik des 17. und 18. Jahrhunderts*. Cologne: Deutscher Ärzte Verlag, 1969.
D'Ester, Karl. *Einflüsse der Deutsche Presse*. Bremen: Internationale Zeitschrift, 1942.
Dodge, Robert K., comp. *A Topical Index of Early U.S. Almanacs, 1776–1800*. Westport, Conn.: Greenwood Press, 1997.
Drake, Stillman. *Galileo at Work: His Scientific Biography*. Chicago: University of Chicago Press, 1978.
Drake, Stillman. *The Unsung Journalist and the Origin of the Telescope*. Los Angeles, Calif.: Zeitlin & Ver Brugge, 1976.
Duchênne, Roger. "Lettres et Gazettes au xviie Siècle." *Revue d'Histoire Moderne et Contemporaine* 18 (1971): 459–502.
Durant, Will, and Ariel Durant. *The Age of Reason Begins*. New York: Simon and Schuster, 1961.
Early Transactions of the American Philosophical Society Published in the American Magazine during 1762. Philadelphia, Pa.: The Society, 1969.
Fabre, Maeleine. "Gazette de Londres." In *Dictionaire des Journaux*. Edited by Jean Sgard. Paris: Universitas, 1991, pp. 470–71.
Ferdinand, C. Y. *Benjamin Collins and the Provincial Newspaper Trade in the Eighteenth Century*. Oxford: Claremont Press, 1997.
Feyel, Gilles. *La Gazette en Province à Travers ses Réimpressions, 1631–1752: Un Récherche Analytique de la Diffusion d'un Ancien Périodique dans tout la France*. Amsterdam: ABA Holland University Press, 1982.
Feyel, Gilles. "La Presse Provinciale au xviiie Siècle: Geographie d'un Reseau." *Revue Historique* 27 (1984): 353–74.
Feyel, Gilles. "Feuille du Bureau d'Adresse (1633–1651)." In *Dictionaire des Journaux*. Edited by Jean Sgard. Paris: Universitas, 1991a, pp. 416–722.
Feyel, Gilles. "Gazette [de France] (1631–1792)." In *Dictionaire des Journaux*. Edited by Jean Sgard. Paris: Universitas, 1991b, pp. 443–49.
Fielding, Henry. *The History of Tom Jones*. New York: Random House, 1964.
Frangsmyer, Tore. "Swedish Science in the Eighteenth Century." *History of Science* 12 (1974): 22–42.

Frearson, Michael. "The Distribution and Readership of London Corantos in the 1620s." *In Serials and Their Problems, 1624–1914.* Edited by Robine Myers and Michael Harris. Newcastle, Del.: Oak Knoll Press, 1993, pp. 1–25.

Grand-Carteret, John. *Les Almanachs Français, Bibliographie Iconographique, 1600–1895.* Paris: Alisse et Cie, 1894.

Groth, Otto. *Die unbekante Kulturmacht: Grundlegung der Zeitungswissenschaft.* 7 vols. Berlin: Walter de Gruyter, 1960–1971.

Guerlac, Henry. *Antoine-Laurent Lavoisier: Chemist and Revolutionary.* New York: Charles Scribner's Sons, 1975.

Handover, P. M. *Printing in London: From 1476 to Modern Times.* Cambridge: Cambridge University Press, 1960.

Harris, Michael. "Newspaper Distribution during Queen Anne's Reign." In *Studies in the Book Trade in Honour of Graham Pollard.* Oxford: Bibliographical Society, 1975, pp. 139–52.

Harris, Michael. "Astrology, Almanacs, and Booksellers." *Publishing History* 8 (1980): 87–104.

Harris, Michael. *London Newspapers in the Age of Walpole.* Rutherford, N.J.: Fairleigh Dickinson University Press, 1987.

Hartmann, Claude. "La revolution verte du siecle des lumiers." *Revue d'Histoire des Sciences* 49 (1996): 5–21.

Hatin Eugène. *Bibliographie Historique et Critique de la Presse Periodique Française.* 1866. Reprint, Paris: Didot. 1965.

Hatin, Èugene. *Histoire Politique de la Presse en France.* 1859. Reprint, Genève: Slatkine, 1967.

Heilbron, John L. *Electricity in the 17th and 18th Centuries: A Study of Early Modern Physics.* Berkeley: University of California Press, 1979.

Hellmann, G. *Neudrucke von Schriften und Karten über Meteorologie und Magnetismus.* No. 1–15. Berlin: Asher & Co., 1893–1904.

Heseltine, G. O., ed. *Kalendar & Compost of Shepherds.* London: Peter Davies, 1930.

Hetherington, Norris S. "Almanacs and the Extent of Knowledge of the New Astronomy in Seventeenth-Century England." *Proceedings of the American Philosophical Society* 119 (1874): 275–79.

Howe, Ellis. "The Stationers' Company Almanac: A Late Eighteenth-Century Printing and Publishing Operation." In *Buch und Buchandel in Europa in Achtzehnten Jahrhudert.* Edited by Giles Barber and Bernhard Fabian. Hamburg: Hauswedell, 1981, pp. 195–209.

Hutt, Allen. *The Changing Newspaper: Typographic Trends in Britain and America, 1622–1972.* London: Gordon Fraser, 1973.

Jager, Patrick. "La Mercure Française (1613–1648)." In *Dictionaire des Journaux.* Edited by Jean Sgard. Paris: Universitas, 1991, pp. 867–69.

Jerome, Lawrence E. *Astrology Disproved.* Amherst, N.Y.: Prometheus Books, 1977.

Johnson, Neal. "Popular French Almanacs of the Eighteenth Century." *British Journal of Eighteenth-Century Studies* 8 (1985): 139–59.

Kirchner, Joachm. *Die Zeitschriften des Deutschen Sprachgebietes von den Anfängen bis 1830.* 4 vols. Stuttgart: Hiersemann, 1969.

Kirchner, Joachim. "Gedanken zur Definition des Zeitschrift." In his *Ausgewählte Aufsätzen*. Stuttgart: Hiersemann, 1976, pp. 96–106.

Klarwill, Victor von, ed. *The Fugger News-Letters: Being a Selection of Correspondents of the House of Fugger during the Years 1568–1605*. Translated by Pauline de Chary. New York: Putnam's Sons, 1924.

Kölving, Ulla. "Les Copistes de la Correspondance Littéraire: Un Première Présentation." In *Edite Diderot*. Edited by Dulac Georges. Oxford: Voltaire Foundation, 1988, pp. 175–93.

Kölving, Ulla, and Jeanne Carriot. *Inventaire de la Correspondance Littéraire de Grimm et Meister*. Oxford: Voltaire Foundation, 1984.

Laboulbene, M. "Histoire de Journalisme Médicale, 1679–1880." *Lancette Française*, 1880 (53): 1057–59.

Lindemann, Margot. *Deutsche Presse bis 1815*. 4 vols. Berlin: Colloqium Verlag, 1969.

Lombard. *Der medizinische Inhalt der Schweizerischen Volkskalendar im 18. und 19. Jahrhundert*. Zurich: Fuesli, 1925.

McClellan, James E., III. *Science Reorganized: Scientific Societies in the Eighteenth Century*. New York: Columbia University Press, 1985.

Marion, Michel. "Dix Ans des Affiches, Annonces et Avis Divers (1752–1761)." In *Regards sur l'Histoire de la Presse et de l'Information*. Edited by Jacques Godechot. Saint-Julienne de Sault: Presses Saltessiennes, 1980, pp. 23–40.

Martin, Henri-Jean. *The History and Power of Writing*. Translated by Lydia G. Cochrane. Chicago: University of Chicago Press, 1994.

Mazauric, Simone. "La Diffusion du Savoir en Dehors des Ciruits Savantes: Le Bureau d'Adresse Theophraste Renaudot." In *Commercium Litterarium: Forms of Communication in the Republic of Letters, 1600–1750*. Amsterdam: APA-Holland University Press, 1994, pp. 151–72.

Morgan, Betty R. *Histoire du Journal des Sçavans Depuis 1665 Jusqu'en 1701*. Paris: Presse Universitaire de France, 1929,

Munck, Thomas. *The Enlightenment: A Comparative Social History 1721–1794*. London: Arnold, 2000.

Nelson, Carolyn, and Matthew Seccombe. *Periodical Publications 1641–1700: A Survey with Illustrations*. London: The Bibliographic Society, 1986.

Nicholson, Majorie. "English Almanacs and the New Astronomy." *Annals of Science* 1 (1939): 1–33.

Oldenburg, Henry. *Correspondence*. Edited by A. Rupert Hall and Marie Boas Hall. Madison: University of Wisconsin Press, vols. 1–9, 1966–1973; London: Mansell, vols. 10–11, 1975–1977; London: Taylor and Francis, vols. 10–11, 1986.

Perkins, Maureen. *Visions of the Future: Almanacs, Time, and Cultural Change*. New York: Oxford University Press, 1996.

Popkin, Jeremy D. "Periodical Publication and the Nature of Knowledge in Eighteenth-Century Europe." In *The Shapes of Knowledge from the Renaissance to the Enlightenment*. Edited by Donald Kelly and Richard H. Popkin. Dordrecht: Kluwer Academic, 1991, pp. 203–13.

Posner, E. *Archives in the Ancient World*. Cambridge: Harvard University Press, 1972.

Raymod, Joad. *The Invention of the Newspaper: English Newsbooks 1641–1649*. Oxford: Clarendon Press, 1996.

Richardson, Lila M. "Conferences of T. Renaudot: An Episode in the Quarrel of the Ancients and Moderns." *Modern Language Notes* 48 (1933): 313–16.
Robinson, Howard. *The British Post Office: A History*. Princeton: Princeton University Press, 1948.
Roche, Daniel. *Le siecle des Lumieres en Province*. Paris: Mouton, 1978.
Römer, J. J., and P. Usteri, eds. *Tagebuch der medicinischen Literature des Jahres 1745 bis 1774*. 3 vols. Hildeshheim: Olms [1789–1791] 1974.
Sarasohn, Lisa T. "Nicolas-Claude de Peiresc and the Patronage of the New Science in the Seventeenth Century." *Isis* 84 (1992): 70–90.
Schöne, Walter. *Die deutsche Zeitung im Ersten Jahrhundert Ihres Bestehens, 1609–1700*. 3 vols. Leipzig: Harrasowitz. 1939–1940.
Sgard, Jean, ed. *Dictionnaire des Journaux 1600–1789*. Paris: Universitas, 1991.
Sinnerstad, Ulf. "Astronomy and the First Observatory." In *The Royal Swedish Academy of Science, 1739–1900*. Edited by Tore Frängsmyr. Canton, Mass.: Science History Public Actions, 1989, pp. 45–71.
Snorrason, Egill. "Danish Physicians and Periodicals of the Seventeenth and Eighteenth Centuries." *Danish Medical Bulletin* 5 (1958): 200–209.
Solomon, Howard M. *Public Welfare, Science, and Propaganda in the Seventeenth Century: The Innovations of Theophraste Renaudot*. Princeton: Princeton University Press, 1972.
Solomon, Ludwig. *Geschichte des deutschen Zeitungswesens von dem Ersten Anfängen*. 3 vols. Oldenburg: Schulzische Hof Buchhandlung, 1972.
Stieda, Wilhellm. "Die Anfänge des periodischen Presse in Mecklenburg." *Geschichte des deutschen Buchhandels* 19 (1897): 65–178.
Sudhoff, Karl. "Das medizinsche Zeitschriften-wesen in Deutschland bis zur Mitte de l9 Jahrhunderts." *Münchner medizinischen Wochenschrift* 50 (1903): 455–63.
Sutherland, James R. "The Circulation of Newspapers and Literary Periodicals, 1700–1730." *The Library* Ser 4, 15 (1934): 110–24.
Synder, Henry L. "Newsletters in England 1689–1715." In *Newsletters to Newspapers: Eighteenth-Century Journalism*. Edited by Donovan H. Bond and W. Reynolds. Morgantown: School of Journalism, West Virginia University, 1977, pp. 3–19.
Tabarroni, Giorgio. "Geminniano Montinari." In *Dictionary of Scientific Biography*. Edited by Charles C. Gillespie. Vol. 9. New York: Charles Scribner's Sons, 1977, pp. 484–87.
Trenard, Louis. "La Presse Française des Origines à 1788." In *Histoire Général de la Presse Française*. Edited by Claude Bellanger et al. Vol. 1. Paris: Presse Universitaire de France, 1969, pp. 27–404.
Vittu, Jean-Pierre. "Diffusion et Réception du Journal des Savants." In *Diffusion et la Lecture des Journaux Française sou l'Ancien Régime. Proceedings of the International Conference at Nijmegen*. Amsterdam: APA Holland University Press, 1988, pp. 167–75.
Weil, Françoise. "Les Gazettes Manuscrites Avant 1750." In *Journalisme d'Ancien Régime: Questions et Propositions*. Edited by Pierre Retat. Lyons: Universitaire de Lyon, 1982, pp. 93–98.
Weil, George. *Le Journal: Origines, Évolution et Rôle de la Press Périodique*. Paris: Renaissance du Livre, 1934.

Weld, Charles R. *A History of the Royal Society*. London: Parker, 1848.

Werkmeister, Lucyle. *Newspaper History of England 1792–1793*. Lincoln: University of Nebraska Press, 1967.

West, John A. *The Case for Astrology*. London: Arkana Books, 1992.

Wiles, R. M. *Freshest Advices: Early Provincial Newspapers in England*. Columbus: Ohio State University Press, 1965.

Wilson, Arthur M. *Diderot*. New York: Oxford University Press, 1972.

Winton, Calhoun. "Richard Steele, Journalist-Journalism." In *Newsletters to Newspapers: Eighteenth-Century Journalism*. Edited by Donovan H. Bond and W. Reynolds. Morgantown: School of Journalism, West Virginia University, 1977, pp. 21–37.

Chapter 15

The Commerce of Letters

Networks and Invisible Colleges in Early Modern Science

In 1712, Isaac Newton anonymously published a collection of letters and other documents to defend his claim to priority against Leibniz for the invention of the calculus. Newton based his claim largely on his correspondence with Barrow, Collins, and Leibniz (Hall 1980:16) in a book titled *Commercium Epistolicum.* This provides evidence of the important role that letters played in the life of science from the earliest times, and in this, one of the most famous (or perhaps infamous) disputes of priority.[1] The term "commercium" is appropriate in this context because letters can be considered a form of commerce in which letters sent confer an obligation of the recipient to respond.

Letters are among the most ancient methods of transmitting scholarly information. Evidence of their existence goes back at least as far as the Babylonian clay tablets that have been found in abundance in archaeological digs. Medical letters found among these tablets date back to the third millennium BC (Sigerist 1951:431). Correspondence played a large role in commerce and other kinds of social and political affairs. Letters survive in significant number from the third century AD onward, but it is the fifth and twelfth centuries that have been said to be the "Golden Age of medieval epistolography" (Constable 1976:31). Letters were also produced in large numbers in the seventeenth and eighteenth centuries in response to stimulus provided by new ideologies and discoveries. The letters that are exchanged between scholars in this period have been called "erudite letters," not because they are always involved with scholarly matters, but because they are exchanged between *erudits.* Pomian says: "One cannot hope to know the thought of the scholars of the seventeenth century without acquiring a knowledge of their correspondence" (Pomian 1974:26).

THE ERUDITE LETTER

Letters as a literary format were well established in classical times, and students who received a classical education in the seventeenth and eighteenth centuries were exposed to letters of the great classical writers. Cicero established a tradition by publishing his collected letters in 68 BC and set a pattern that was followed for centuries thereafter (Howland 1991:15). These formulas and styles of letter writing were carried over into the Middle Ages and found a place in the manuals on letter writing that were produced in increasing numbers well into the eighteenth century. Some of the manuals prescribed formulas that are reminiscent of the formulas recommended for writing scientific papers in the twentieth century: *Exordium* (introduction), *narratio* (narrative), *proposito* (principle theme), *confirmatio* (confirmation), *conjuratio* (agreement), and *peroratio* (conclusion) (Robertson [1942]1973:58). In Germany, in the last third of the seventeenth century, there were about sixty manuals[2] on letter writing published; by the second half of the eighteenth century there were at least a hundred and fifty on the market (Oellers 1989:9, 22). Not all the works that are designated *epistola* in this period can be considered letters. The term is found frequently in the titles of publications, particularly in the sixteenth and seventeenth centuries, but my impression is that they have the same kind of interpretation as such words as *observationes, disputationes,* and *disserationes* to describe collections of texts. They are sometimes used in combination as in Marcello Malpighi's *Dissertatio Epistolica de Bombyce,* which was printed under the auspices of the Royal Society of London by the society's printers Martyn and Allestry in 1669, as well as his equally famous *Dissertatio Epistolica de Formatione Pulli in Ovo,* printed under the same auspices in 1673. The words *epistolarum* or *epistolae* are frequently found in the titles of collections of texts printed in the sixteenth century, particularly of medical works.

Letters of some Humanist and Renaissance scholars are of interest to science historians as well as general historians because many of them, particularly those written by physicians, were concerned with natural history as well as philosophy. Collections of letters by these writers were frequently published in such works as *Epistolae medicinales diversum authorum;* many of the authors included were still alive when their books were published in Lyon. Renaissance physicians such as Giovanni Manardi (1462–1536), Johannes Lange (1485–1565), and Pietro Andrea Mattioli (1501–1577) published collections of their own letters at various times while they were still alive. Marsilio Ficino (1433–1499) began to collect his letters for publication at the end of 1473, including those letters on medicine and astrology. The letters previously had been disseminated among his friends but not published, a common practice at that time. The editor of his letters suggests: "He may have been prompted to start this collection by the fact that letters falsely purporting to be his were being circulated." Twelve books of his letters were first printed in Venice in 1495 four years before his death (Ficino 1981:27).

The great humanist Desiderius Erasmus (1466–1536) did not contribute much to the advancement of science,[3] but was one of the most prodigious letter writers in his period. A cleric, like many of his humanist colleagues, he acquired a wide range of scholars in his peregrinations around Europe and England and wrote to many of them. Letters constituted a form of publication for these scholars, and even though they may have been distributed from hand to hand, they were also printed to assure a wider audience. Erasmus published several series of his letters during his lifetime. In the period 1515 to 1536, he edited twenty-one collections for the press (Darriceau 1985:433). It has been suggested that Renaissance scholars "published their correspondence with other scholars as a university professor now submit articles to scholarly journals . . . [and thus] . . . his collected letters became his dossier or portfolio" (Henderson 1983:94). Erasmus's contribution included his widely circulated manual of correspondence *De Consribendis Epistolae*, published in 1522. It went through twenty-eight editions in his lifetime and is said to have exercised a profound influence on epistolary theory (Nellen 1993:88).

Athanasius Kircher (1602–1680) was a transitional figure between the humanists and the scientific virtuosi of the seventeenth century. Writing to Kircher in 1652, Marclaurelio Severini, a professor of anatomy, called him "a resuscitated Pythagorus, an immortal encyclopedist." Fletcher regards this as a typical statement in the accepted etiquette of the seventeenth-century correspondent. He, nevertheless, asserts that in Kircher "we find a man who on occasion deserved more than most, extravagant praise" (Fletcher 1970:52). The large volume of his collected works has been described "as a vast storehouse and quarry of medieval learning [which] frightened the timid of the seventeenth century and were deplored by Kircher's posterity" (Fletcher 1969:259). History delivers a mixed appraisal. He is said to have provided a clearinghouse for reports on eclipses (Fletcher 1970:61). Other scholars call attention to the fact he was "considered credulous to any story brought to him if it seemed plausible" (Thorndike 1923, vol. 8:177). Kircher corresponded with many of the well-known scientists of his period, including Christiaan Huygens, Leibniz, Hevelius, Pieresc, and Gassendi. Huygens characterized his writings as "nothing but a heap of idle unreasonable stuff," but nevertheless addressed him in flattering terms in his letters (Fletcher 1970:59). Much of this correspondence remains unpublished, except for those letters that are included in the collected letters of some of Kircher's contemporaries. Fourteen folio volumes of his unpublished correspondence are to be found in the archives of the Pontificia Università Gregoriana in Rome. On the other hand, he published over thirty-eight volumes in which his work primarily is preserved (Fletcher 1970:52). The correspondence of Evangelista Torricelli (1608–1647), whose work was well known in the West, shared a similar fate. He published only one book in his lifetime; his accomplishments were disseminated largely through correspondence, his own or that of other contemporaries. Torricelli began to organize his letters and made plans to publish them, but died before it could be accomplished (Gliozzi 1976:433–40). His letters were included in conjunction with an edition of his works published between 1919 and 1944 (Torricelli 1919).

In the first half of the seventeenth century, the primary method scholars used to disseminate or receive information or put their ideas or discoveries on the record was by means of writing letters. In his *History of Non-Printed Science in the Waller Collection*, Marco Beretta emphasizes the importance of correspondence as a historical resource and provides examples from letters in the collection.

> The letter in fact not only constituted for scientists the fastest means of communication and collaboration but it also represented a form of assessing in a public way, even if not officially, the priority of a discovery or of an experiment . . . following a very common routine, relevant scientific letters were copied and made to circulate throughout Europe. Galileo transcribed a letter by Kepler and had it circulated. Like-wise his pupil Evangelista Torricelli explicitly asked his addressee to copy the letter in which he made an original geometrical demonstration and to make it known to other scientists in order to assess the priority of his discovery. (Beretta 1993:31)

Thus a letter represented a form in which a scientific article could be disseminated for comment and may be considered equivalent to reading a paper at a scientific meeting before submitting it to a publisher or editor for peer review.

There were other options, of course, such as publishing a pamphlet or a book, but these methods were costly and did not guarantee an audience. There were, in addition, certain advantages in circulating one's ideas in the form of letters rather than presenting them in the open forum provided by print. Frequently, letters represented work in progress rather than work completed and thus could benefit by responses from those who were engaged in similar endeavors. In addition, correspondence was not faced with the kind of restraints that are inherent in publication, where works were subject to authoritarian review. They also did not risk offending anyone whose opinions or work was being criticized or repudiated. The erudite letter provided a freedom of information exchange and expression of ideas because it existed for the most part outside the scrutiny of governmental censorship systems, and the restraints of religious and national controversy (Rochot 1966). Even after his condemnation by the Inquisition in 1633, Galileo was able to maintain his correspondence with his French admirers such as Elia Diodati and Nicolas Fabri de Pieresc. This correspondence was mutually beneficial because it not only stimulated the interest of French astronomers, but also provided Galileo with information about the astronomical efforts of such scientists as Gassendi (Baumgartner 1988).

A sampling of the vast published correspondence of the scientists of this period makes it clear that few generalizations can be made about the character of the erudite letter as a literary genre. There are too many variables to consider in characterizing the nature of any particular correspondence. Among them are the personality of the writers and their relationships with their correspondents, which may change during the course of the exchange. Such a change can be seen in a sequence of letters between Alexander Garden, a physician in South Carolina, and William Ellis, an English naturalist who held administrative posts in the colonies. Garden began his letter to Ellis on March 25, 1755, by addressing him as "Dear Sir," which on November 19, 1764, had become "My first and still dearest botanical friend." On December 16, 1765, Garden began his letter to Ellis with the news that "The fatal Stamp Act is likely to put an end to our intercourse. You have imposed a taxation in America, which Americans say they will not receive." After commenting for another page or so on the political situation he adds: "Alas!—enough of this!—Now for Natural History." (Smith 1821:485, passim).

In many cases, letters provided the only form of communication that existed between correspondents who may never had met in person. It is not always clear what the implicit or explicit purposes were of a particular letter. As in the writing of diaries (journals), in correspondence there is always the presence of an unconscious or conscious audience; first, there are the authors themselves who wished to reflect a favorable image, and then there are the known or unknown readers, present or future, whom the writer is addressing. Mireille Bossis makes a distinction between what she calls "literary correspondence" and "natural correspondence," the first written with an immediate or distant object of publication, and the second written for other purposes. She, nevertheless, finds it necessary to add that the distinction is sometimes difficult to make (Bossis 1966:64). The editor of the Newton correspondence reflects this ambivalence when he says:

> It would appear that the present distinctions between the personal and official letters, or again between a letter and a contribution to a scientific periodical,

> had little or no significance in the seventeenth century. Thus the great communication in February 1671/2 upon the refrangiability of coloured light, is regarded as a letter because it is addressed personally to Oldenburg, yet it has a dignity and breath that makes it stand entirely suited to its place as a paper in the *Philosophical Transactions*. (Turnbull 1959, vol. 1:xxviii)

One characteristic that applies to all letters, however, is that, unlike speech, letters (that is, when they are not recorded) are irrevocable and therefore impose some restraints on the author. It is "a document whose purpose is the transmission of information when oral communication is not possible" (Nellen 1993:87).

Correspondence also served other functions in addition to the dissemination of information. It provided a means of achieving and maintaining a position in the international company of scholars and of staying in touch with new developments in scholarship and science. In their letters, scholars often requested assistance in acquiring books, particularly when they were difficult to acquire through the booktrade. The efforts of scholars to procure books from distant places with the assistance of their correspondents occur frequently in the letters that circulated in this period. Much of the published erudite correspondence, or rather the correspondence of *erudits*, is personal and not always scholarly, but others are of great historical importance.[4] An early example is the letter that Galileo received in 1610 from his disciple Benedetto Castelli. It raised an astronomical question regarding the heliocentric theory and stimulated Galileo to look upon the telescope as an important instrument in studying the problem (Westfall 1985:11). In another example, Carl Boyer tells us:

> While Pascal in 1654 was working on his *conici*, his friend, the Chevalier de Méré raised with him questions such as the following. In eight throws of a die a player is to attempt to throw a one, but after three unsuccessful trials the game is interrupted. How should he be indemnified? Pascal wrote to Fermat on this, and their resulting correspondence became the effective starting point for the modern theory of probability. (Boyer 1968:397)

An example of the function that correspondence played in resolving controversies without making them public can be seen in an unpublished letter of Robert Hooke, which he wrote on December 6, 1679, to Newton to make amends for having wounded his sensibilities.[5] At a meeting of the Royal Society, Hooke had not only criticized Newton's work on optics but also claimed priority for most of the results. In his letter he writes:

> First, that I doe noe ways approve of contention, or feuding or proving in print, and shall be very unwillingly drawn to such kind of warre. . . . If, therefore, you will please to correspond about such matters by private letters, I shall gladly embrace it; and when I shall have the happiness to peruse your

excellent discourse, (which I can as yet understand nothing more of by hearing it cursorily read), I shall, if it be not ungrateful to you, send you freely my objections, if I have any, or my concurrence, if I am convinced, which is the more likely. (Koyré 1952:314)

Newton responded in a friendly manner in which he used the well-known expression, which is sometimes attributed to him: "If I have seen farther, it is by standing on the shoulders of giants."[6] Koyré comments: "Neither, of course, availed himself of the magnanimous and high sounding invitation" (Koyré 1952:314–15). In the eighteenth century, Joseph Black could try out his ideas by letters before presenting them publicly, particularly if he thought them speculative. Furthermore, he could discuss issues relating to securing rights in his discoveries without risking being considered a "projector"[7] (Golinski 1992:39–40).

A study of the correspondence of this period reveals a measure of ambivalence about the confidentiality of letters written to individuals. Correspondents had to be circumspect, particularly when they feared their letters might fall into the wrong hands. Erasmus, for example, could express a point of view about Luther in some letters that he did not wish would be seen in Rome and another point of view in those that he had no reservations being circulated. Erasmus edited most of his correspondence published during his lifetime. Bietenholz reports that: "Erasmus generally assumed that he was at liberty to publish the letters he received from others along with his answers, and . . . most correspondents seemed to concede this privilege to him without demanding reciprocity" (Bietenholz 1977:64). However, when his private views regarding Luther were published in collections of other authors, he was not very happy about it (Bietenholz 1977:64–70). In Italy, the Medici librarian Magliabéchi (1633–1714) complained in a letter to Jean Mabillon (1632–1707) that the publication of certain of his letters in Germany without his permission had caused him great pain (Pomian 1974:27).

In general, letters rarely required any confidentiality. They were diffused, sometimes in extracts and sometimes in their entirety among the recipients' friends. Correspondents frequently received letters for transmission to a third party. Dibon cites as an example a letter from Grotius from Amsterdam to Pierre Dupuy,[8] which contained an observation of a lunar eclipse seen in Amsterdam. The letter was destined to be sent to W. Schickard in Tübingen as well as to Gassendi (Dibon 1976:41). In a letter to Mersenne on December 4, 1634, Peiresc asked him not to spread the news about Galileo's sentence because the Roman curia were attempting to keep it secret. If it was a secret, it was not very well kept because Theophraste Renaudot carried the news in the December 1633 issue of his *Gazette* (Sarasohn 1993: 84).

In a letter written to one of his correspondents in 1641, Christiaan Huygens's father Constantijn writes: "This letter must in no case leave your cabinet" (Nellen 1993:95). The presumption seems to be that openness was assumed unless it was specifically indicated that they were to be held confidential. In practice, however, this presumption is not always respected. Marie Boas Hall tells us: "Everyone knew the conventions; everyone knew that letters to Oldenburg were forms of publication and expected him to show them to appropriate colleagues" (Hall 1965:287).

In Oldenburg's case, as secretary of the Royal Society of London, he was obliged to share letters received in this capacity with other members or to the society as a whole. Nevertheless the astronomer Auzout wrote to Oldenburg on October 3, 1665, from Paris:

> I am very sorry that you are displeased at my having printed (at the request of my friends) the letter you wrote me setting forth the opinions of Mr. Hooke. I did not regard this letter as being altogether your own work so much as the reply of Mr. Hooke and since both of us had already begun by printing this material saw no inconvenience in printing the rest if our friends wished to see the sequel of the dispute. For I see little difference between printing scientific matters contained in letters and showing these same letters to those learned in these matters who can copy them out when they have them on loan, and everyone knows perfectly well that when one exchanges ideas by letter one does not look for eloquent and polished but plain and simple language and that there is a great difference between an eloquent discourse and some treatise which has been written on some occasion for oneself alone which one would never print without the author's permission or a letter which one writes well aware of the fact that it will be shown to many learned men. . . . It is not the case that, if exchange of letters were not so uncertain just now, and if I had not been afraid of not receiving a reply soon enough, I should have failed to warn you, but this, as well as a little journey I had to take and which caused me to hurry the printing before I undertook it, led me to give it all [to the printer] without warning you, and since there were several words which were not very colloquial I took the liberty of changing them as I have already told you and I did not do that without consulting Mr. Justel who found nothing to say against it. However, Sir, I apologize and shall not again have anything of yours printed unless you authorize it. (Oldenburg 1966, vol. 2:518)

This goes beyond the simple act of sharing letters and provides us with some idea of the conventions of the period. Statements requesting confidentiality are to be found all through the collected correspondence of the seventeenth and eighteenth centuries. Albrecht von Haller (1708–1777), an industrious correspondent in the eighteenth century, was also held to be a transgressor. When the Swedish astronomer and longtime secretary of the Royal Swedish Academy of Science, Pehr Wilhelm Wargentin (1717–1774), found to his surprise that Haller had printed a letter that he had sent him in the former's *Epistolae* (1774), he regarded it "a compliment that he tactfully hoped wold not be repeated." Frangsmyr adds: "Enthusiasm for spreading the news from the field of science could sometime overstep the bounds of propriety" (Frangsmyr 1989:10).

I. Bernard Cohen's statement in the preface to his 1941 book on Benjamin Franklin's experiments sums up the role of the erudite letter in the eighteenth century as follows:

> Some commentators go so far as to suggest that Franklin is not to be considered a scientist at all, since he never wrote a "treatise" or an "article" intended for publication in a scientific journal. Actually Franklin's scientific writing took the form of letters addressed to Peter Collinson and others. Collinson was a Fellow of the Royal Society and in that capacity introduced Franklin's letters in the Society meetings. There they were discussed and eventually found their way into the *Philosophical Transactions*, the scientific journal published by the Society. Franklin knew that his letters were being made public and was very much interested in the criticism which they occasioned. The greater proportion by far of the "articles" published in the *Transactions* were letters addressed to Fellows of the Society or to the President. The scientific letter was one of the regular forms of scientific communication in the eighteenth century.

On the other hand Cohen adds:

> The public character of Franklin's scientific letters is further attested by the fact that his personal or private letters to Collinson were neither read to the members of the Royal Society nor published. In some cases when Franklin wrote a scientific letter in which he indulged a hypothesis which had not been confirmed by experiment, he would stress the fact that this letter was not to be made public but was only for the eyes of his friend and was to be returned. (Cohen 1941:viii–ix)

There are two issues about the erudite letter that have not yet entered the discussion. One relates to ownership of the physical letter. Is it the author or the recipient, or in many cases, those who have purchased or inherited the letters? Letters written to individuals today are the physical property of the recipients to dispose of by sale, auction, or gift as they wish. Nevertheless, the right to publish remains with the author if not otherwise assigned and was so regarded even before it was recognized by law. The other issue, copyright (the right to reproduce the text) belongs to the author. These issues may be related, either directly or tangentially, to some of the examples cited above, but copyright was rarely an issue in the seventeenth and eighteenth centuries. Copyright (that is, if any related legislation was in place) was usually assigned to the printer or publisher who paid for the manuscript. However, the principle could create difficulties when the author of the letters wished to publish them when they were in the possession of someone who did not wish to relinquish them for this purpose. This was one of the reasons why some authors retained drafts of the letters they sent.

I cannot discuss these issues in depth here except to cite a few instances in which they were considered in the past. Constable cites a source that says: "In Antiquity when it was a question of epistolary correspondence, the propriety right, the dominium [right of ownership] passed to the addressee, unless the author stipulated to the contrary" (Constable 1976:16). In his book on "the invention of copyright," Mark Rose describes in detail a celebrated case in the eighteenth century on this subject. Alexander Pope filed a suit against the publisher Edmund Curll over the publication of some of his letters. It is said that this case established the rule that copyright in a letter belongs to the writer. "[It] remains a foundational case in English and American copyright law." The case involved the publication of some of Jonathan Swift's letters in which letters to and from Pope were included. Rose comments: "In 1735, as is well known, Pope tricked Curll[9] into publishing his correspondence. For a gentleman to publish his own letters would have seemed inexcusably vain, but a prior unauthorized publication by Curll would open the way for Pope to publish his own edition as a way of setting the record straight" (Rose 1993:60).

Dena Goodman reports a somewhat variant view put forward by Michel de Servan, a French lawyer: "In 1782 Servan published *Réflexions sur la Publications des Lettres de Rousseau et des Lettres en Général*, a tract in which he urged that letters be considered the legal property of both sender and receiver and thus unpublishable without the express permission of both" (Goodman 1994:151). This seems to be an extreme case in which the sender had not retained drafts of his letters or was responsive to the feelings of a recipient who had some reason not to see them in print. In a lengthy conference in Paris in 1983 on the preservation and publication of correspondence, one of the participants reported on what he judged to be an international consensus on this issue. The consensus, as far as I am able to determine it, was that the content of a letter belongs to the sender (i.e., the author), while letters as physical property belong to the receiver, who can buy and sell them as property, but cannot publish them without the consent of the author, as is the case with other manuscripts (Koumantos 1984:201–209) Elsewhere it has been indicated that letters held by institutions, such as archives, libraries, or museums, may be copied for research even if only the original is privately held, at least fifty years after the author's death (Cavendish 1984:61).

By the end of the seventeenth century, the erudite letter was widely accepted as a publications medium. This conclusion is supported by the frequency in which terms like "letters," "correspondence" and their cognates in other languages appeared in the titles of periodicals and other kinds of publications (Howland 1991:5). In his book *Polyhistor literarivs, philosophicvs et practicvs,* Daniel Morhoff (1708) devoted two chapters to published correspondence and acknowledged that, if he had not been selective, the entries would have required a large volume itself (Wacquet 1993:105). The section headed *epistolae* in the catalog of the library of Joseph Banks published in five volumes covers five pages (Banks 1798). The index in the catalog of the French periodical press published by Eugène Hatin lists over seventy titles beginning with the word *correspondances* and over fifty that begin with the word *lettres* (Hatin 1866:617, 639). The German librarian Joachim Kirchner, in his bibliography of German periodicals, lists at least twenty titles published before 1800 that begin with the word *Briefe* or *Korrespondenz* (Kirchner 1969). The vogue of letters as a literary format is also demonstrated by use of the "form of fictional correspondence" (Beretta 1993:32) or "epistolary novels" (Goodman 1994:14) utilized by many eighteenth-century authors such as Diderot, Montesquieu, and Rousseau as well as scientists such as Euler, Cassini, Fontenelle, Magalotti, and Volta (Beretta 1993:32).

Like the scientific article, the letter is limited in coverage, stands alone, and reflects a singular unity of time and substance. Also, like the scientific article, it is a personal communication that requires analysis and study on the basis of its historical setting, the personality and life history of the author, and the intellectual consensus and its deviations from which it is derived. It is difficult, therefore, to make generalizations regarding the correspondence of any individual or any period, which depends in the first place on what has been preserved in published or manuscript form. Study of individual correspondences, nevertheless, are informative. We can see the transition between correspondence and the journal in the work of Thomas Bartholin (1616–1680), who is credited with publishing the "first Danish scientific journal" (O'Malley 1970:493). Bartholin, a member of a distinguished Danish medical family, returned to Copenhagen in 1646 after traveling widely to Leyden, Paris, Orleons, Montpellier, Padua, Rome, and Naples, and receiving his medical degree in Basel. Like his father, he became a member of the university faculty and achieved renown as a pedagogue and medical investigator. As an appendix to the first edition in 1641, Bartholin included his father's *Institutiones Medica,* letters that he had received from Johannes Waleus (1604–1649) in defense of William Harvey's theories on the circulation of the blood (Schouten 1974). He had published a collection of his letters in Amsterdam in 1654 with the title *Epistolarum Anatomicorum Rariorum.* It was not until 1662 that he published another collection of his own letters under the title *Cista Medica Hafniensis,* to which he added letters of another famous Danish physician, Oluff Worm (1588–1654) and the astronomer, Tycho Brahe (1546–1601), along with "medico-legal dissertations from the medical faculty and articles which he had written himself" (Snorrasen 1958:202). The favorable reception of *Cista Medica* encouraged Bartholin to begin a series of collections of letters under the title *Epistolarum Medicinalium a Doctis vel ad Doctis Scriptorum* between 1663 and 1667. Another stimulus was provided by the appearance of the first annual volume of *Miscellanea Curiosa Medico Physica*, published by a German scientific society that predated the Royal Society of London by at least ten years. The contributors were primarily German, but the first volume contained contributions from other countries in northern and central Europe as well as Denmark, Switzerland, and Italy (Jarcho 1972:65). Bartholin made the final step to a periodical by publishing *Acta Medica et Philosophica Hafniensis*, which was published in Copenhagen from 1673 to 1680 and, Thorndike says, had the same character and contents as the *Miscellanea*. Thorndike commented: "In the case of many of their observations and *scholia* the interest is rather odd and curious, the astounding and marvelous, the unnatural and the abnormal." He also reminds us that *Philosophical Transactions*, which had been published in London some eight years earlier, had many similar contributions (Thorndike 1923, vol. 8:236).

There are other collections of correspondence that can help reveal some of the problems of the epistle and its relationship to the journal. We can explore them in some detail thanks to the scholarship of Howard Adelmann, who has edited the correspondence of Marcello Malpighi (1628–1694) and has written a compendious history of Malpighi's contribution to embryology (Adelmann 1966, 1975).[10] Through the correspondence, we can trace the history of the letter on the conglobate glands that Malpighi had written with the intention of submitting it to the Royal Society of London. Malpighi's reputation was well established by this time and the Royal Society had already authorized the printing of several of his earlier works. Malpighi indicated his continuing interest in the conglobate glands in a letter (AC:642) he wrote on July 19, 1687, to his colleague in Pisa, Lorenzo Bellini (1643–1704), who Castiglione includes among the "eminent Italian anatomists" of the period (Castiglioni 1941:525).

In the same year, Malpighi wrote to Prince Marcantonio Borghese, sending his greetings for the coming Christmas holidays, reporting that he is in satisfactory health, and that he is working on the conglobate glands. "If God granted him the good fortune to compile these studies, he would send them to the Royal Society and to the Prince at the same time" (AE:520; AC:652). On June 23, 1688, he wrote to John Wallis [11] saying that he is sending him an epistle on the glands and lymphatics (AC:684). Malpighi's letter, dated August 22, 1688, was sent to Antonio Ferrarini, who was the physician to the Duke of Modena, Francescso II d'Este, brother of Queen Mary, the consort of the English King James II. In this letter, Malpighi asked Ferrarini to enlist the Duke's help in seeing that the package containing the conglobate gland letter, which he had entrusted to a merchant traveling to England, was delivered to the Royal Society. The merchant had been instructed to deliver it to a Signor Francisco Riva, who was a Bolognese on Queen Mary's staff, to present it to the Royal Society. Riva had received it, Malpighi wrote, but refused to deliver it because he thought it might harm him in some way. Would Ferrarini please request the Duke, Malpighi adds, to get his representatives in England to get the copy from Riva and deliver it to the society (AC:695)? In his letter to Wallis in the same month, Malpighi told him that in case Riva had not delivered the promised package to the Duke, he had given the Duke a second copy to be sent as soon as possible (AC:698).

Further complications occurred when Malpighi discovered he wished to make corrections to the original manuscript. In his next letter to Ferrarini on September 5, 1688, Malpighi asked him to make those corrections to the copy he had sent and that he had been told the Marchese Ragoni would deliver in London (AC:700). Malpighi wrote him again on September 8, 1688, saying he was returning a corrected copy of the gland epistle back to him. Adelmann notes: "The copy which Malpighi had sent to Modena for transmittal to Rangoni (and which he had asked Ferrarini to correct) had evidentially been returned to him. . . . Malpighi was now sending back the note listing his corrections, so that they could be made on the manuscript copy of his work on glands which he had sent to Modena for the Duke" (AC:701). Malpighi wrote to the physician again on November 17, 1688, informing him that he had been told that Rangoni had been robbed after leaving Paris, and although he regretted the Marchese's losses, that as far as his own epistle is concerned, he fears that fate has destined it for the fire (AC:711).

On November 17, 1688, Malpighi received a letter from Bellini acknowledging the receipt of a copy of the epistle on the glands. He extolled the work and asks permission to have it printed (AC:711). On November 21, 1688, Malpighi wrote to Ferrarini and told him that he was glad to "hear that Rangoni has reached London and hopes that his epistle on the glands has already been delivered" (AC:712). On December 13, 1688, he wrote to the physician again and told him that both the Riva and the Ragoni copies had been delivered to the Royal Society (AC:716). Indeed, in a letter dated December 15, 1688, Wallis wrote to Malpighi that both copies had been received and that the society had ordered the epistle to be printed (AC:719). On the same day, the Duke Borghesi wrote Malpighi to tell him he had received a copy of the epistle on the glands through his physician Lucantonio Gestaldo (AC:718). On January 19, 1689, the Count Rinieri Merecotti also acknowledged receipt of a copy of the manuscript (AC:728). Finally, in March 1689,[12] Wallis wrote that fifty copies of the publication on the gland were "ordered to be sticht in marbled Paper and to be sent to the author" (AC:736). Almost one year later (April 21, 1690) Malpighi received a letter from Gaudenzio Roberti, one of the editors of the *Giornale de' Letterati* published in Parma, that the gland epistle would be printed "when the printers are at liberty (AC:821).

About those corrections for which Malpighi had shown concern, Adelmann reports from the record of the Royal Society's meeting of December 31, 1690, at which Richard Waller, the Society's secretary, read a letter he had written to Malpighi. Adelmann says:

> As for the mistakes in the epistle on glands, Waller regretted that they had been made; but the world would now see them corrected as carefully as possible, for they would be listed on a sheet to be pasted at the end of each copy. . . . Waller's promise to print a leaf of errata to be pasted in the copies already printed may never have been kept. Richard Chisell, the printer of the original edition of 1689, failed even in his reprint of 1697 to make the changes Malpighi had requested, and none of them is made in the reprinting of 1690 and 1698 or in the manuscript in the *Biblioteca Estense* at Modena. Three of the four desired corrections, however, have been made in the manuscript in the library of the Royal Society, and the same three are found in the reprint of the epistle contained in the second edition of the LeClerc and Manget's *Bibliotheca Anatomica.* All four appear in the reprintings of the epistle contained in the *Giornale de' Letterati* for 1689 and in the *Galleria di Minerva* for 1696, as they do at Bologna in the manuscript in the *Biblioteca Universitaria.* (AE:705–706)

Malpighi became aware, when he saw a copy of the book while visiting a German professor in 1691, that the epistle he had sent to the Royal Society had gone to the printer before his corrections could be made (AC:813). At least six separate copies of the manuscript letter can be identified in this account. This means it must have been recopied at least that many times with the attendant possibilities of the errors that might occur. While it probably does not represent the history of the typical erudite letter, it reveals some of the hazards of the modes of transmission by which they were served in this period.

THE REPUBLIC OF LETTERS

Writers of scholarly letters in the seventeenth and eighteenth centuries, by virtue of their activities, were regarded as members of a Republic of Letters. It was a Republic of Letters not only in the sense that orthography, the ability to put letters of the alphabet together in a meaningful way, was involved, but also in that epistles were the mode in which the letters were transmitted. Letters were the glue that held the Republic of Letters together. The republic was inhabited by "gens de letters," or "gentlemen of letters," who are frequently found cited in title pages and prefaces as responsible parties in seventeenth- and eighteenth-century journals. The republic was sustained by individuals whose contacts frequently were only by letter and who may have had little in common in national origin, religion, or political persuasion. The word "republic" has a definitive meaning in a political sense that does not strictly apply to the Republic of Letters. However, it is useful in an evocative sense in which the members found in the term a means of defining themselves. Though it had no overt leadership or organization, the republic was maintained by mutuality of interest. Although responses to letters could not be commanded, they required an answer as a matter of courtesy and civility. As Dena Goodman says: "Reciprocity was the fundamental virtue of the republic" (Goodman 1994:18).

Before the emergence of the scientific journal, letters were the primary modes by which scholars could exchange ideas and establish their property rights in their ideas. "Seventeenth-century scientists," says Marie Boas Hall, "were obsessed with the necessity for communication and the interchange of information to a degree hardly known in earlier centuries" (Hall 1981:179). Saskia Stegman amplifies this idea:

> It was generally acknowledged that no scholar aspiring to play a role of any significance in international learned circles could afford not to have an extensive network of correspondents. Witness the testimony of the enormous collections of letters amassed by many seventeenth-century scholars such as Leibniz, Mersenne or Rivet. A considerable portion of their working day was taken up by writing letters and nurturing their contacts.

Stegman amplifies the statement by showing how a "relatively unknown scholar," Theodorus Janssonius van Almeloveen (1657–1712), was able to have correspondents in France, Germany, and Italy, and leave behind a collection of some three thousand letters (Stegman 1993:227).

Many scholars have been engaged in discussing the Republic of Letters as a concept. In his survey of the subject, Ultee cites what he calls a "veritable academic industry that has arisen over the past half century" (Ultee 1987:96). Although it has been suggested that the idea dates back to antiquity (Daston 1991:370),[13] it is generally agreed that it did not begin to flourish until the seventeenth century when the impact and wide distribution of printed materials became more readily available and scholars began to gain greater recognition. In the seventeenth and eighteenth centuries, the Republic was identified as an idealized utopian state. A definition that is frequently cited is that of the philosopher Pierre Bayle (1647–1706), who in his *Dictionnaire Historique et Critique* (1697), states: "Liberty reigns over the Republic of Letters. This republic is an extremely liberal state. It recognizes only the rule of truth and reason" (Goodman 1994:12).

Bayle incorporated the term into the title of the important review journal that he edited from March 1684 to February 1687, *Nouvelles de la République des Lettres*. It was cited as a model by the anonymous editors in the preface to the first volume of their *Bibliotheque Raisonnée des Ouvrages de l'Europe*, which appeared in 1728 and, like the *Nouvelles*, was published in Amsterdam. The "republic" is described rhapsodically in 1700 by Vigneul-Marville in his *Melanges d'Histoire et de litterature*: "Never has the Republic been so great, or more populous, freer or more glorious. It extends over the whole earth and is composed of people of all the nations, of all positions, ages and sexes; not even women or children are excluded. All sorts of languages are spoken there, both living and dead. Art, literature and crafts all have a place there" (Wacquet 1989:485).

Voltaire is equally flattering in this passage from his *Le Siècle de Louis XIV* (1751): "A republic of letters was gradually established in Europe, in spite of wars and different religions. Every science, every art, was mutually assisted in this way, and it was the academies which formed this republic . . . true scholars in every branch drew closer the bonds of this great fellowship of intellect, spread everywhere and everywhere independent" (Ultee 1987:97).

The idea was widespread. In its annual volume for 1780 (published in Amsterdam in 1781), the editor of *Histoire de la République des Lettres* reports: "There exists a certain empire which holds sway over only the mind, that we honor with the name Republic, because it preserves a measure of independence and because it is almost its essence to be free. It is the empire of thought" (Daston 1991:367–68).

A description of the republic seen in the literature frequently is that of Annie Barnes:

> The idea of a utopian state is ancient but the concept of a Republic of Letters does not seem to have arisen until the first part of the Seventeenth century. . . . It was a state that was strongly democratic, in which birth had no part and only knowledge placed each citizen in his appropriate rank. Differences of nationality as well as religion was effaced; an orthodox prince could correspond with an Anglican Bishop and a pastor in Geneva. It had a language which was international: Latin—and later French. The first duty of each citizen was to serve "les lettres" and the best way to do so was to participate in a system of exchanges. This was accomplished by a vast correspondence which covered the entire continent. (Barnes 1938:13–14)

These quotations also could be applied not only to the correspondence in this period, but ideally move beyond it to the whole system of scientific and scholarly communication. On ceremonial occasions, which are characterized more by sentiment than strict adherence to the truth, they could be used today to describe these systems as they exist today.

The Republic of Letters, like other lofty conceptions, was not quite as utopian as its celebrants proclaim. Ultee cites an anonymous contributor to the *Bibliothèque Raisonée des Ouvrages des Savans l'Europe* in 1747, who said: "The Republic of letters does not lack people who work in its service. But it is in a state of anarchy, having neither a leader, an archive, a chancellor or a unified center." The article appears under the title: "A project for the establishment of a general office for the republic of letters." Ultee adds that the anonymous author fails to note that there had been efforts of this kind in the past, and that there would be others in the future (Ultee 1987:104, 111 n. 33). Class did indeed play a part, as we have seen in the copies of his conglobate gland letter that Malpighi sent to his noble Italian patrons. It was not always easy to gain entry into or to establish links to the republic, as Alexander Garden discovered when he wrote to Linnaeus on March 15, 1755, commenting of Linnaeus' *Fundamenta Botanica*: "Such neatness! such regularity! so clear and supremely ingenious system, never appeared before in the Botanical world" (Smith 1821:284). Garden wrote Linnaeus three more letters over a period of four years before receiving a response. The reasons for the tardiness of Linnaeus' response are reflected in a letter he wrote on January 1, 1767, to John Ellis, an amateur but productive English botanist, to whom Garden had also been writing: "I am every day receiving letters, for the most part, if not altogether, to little purpose; filled with compliments indeed; but when I review the accumulation of last year, I derive, on the whole, but little real benefit. Yours indeed are always rich in agreeable and learned information" (Smith 1821:194).

The Republic also was not entirely devoid of national differences, particularly in the case of priority disputes in which scientists tended to champion their compatriots. Correspondence played a large part when letters were freely circulated and a claimant could argue that his opponent had an opportunity to see letters in which his ideas had been disseminated. In the seventeenth century, both Newton and Leibniz, by different routes and at separate times, arrived at the calculus, a significant development in the history of mathematics. It resulted in what A. Rupert Hall has called a "bitter quarrel between two of the greatest men in the history of thought the most notorious of all priority disputes" (Hall, A. 1980:ix). Newton made his case in a collection of letters, which were published under the title *Commercium Epistolicum*, in which he claimed the discovery of the calculus largely on his correspondence with Barrow and Collins, and to which Leibniz had access as well two letters he had written to Leibniz himself.

Another famous dispute occurred at the end of the eighteenth century, when it was possible to speak of a French chemistry and a German chemistry. The dispute was based not on a question of priority, but on a difference of opinion on the existence of a hypothetical chemical element, phlogiston, which had been identified by the German chemist Georg Ernst Stahl (1660–1734). The French chemist, Lavoisier, proved that phlogiston did not exist, and in the process revolutionized chemistry. His work, however, was disputed by German chemists for many years.[14] The controversy was played out in the scientific journals that had developed in the eighteenth century, but correspondence also had a significant role (Goupil 1990).

The importance of the republic of letters to the development of the scientific journal is reflected in those periodicals that incorporated the word republic into their titles. First and foremost, as we have seen, is *Nouvelles de la République des Lettres*, published in Amsterdam (1684–1718), edited by the French philosopher Pierre Bayle from 1684 to 1687. It provided a vehicle for scientific news, original reports, and reviews. There were two journals in the eighteenth century that paid tribute to Bayle by assuming the same title. Another *Nouvelles de la République des Lettres* was published in Laussane between 1775 and 1777. It was continued for one issue under a new title, which more adequately expresses its scope: *Observations Periodique sur la Philosophie et des Belles-Lettres*. In 1778, Pahin de Champlain de la Blancherie expanded the title to read *Nouvelles de la Republique des Lettres et des Arts*. The journal was designed to supplement a society established to exchange ideas by letters and meetings. It had an ambitious but perilous existence from 1779 to 1788. Two other journals that incorporate the republic in their titles are *Histoire Critique de la Republique des Lettres*, published in Utrecht and Amsterdam between 1712 and 1718, and the annual review issued under the title *Histoire de la Republique des Lettres en France*, a review journal covering the years 1779 to 1782 published in Amsterdam. The Dutch also paid homage to the republic with the titles *Republyk der Geleerden*, published in Amsterdam from 1710 to 1711 (based largely on material extracted from the *Journal des Sçavans*), and *Het Republic der Geleerden of Boekzaal van Europa* (Wacquet 1989:482). Despite the success of the scholarly and scientific journal in creating an effective communication channel, correspondence between scientists and scholars continued to provide a "more personal form of scientific collaboration than journals" (Dooley 1991:118).

NETWORKS IN THE REPUBLIC OF LETTERS

The word "network" has also been used frequently to describe the dissemination of correspondence by scholars in the seventeenth and eighteenth centuries. Lines of correspondence were opening up between a great many of the scholars in this period, but as William Eamon indicates: "All these channels of communication, however, were more or less informal and restricted to small circles of intimates" (Eamon 1990:353). Strictly, the term "network" should be used only with organized systems. Networks can appear in many configurations, centralized or distributive, radial or sequenced, but usually in a prefigured arrangement. Few systems of this kind seem to have existed in the widespread distribution of letters in this period. The term "network," therefore, like the term "republic," can only be regarded as a metaphor when applied to the distribution of letters at this time. One network, which can be said to have been organized, was established in 1763 by François Richard de Hautesierck to gather morbidity, mortality, meteorological, and clinical data from physicians and surgeons in French military hospitals (Cilleuls 1972). The results were issued under the title *Recueil d'Observations de Médecine* (Paris, 1776), having been originally published as a part of the *Journal de Médecine, Chirurgie et de Pharmacie*. However, for Cilleuls to refer to *Recueil* as a precursor of the medical press in France is to ignore not only *Journal de Médecine* (Paris, 1754–1794), but especially the very first medical journal of the medical entrepreneur Nicolas de Blegny, whose *Nouvelles Découvertes sur Toutes les Parties de la Médecine* appeared in Paris in 1679 (Kronick 1960).

Another example, also French, is that of the physician Felix Vicq d'Azyr, who established a "medical Commission in Paris for correspondence with physicians in the provinces for everything relative to epidemic and animal diseases" in 1776 (Hannaway 1974:81ff). Rusnock states that: "During the eighteenth century, correspondence networks were frequently organized by physicians and generally were designed to gather medical intelligence, for example, on the incidence and severity of diseases within different communities" (Rusnock 1999:159).

A more appropriate word, perhaps, to describe major participants in the dissemination of information in this period is the term "gatekeeper." It was introduced in communications research in the 1940s to designate individuals who acted as information sources for others. The concept has been investigated in a wide variety of environments, of which the distribution of letters could be an example. In that sense, there were a number of "gatekeepers" in the seventeenth and eighteenth centuries, of which only a few can be cited here. Henry Oldenburg is, of course, the archetypical "gatekeeper" of this period (Hall, M. 2002), but his contributions took place in a more structured environment than most of the other correspondents. Not every correspondent can be regarded as a gatekeeper; a readiness to share information and a sustained interest in doing so is a requirement. Throughout this early period there were several individuals who took on the role voluntarily, by reason of their wide acquaintance and interest. One of the most active gatekeepers in England was John Collins (1625–1683). He has been described as "an amateur scholar so eager and enthusiastic in the cultivation of correspondence that he became the scientific gazette of his time" (More 1934:69); as "the mathematical 'intelligencer' for the Royal Society" (Scriba 1964:50); as "a broker between eminent scholars" (Hunter 1981:96); and as "a mathematical impresario who had made it his business to foster his favorite subject." To these characterizations, Westfall adds: "To that end, he served as a clearinghouse for information, attempting by his correspondence to keep the growing mathematical community of England and Europe abreast of the latest developments" (Westfall 1993:67–8). In this capacity he also played an important part in the Leibniz/Newton dispute on the invention of the calculus, as we have seen.

Peter Collinson's (1693–1768) involvement as a gatekeeper was in some ways similar to that of Collins. Collinson was a successful merchant rather than a civil servant like Collins, but both of them were fellows of the Royal Society of London and, in that capacity, could submit letters they received from their correspondents to be read to the society and to be considered for publication in *Philosophical Transactions*. Collinson was largely self-taught as a botanist but had acquired enough wealth to pursue his interest in gardening and botany. He is credited with having introduced a hundred and eighty new species of plants into England, of which fifty were from his American correspondents. He acted as an agent for the Library Company of Philadelphia for the purchase of books and, in that way, began to correspond with Benjamin Franklin. Franklin sent him accounts of his electrical experiments, some of which were read before the society. George Frick, in his biography of Collinson in the *Dictionary of Scientific Biography*, says: "In the period of his greatest activity, from 1730s to near the time of his death, Collinson was at the center of a network of scientific intelligence which reached from Peking to Philadelphia, but which was concerned principally with the communications of American with European and English savants" (Frick 1971:349–51). His correspondence with Franklin as we have seen, was of both a private and a public nature. In the letters in which Franklin described a hypothesis for which he had not yet tested in an experiment he asked Collinson to return the letters after he had read them (Cohen 1941:viii).

Collinson is also described as "one of the earliest and most constant correspondents of Linnaeus" (Smith 1821, vol. 1:3). In a strange but indicative instance, he involved Linnaeus in a dispute he had with a German functionary who had published a statement avowing that swallows do not migrate but instead hibernate under water. Collinson refuted this "theory" in a paper he read before the Royal Society on March 9, 1758, in which he called the theory "contrary to nature and reason." He sent a copy of the paper to the German author who he addressed as "my learned friend Klein." He also sent a copy to Linnaeus in the same year asking him for his opinion on the issue. Linnaeus apparently did not reply because Collinson wrote him again on May 20, 1762, observing that "Some years have passed since I sent my dear Linnaeus my dissertation on the migration of swallows, but I have received no satisfactory response." Linnaeus apparently responded to this reminder since when Collinson wrote him letter again on September 15, 1763, in which he said: "After reading my dear Linnaeus's letter of the 23d Nov. 1762, how can I any longer doubt that Swallows live under water all winter" (Smith 1821, vol. 1:45–59).

There were also a number of writers of erudite letters in seventeenth-century France who qualify as gatekeepers. Among these the names of Peiresc and Mersenne are paramount. Along with Thevenot, Montmor, Bourdelot, and the Dupuys, they were also successful in bringing groups together in informal academies that preceded the foundation of the Académie des Science in Paris. Nicolas Claude Fabri de Peiresc (1580–1637), as a man of wealth, a magistrate, and member of a provincial parliament, was able to apply himself to the scientific interests of his time (Tolbert 1992). He is credited by his colleague and biographer, Pierre Gassendi (1592–1655), with establishing certain norms of relationship among scientists by appealing to them to avoid quarrels, to live in concord, and to be careful in making judgments (Pomian 1974:28). He had spent some time as a student in Padua and as a visitor of one of the earliest scientific academies, the Academia dei Lincei, of which he became a member. As a result of his wide connections, he became a principal channel for information about Italian science in France (Caton 1988:77). Although he never published any of his own work, he was known widely through his correspondence (Bernhardt 1957:165) His letters over a period of forty years number more than ten thousand. In his biography of Peiresc, Gassendi says that he was so busy with his correspondence in his later years that he had no time to read books (Sarasohn 1993:77). Despite being based in the provinces, he was able to get scientists in several countries to work together. By means of correspondence, he was able to mobilize astronomers in France, Italy, Germany, and North Africa to observe the eclipses of the moon on March 3 and August 27, 1635 (Nellen 1993, 91). Pierre Bayle said of him: "Never has any man rendered greater services to the Republic of Letters" (Pomian 1974:30). Several collections of his letters have been published, including a definitive edition by Tamizey de Larroque, who noted that he "like most men of his time who were celebrated in literature and in science, had the habit of keeping copies of all the letters which they sent (Tamizey 1888, vol. 1:ii).

Marin Mersenne (1588–1648), a member of the Franciscan order of Minims, has also been given a central position in the scientific communication of the seventeenth century. Despite the fact that he was only eight years older than Mersenne, Peiresc can be considered his mentor as well as one of his correspondents. A. C. Crombie's reference to the systematic way in which Mersenne went about making his epistolary connections would seem to refute the notion that organized networks did not exist in the seventeenth century: "From about 1623 Mersenne began to make careful selection of savants who met at his convent in Paris or corresponded with him from all over Europe and as far afield as Tunsia, Syria and Constantinople" (Crombie 1974:316).

In another reference, Mersenne is described as functioning as the "secretary of scientific Europe" (Dolby 1977:8). There is a suggestion of the existence of a network in Aarslef's description of the channel, which Mersenne opened up between France and Holland through Constantijn Huygens (Christiaans's father), who is said to have regularly transmitted mail from Mersenne in Paris to recipients in Holland, including Descartes (Aarslef 1976:368). Mersenne conducted a correspondence with many of the scholars of his period concerning not only questions of science, but also theology. C. DeWaard, in his article on Mersenne, reproduces an account of his correspondents published by one of his confreres in the order that covers three pages of text and includes many of the outstanding scientists of the seventeenth century (DeWaard 1948:14–17). The memory of Mersenne was still alive in the last third of the eighteenth century when the Abbé Rozier said in the preface to the first volume of the new series of his journal *Observations sur la Physique*:

> It is known that the cabinet of Father Mersenne was established as a Bureau of correspondence where one could address foreign scholars, and the place where those French scholars who occupied themselves with physics could meet. It was there that Pascal brought the results of his experiments on the Puy de Dom. (1756)

In the same preface, Rozier mentioned Mersenne again, fully acknowledging the journal as the heir to the exchanges of letters that preceded it (and which showed no signs of diminishing in Rozier's time), that: "Genuine scholars do not have a mania for keeping their discoveries secret; friends of humanity derive glory from being useful, thus it is to these to whom we offer this magazine, as a depository where they can put their discoveries into action. We invite them to regard our assembly, as that of Father Mersenne" (1756).

The gatekeepers described above represent only a small sample taken from the large exchange of letters that took place in this period. Many other correspondents could be said to have played similar roles. Among them are Hans Sloane and Joseph Banks, both of whom, a generation apart, served as presidents of the Royal Society of London, and the brothers Dupuy in France, Pierre (1582–1651) and Jacques (1586–1656), who not only corresponded widely with many of the scholars of the day, but also hosted one of the informal scientific academies in Paris. Sir Hans Sloane

(1660–1753), a physician, was appointed secretary of the Royal Society in 1693. In this capacity, as other secretaries who had served before him, he had responsibility for foreign correspondence and soliciting papers for the society's journal, *Philosophical Transactions*. One of his principal correspondents was the Abbé Jean Paul Bignon (1662–1743) who served as president of the Académie des Science in Paris and played a major role in the reorganization of the society in 1699, and later as editor of the *Journal des Sçavans* in 1702–1714. They were both interested in collecting books, Sloane for his own large library and Bignon as a representative of the king's Library. Sloane built an impressive library for his time of over four thousand volumes, which eventually became the nucleus of the Library of the British Museum (Clarke 1980).

Sir Joseph Banks (1743–1820), who was president of the Royal Society from 1778 to 1820, was a wealthy land owner. Banks' letters, which number over six thousand (Dawson 1958:viii), deal with the various interests that engaged him. Most of Banks' papers were scattered over the world at sale in 1886. The acknowledgements in the preface lists twenty institutions in Great Britain, fourteen academic libraries and learned societies, and fifteen private collections that contributed letters (Dawson 1958:54). In addition to his duties as president of the Royal Society, he was also involved with the Board of Longitude, which was established by the crown in 1714 to encourage development of a method to determined the longitude at sea by offering prizes equivalent of millions of dollars today (Sobel 1995:viii). He served with the Royal Greenwich Observatory and the Board of Agriculture as well. His correspondence is characterized as "probably totaling between fifty and one hundred thousand items (though many of these are no longer extant)—is very largely that of a public man" (Gascoigne 1994:15). Most of his correspondence was with scientists, but they also included exchange of letters with engineers, agriculturalists, instrument-makers, and other craftsmen (Musson 1969:124). During the ballooning trials in France in the 1780s, he was kept informed by letters from both English and French correspondents (Gillispie 1984:261). The large collection of Banks' correspondence was widely dispersed by sale in 1886. Gascoigne comments:

> Some of his descendants regarded this carefully constructed and catalogued collection as little more than a source of spare cash, with the result that much of it was scattered around the globe and, in all too many cases, lost to scholarly scrutiny. (Gascoigne 1994:2)

Fortunately a devoted search was made by Warren Dawson, who was preparing to write a biography of Banks and was able to make copies of a great many which are now preserved at the British Museum (Natural History) in twenty volumes. Publication was considered "but proved to be an undertaking of so vast a magnitude that financial consideration made it impossible." Dawson decided to prepare a calendar or chronologically ordered list instead (Dawson 1958:viii).

The brothers Pierre and Jacques Dupuy corresponded widely with many of the scholars of their day. A large part of Peiresc's correspondence between the years 1617 and 1637 was with the Dupuys (Brown 1934:12). Melchisédech Thévenot (1620–1692), who also gathered scientists for meetings in Paris in the 1660s, has been called "one of the important correspondents linking Paris to the European scientific world" (Gillmor 1976:334–37). Among the many important correspondents in other European countries, the Dutch physician Hermann Boerhaave (1668–1738) played an important role. He is noted as a botanist, chemist, and clinician (Löfstedt 1995:161). He accumulated a large collection of letters from his students and from other scholars throughout Europe and Great Britain (Lindeboom 1962:vii). Lindeboom used letters from five institutions as far apart as London and Montreal. He estimated that "several hundred letters by Boerhaave have not yet been published." In Italy Evangelista Torricelli (1608–1647) was able to publish only one work during his brief life (Gliozzi 1976:433–40). Nevertheless, his work was known "from Rome, where it began, eastward to Warsaw and northwards to Sweden, and had been extensively discussed in France and England" (Hall, M. 1974–1975:176). The Medici librarian Antonio Magliabechi (1633–1714) had an international reputation as a bibliographer (Taylor 1951:111) and is credited with supplying the editors of *Acta Eruditorum*, one of the most important scholarly journals of the seventeenth and eighteenth centuries, with news about scholarly activities in Italy (Dooley 1991:19). Though not particularly involved with science, he numbered such scientists as Niels Stenson (1638–1686) and Leeuwenhook among his correspondents (Scherz 1961; Palm 1983). At the time of death, Magliabechi's correspondence filled 160 volumes (Scherz 1961:25) estimated to contain 20,000 letters, which are preserved in the Biblioteca Nazionale Centrale in Florence (Laeven 1990:177). Besterman claims the same quantity for Voltaire, who he says he had over 1,700 correspondents, from whom there are 20,000 surviving letters. In case we are awed by these figures, Besterman reminds us that over a period of 36 years, Lewis Carroll wrote and received 98,721 letters (Besterman 1969:13). These few examples are merely a token of the traffic that took place in letters among scientists in the seventeenth and eighteenth centuries. The correspondences of many of them, such as Newton, Haller, Lavoisier, Bonnet, Jenner, and Leibniz have been published and some are still in progress, though there are still many letters preserved in archives and other collections that appear only in the published correspondence of other scientists.

SOCIETY CORRESPONDENCE AND CORRESPONDENCE SOCIETIES

There appears to be no evidence to support Abbé Rozier's statement in the preface to the first volume of his *Observations sur la Physique* in 1773: "It seems that in the degree that the number of scholars has increased, the correspondence among those of different nations has decreased" (Kronick 1976:108). I do not know on what data Rozier based this generalization, nor do I have any except my own perceptions to contradict him, but, it appears to me that, on the contrary, correspondence between scholars, foreign as well as domestic, increased with the growth of interest in the new natural philosophy as well as with the increase in the number of societies in the seventeenth and eighteenth centuries as a result of these interests. Scientific academies and societies were founded to meet some of the same objectives that stimulated the development of correspondence and periodical publications like Rozier's, the desire to facilitate communication and discussion among those who had similar scholarly and scientific interests. The scientific revolution and the Enlightenment and even the proliferation of scholarly journals did not stifle the flow of correspondence. It was in fact enhanced by the new channels, the media that became available, and by the creation of scientific and scholarly societies that made correspondence an essential part of their mission. These factors were all reciprocal, as we can see from some of the individuals who have figured in the discussion of the erudite letter, like the brothers Dupuy and Mersenne, who were also involved with the development of scientific societies and salons.

The interaction of these three elements, correspondence, organized societies, and periodical publications, can be seen both in the seventeenth and eighteenth centuries in efforts to integrate them into a single agency. An early example is that of the innovative French physician Theophraste Renaudot (1584–1653) who Castiglioni has called "physician to the King and commissary general of the poor of the Kingdom" (Castiglioni 1941:55). He not only organized an information center he called the Bureau d'Adresse for the exchange of information of all kinds and published the *Gazette*, the first French newspaper; he also conducted a series of public meetings on scientific and other subjects, which gave rise to a series of publications in Paris from 1634 to 1642 under the title *Conferences du Bureau d'Adresss* (Solomon 1972:256–57).[15] Reports of the conferences were published each week. It has been suggested that they might be considered a precursor of the scientific journal (Brown 1934:21). In 1643, he published a work describing a method of treating patients by correspondence, a technique with an ancient history (Solomon 1972:248).[16]

Renaudot's influence can be clearly seen in the efforts of Samuel Hartlib (d. 1662) who came to England from Germany as a young man and established residence there. He became a member of a group of social and educational reformers and attended the informal gatherings of many of the individuals whose efforts led to the formation of the Royal Society of London. He was aware of Renaudot's efforts as early as 1639, but it was not until 1647 that he published his proposal to establish an Office of Publicke Adresse, which would provide information on social services and new developments in science and technology. He also planned to issue a newspaper to disseminate this information as well as information on advances in science and technology (Jones [1961] 1982:151). Hartlib's plan also called for the employment of two secretaries to manage an international correspondence, one for Latin and the second for communications in the vernaculars. He succeeded in receiving a government stipend of £100 per annum to serve as the "Agent for Universal Learning." Although the plan gained interest under the Commonwealth, Webster concludes: "By 1659, a state-endowed institution having ambitious and long term goals for the advancement of learning and its application to the public good was entirely out of place" (Webster 1975:71, 76).

Another attempt along the same lines, pursued by Renaudot and Hartlib more than a century later, is provided by Claude Mammis Pahin de la Blancherie (1751–1811), who came to Paris from the provinces to seek his fortune in his early twenties. In a period in which *lycées* and *musées* flourished in Paris (Guenot 1986) he established the *Correspondance Générale Pour les Sciences et des Arts* in 1781. His mission, however, was much more ambitious than the *lycées* and *musées*; in addition to a society of arts and sciences, he sought to combine the function of promoting correspondence and the publication of a periodical in which these other functions would be reflected. For the periodical, he adapted the title of Pierre Bayle's famous *Nouvelles de la République des lettres* (1684–1718) by adding the words *et des arts*. His program was published in 1779 in *Journal de Litterature, des Sciences et des Arts* (vol. 6:61–67) in the section titled "News of the Republic of Letters," he stated that a "Correspondance générale" was being established as part of a society to establish a rendezvous for authors, artists, and scientists, and particularly for distinguished travelers to the capital. Meetings would be held every Wednesday *chez* Blancherie, except on holidays when they would take place on the following day. The assembly would take place in the morning and last until noon, after which ladies could be admitted. Musicians who wished to exhibit their talents could be given permission to perform either in the morning or the evening of the assembly day. These meetings would not by any means be an open forum; only "those who are distinguished by rank position, or the public profession of science will be admitted."

Blancherie, who established himself as the Agent Générale de Correspondance for the organization, offered to act as a medium for all artists and writers, both domestic and foreign, who wished to obtain information relative to their work or to be put in touch with others who might be interested in their work. He proposed to finance the project by publishing a weekly journal under the title *Nouvelles de la Republique des Lettres et des Arts*, which would appear after each assembly, and would include notices of new works in all parts of the world, discoveries of interest in the arts, notices of the proceedings of academies, and finally a resumé of the activities of the preceding assembly. All those who regularly sent correspondence to the journal would receive free copies, while others could obtain the journal by subscription of 24 livres in Paris and 30 livres elsewhere. The king and queen, he added, and other members of the court, had authorized and encouraged this venture and had subscribed to the journal (*Extrait* 1785). The venture met with success initially, and subscriptions at first apparently provided sufficient funds to support the entire enterprise. Among those who attended the assembly were such distinguished visitors as Volta, Franklin, and D'Alembert (Guenot 1985:60). The journal continued to appear weekly from 1779 to 1787 with the exception of 1784, when financial difficulties caused it to be suspended. It resumed publication in 1785, but eventually the burden of debts became too heavy and Blancherie had to leave for London where he tried without success to establish himself anew (Guenot 1985:49, n. 1).

The development of academies and societies did not serve as a deterrent to correspondence as Rozier suggested; it was rather enhanced by the efforts of the societies to provide continuity and stimulation, and by assigning responsibility to designated members. The enhancement took two general forms: first, in the formation of corresponding societies, that is, societies that existed largely through the exchange of letters, and second, in their efforts to initiate and foster correspondence by incorporating the activity into their rules and regulations. Both forms existed from the development of the earliest societies until well into the eighteenth century. An early example is Accademia dei Lincei[17] founded in Rome by the young Prince Frederico Cesi (1585–1630) in 1603. After it was disbanded by parental and official disapproval in 1604, it was sustained primarily by correspondence until it was able to resume activities in 1610 (Kidwell 1970:22). Another manifestation of what was largely a correspondence society was Societas Christiana, a utopian society that had scientific interests, founded by Johann Valentin Andreae (1586–1654). It maintained its existence primarily by correspondence because its members were widely scattered geographically (Webster 1975:61). A more important and sustained example is that of Academia Naturae Curiosorum founded in Schweinfurt in 1651 by the German physician Johann Laurentius Bausch (1605–1665). After going through many changes in location and in name, it remains in existence today as Deutsche Akademie der Naturforscher

(Leopoldina). At the beginning, its seat changed from one city to another with the change of presidents, and was coordinated by correspondence. It provided the basis for one of the earliest scientific journals, *Miscellane Curiosa Medico-Physica*, which was published in a series of volumes, collected at ten-year intervals from 1670 to 1706 (Herrlinger 1968).

A similar effort to create a corresponding society over a century later was made by a group of Swiss physicians who convened in Zurich 1789. It included Johann Heinrich Rahn (1749–1812), Johann Jacob Römer (1760–1819) and Paul Usteri (1768–1831), each of whom were individually involved with several medical periodicals. Invitations were sent to the physicians of all the Swiss cantons to join a Helvetische Gesellschaft correspondiriende Aerzte und Wundaerzte. The preface to the first volume of *Museum der Heilkunde*, which was issued under the auspices of the society in 1792, included a brief history of the organization along with the regulations under which it was to function. The principal requirement for membership beside payment of yearly dues was stated in the ninth regulation. It specified that any member who did not submit either a useful observation or an abstract from a recent medical or surgical publication in any three-year period would be stricken from the membership. The membership list published in the same volume (p. li–liv) included over a hundred names. Four volumes of the journal were published between 1792 and 1795.

There were two other efforts in the eighteenth century by societies to establish what might genuinely be called correspondence networks. McClellan cites the efforts of Dubois de Fossseux who, as secretary of Académie Royale des Belles Lettres in Arras, France, attempted to coordinate the efforts of the provincial academies that had proliferated in France in the last half of the eighteenth century. Between 1785 and 1792, he sent out nearly fourteen thousand letters to prospective correspondents. By this means, he succeeded in establishing a network that included "some fifty-four learned societies and hundreds of individual correspondents." McClellan attributes the success of the effort to the free postal service that was provided by the French government, and to Fosseux's use of a "collective letter" that could be individualized "so that each correspondent received the individualized communication that best matched his own interests" (McClellan 1985:186).

Another ambitious attempt to set up such a network was made by a "patriotic" society in a small principality near Frankfurt in 1775. This was one of a large number of such societies that were modeled loosely on the flourishing Society for the Encouragement of Arts, Manufactures, and Commerce founded in London in 1754. The Hesse-Homburg society is listed as ninety-third in a chronological list of 117 such societies established between 1731 and 1789 (Im Hof 1982:259–63). The Patriotic Society of Hesse-Homburg[18] was established by the prince under the influence of the French journalist Nicolaus Hyacinthe Paradis (b. 1732) who was then working in Frankfurt. The statutes of the society stated: "The distances between places, the diversity of languages, often the lack of mutual understanding, and, shameful as it is, national prejudices create barriers for the exchange of useful matters, which are always difficult to breach" (Hubrig 1957:62). One of the objectives of the society as stated by the statutes was to facilitate the circulation of information relating to science, economics, and the arts (Hubrig 1957:62–63) The society established contacts with many of the scientific and economic societies in Europe and formed committees in various locations to serve as centers of communications. The initial response to the Homburg society's invitation was widespread. Correspondence committees were initially established in Nancy and Paris. By April of 1777, the society was able to name in their program 143 members from 45 cities in nine European countries. As Hamburg had no printer, the *Bibliothèque du Nord*[19] was published in Paris between January 1778 and January 1780 under auspices of the Hesse-Homburg Society. It appeared irregularly and covered science as well as many other subjects. It probably numbered among its subscribers at least the six hundred or so members of the society and network participants, but does not seem to have left any other mark in the history of scientific communication. Dena Goodman cites two similar attempts to establish correspondence centers for European societies in the eighteenth century by the Société Patriotique de Bretagne and by the Academy at Arras (cited above), both in 1785 (Goodman 1994:26–27).

It seems that the only long-lived efforts to manage scientific correspondence occurred where organized societies were involved. Correspondence was recognized as an essential aspect of a society's activities even before the official societies were established. The group of scientists that met in 1654 at the home of Henri-Louis Habert, seigneur de Montmor, formalized this objective in the seventh of their regulations, which stated: "That the Assembly requests those who have the opportunity should engage in correspondence with scholars in France and foreign countries in order to learn from them what projects they have completed in the arts and sciences, and those which they have discovered or intend to publish" (Bigourdan 1917:132). This suggestion apparently was followed when the Académie des Sciences was established in Paris in 1666. Du Hamel, one of the founding members of the Académie, served as secretary until Fontenelle was named the permanent secretary in 1697. One of principal functions of the secretary was "to see to it that papers or proposals presented at meetings were copied into the minutes" (Stroup 1987:37). Marie Boas Hall says that the Académie des Sciences, in the beginning, did not have "any regular means of communication or publication; its members either published as private individuals, or they published cooperatively, or anonymously and in the name of the Académie." She contrasts this with the practice of the Royal Society of London, whose secretary, Henry Oldenburg, described one of his responsibilities when he informed the Council that he "writes all Letters abroad and answers the returns made to them, entertaining a correspondence with at least 30 persons," a number which increased as the times passed (Hall, M. 1991:55–56).

After Fontenelle was appointed *sécretaire perpétuel* in 1697, and the Académie was reorganized in 1699, a new set of regulations was adopted. Number twenty-seven of the fifty rules and regulations at that time stated:

> The Academy will take care to entertain commerce [presumably in person or in writing] with various scholars, whether in Paris, the provinces of the realm or in foreign countries, in order to be promptly informed about what occurs there of interest in mathematics, and physics; in the election to fill places among the academicians preference will be given to scholars who have been exacting in this kind of commerce. (Fontenelle [1709], 1969:37)

The Académie attempted to accomplish this goal with the appointment of nonresident corresponding members. Hahn describes this function as follows: "Each of these corresponding members resided outside Paris, in France or abroad, and communicated with the society through one of the regular members. No limit to the number of correspondents 'attached' to a regular academician was set, but in practice they rarely numbered as a group more than one hundred, and only a small percentage was regularly engaged in significant scientific correspondence" (Hahn 1971:77). The Académie had correspondents who reported to them on a regular basis (Perrin 1973:426). In his statistical analysis of the membership of the Académie in the period 1899 to 1793, when it was temporarily suspended, McClellan found that the largest category of membership was that of corresponding members, who numbered 389 or nearly 55 percent of the members. Of these, only 32 advanced to other positions (McClellan 1981:547).

Early in its history, the Royal Society of London established a mechanism for creating correspondence linkages and a means for managing them. Among the eight committees established by the society was a committee for correspondence of twenty members that met for the first time on August 16, 1661. The members defined their objective to be the use of correspondence as a means of collecting information from wherever it was available. The committee did not achieve anything, particularly as it lasted little more than a month (Hunter 1989:93). Individual fellows, however, communicated with savants on the continent and shared the information they received with the other members of the society. This activity was not firmly established until Henry Oldenburg became one of the first two secretaries for the society. The society inaugurated a "letterbook" (Altman 1986) as soon as their existence was formally established in 1661. It was one of the responsibilities of the secretary to enter the texts, or at least the substance of all the letters read at the meetings as well as the replies (Hall, M. 1974–1975:178). Many of the letters in the correspondence books remain unpublished. William Wilson submitted a paper in 1746 to the Royal Society of London that was never published. Nevertheless, ten years later he sent the society a retraction that was published in their *Philosophical Transactions* (Singer 1959:167).

We probably have more information about Oldenburg in his role as secretary of a scientific society than any other secretary in the early modern period who has served in a similar capacity.[20] Marie Boas Hall said: "It is not too much to say that he invented the profession of scientific administrator and scientific journalism" (Hall, M. 1965:277). The latter role was intimately related to his skill and industry as a manager of the society's correspondence, which in turn provided him with the inspiration and opportunity to give birth to the first scientific journal, *Philosophical Transactions.* As one of the founding fellows of the Royal Society, he came well-equipped for the role. Before establishing himself in England, he had served as a diplomat for the government of Bremen. As the tutor who accompanied Robert Boyle's nephew on an extended tour of Europe, he had made the acquaintance of many of the leading intellectuals on the continent as well as in England, before the Royal Society was founded in 1660. John Milton, Hobbes, and Spinoza are numbered among his many correspondents before he became secretary. He responded to letters in French, German, and Latin, all languages in which he was fluent. By 1665, most of the society's correspondence was going through Oldenburg. The number of letters reached 319 by 1669, and there were few years when the number was less than 250 (Hunter 1988:168). A. R. Hall, one of the two editors of the Oldenburg correspondence, reported that the number of Oldenburg's correspondents were over two hundred, with only eighty who lived in Great Britain, and that they ranged "from India to Iceland and from the Bahamas to Stockholm" (Hall, A., 1971:112).

After Oldenburg's death in 1677, those fellows appointed to his position had difficulty maintaining the tradition he had established. It was not until 1723, however, that the Council of the Society decided to employ an "Assistant to the Secretaries for Foreign Correspondence." The first appointee to the position was Philip Henry Zollman, who was chosen from among seven candidates. The qualifications were stated as: "1st In their handwriting, 2dly their knowledge in languages. 3dly Their Conversation in Library & Books. 4thly The Sufficiency of their Security they were able to give for their trust" (Massarella 1992:220). Robert Hooke, who was a secretary from 1677 to 1682, indicated what he thought was the scope of the position in his "Designe of the Royall Society" in a document in the Society's "Classified Papers" attributed to him. Among the several "Curators" he proposed:

> The business of the 4th to hold correspondence with as many correspondents as can be procured here in England or in any of our plantations as Newfoundland, New England, Virginia, Barbadoes Jamaica &c. to send to them what is ordered by the Society & to bring in whatever is Returned back againe.
>
> The business of the 5th to hold correspondence with Italy France Spaine, Portugall or any forrein plantation; belonging to them in the east or west indies. The business of the 6th to hold correspondence with the lower & higher Germany Denmark Sweden Poland Hungaria &c. (Hunter 1989:236)

These "Curators," along with the other personnel described by Hooke are, of course, reminiscent of Bacon's ideas about the gathering and organization of information in "Solomon's House" in his *New Atlantis* (Kronick 1976:41–42) The Royal Society also had other methods of maintaining espistolary contacts. Letters could also have been submitted to the society through any fellow, as we have seen in the case of Collinson and Franklin. After Oldenburg's death, the society decided to send a letter to ten of its favorite correspondents, including Malpighi, Hevelius, Leibniz, William Petty, and Isaac Newton (Hall, M. 1978:161). At the Council Meeting of December 10, 1679, the following orders were issued:

> That a particular book be provided for entering all such letters, as are written by the secretaries to the several correspondents at home and abroad; and another for copying all such answers and returns, as come to the Society or secretaries, from abroad or from any part of England, Scotland, or Ireland.
>
> That a committee be appointed to view the former Letter-books, to see how perfect the entries are; and to take an account of all loose letters recovered from the administration of Mr. Oldenburg, and of those, that have come to the secretaries since; and to take care, that those, that are not yet entered, be entered forthwith, and that the originals be with all care got together, in order to their being preserved in such a way as the Society shall direct. Mr. Hill, Sir John Hoskyns, Dr. Gale, and Dr. Croune, or any two of them, were appointed of this committee. (Birch [1756] 1968, vol. 3:515)

Like the Académie des Sciences, the Royal Society also attempted to expand its information network by electing foreign fellows. Between 1622 and 1800, the foreign fellows numbered 571, which was 21 percent of all the fellows that had been elected in that period. The group included some of the most distinguished names in science of that period (Lamoine 1993:169–70).

These two national scientific societies provided a model for other societies that were established in the seventeenth and eighteenth centuries.

Many of them also made provision for correspondence as a means of participating in exchange of information. Among the examples of this practice is the Dublin Society for Improving Husbandry, Manufacture, and other Useful Arts that, at a meeting on December 18, 1731, some six months after it was founded, inserted into their rules the requirement "that the society hold a correspondence with other societies and private persons" (Berry 1915:16). Another example is found in the regulations of the newly formed Académie de Chirurgie. In 1732, it specified that the secretary would give the membership an exact rendering of all the letters that had been received (Ribier 1905:6). It included the following statement: "Since it is important that the Academy be accurately informed about the most interesting developments in surgery, as well as the new books which are related, a number of academicians will be charged with forming and entertaining correspondences with capable surgeons of the realm and also with those of foreign countries" (*Reglement* 1738:2).

Many of the members of other societies in which the regulations did not specifically mention correspondence were also actively engaged in writing letters. An example is the Akademie der Wissenschaften in Göttingen, an offshoot of the University at Göttingen founded in 1734. Relatively isolated as they were from large urban centers, members found correspondence particularly important. The correspondence of such members as Georg Lichtenberg, Albrecht von Haller, Johann David Michaelis, and Christian Heyne kept them abreast of recent developments in science, literature, and the arts (Dougherty 1988:82). Much of this information found its way into a periodical inaugurated by the university, *Göttingische Anzeigen von gelehrten Anzeigen*, which, like the society, still exists.

Other societies in this period made specific arrangements for managing their correspondence. The Academia de Sciencias da Lisboa in 1781 published a brief instruction to guide their correspondents in communicating their observations. In the preface to the first volume of the *Berlinische Schriften* in 1780, the editor informed his readers:

> Since the fourth volume of the Transactions [*Beschaftigungen*] there have been many changes in the society. The growth in the numbers of correspondents have made it impossible for a single individual to take responsibility for the correspondence and to tend to the society's affairs in his leisure hours. The society, therefore, has changed its structure by dividing the various parts of natural history and the correspondence which relate to it among its members. The responses will simply be signed with the society's name to indicate a collective judgement. We also repeat our request to correspondents not to address individual members, but to send their letters to the society. This will enable us to deal with the correspondence more expeditiously. We also ask correspondents to identify the geographic area from which they are writing, particularly as it relates to the nomenclature used. (*Schriften* 1780:vii).

In the same preface the editor complains:

> The management of the *Gesellschaft der naturforschenden Freunde* (Berlin), exasperated by a great increase in their scientific correspondence, decided to drop the usual flowery salutations and compliments, and begged their correspondents to do the same and quickly come to the point. (Heilbron 1979:78)

And finally, there is the medical society that was established in Paris in 1776 in defiance of the venerable Faculty of Medicine, which can serve as an example of both a corresponding society and a network. It began as a "Medical Commission in Paris to Maintain a Correspondence with Physicians in the Provinces on Matters Relating to Epidemic and Epizotic Diseases." The commission led to the formation of the Société Royale de Médecine, which acquired legal status to systematically collect data from 66 correspondents in the provinces, a number that grew to 125, representing most of the provinces in France (Hannaway 1974).

PRESERVATION AND PUBLICATION OF CORRESPONDENCE

A symposium on the importance of correspondence to historians of science and philosophy was held at Chantilly in 1975. The proceedings were published in *Revue de Synthèse* (vol. 97:1–187). Several editors reported on collections that had been published or that were in process and the problems associated with assembling, editing, and publishing the letters of seventeenth- and eighteenth-century scholars. Much has already been done but much remains to be done in filling some of the gaps in published correspondence and in making the unpublished correspondence of the key scientific figures in this period available. Jacques Roger lists some important gaps in publishing correspondence in this period. He includes such important figures as Fontenelle, Bignon, Buffon, Cuvier, and Lamarck (Roger 1976:144–46). Unlike the correspondence of early humanists whose letters were published during their lifetimes, the letters of many of the participants in seventeenth- and eighteenth-century science were not published until long after their deaths. Because of an intense interest in his philosophy, an edition of the letters of René Descartes appeared shortly after his death in 1650: *Lettres de M. Descartes où sont traitées les plus belles questions de la morale, physique, medécine et des mathematiques* (Paris: LeGras, 1657). There were additional editions in 1659, 1663, 1666, and 1668. There were several other collections of contemporary correspondence. In his catalog of the library of Joseph Banks, Dryander lists over twenty collections of this type beginning with *Epistolarum medicinalium* of Joannis Mardi in 1542 and ending with Linnaeus's collection *Epistolarum Que ad Viros Illustres et Clarissimos Scripit Carolus Linné* (Hamburg:

Hoffmann, 1792). Charles Bonnet (1720–1793) encouraged Albrecht von Haller to publish his letters: "The letters which you have received from various scholars will be a bit more interesting for the Public; ordinarily scholars are in their shirtsleeves in their letters; it is the kind of conversation in which the reader loves to be engaged" (Sonntag 1983:8).

Haller did not need much persuasion. Between 1773 and 1775, he published a collection of his Latin letters in six volumes as well as several collections of his correspondence with other individuals.[21] There are numerous occurrences in which letters were published in Haller's lifetime in the work of other authors. Robert Boyle's first literary contribution, for example, was in the form of a letter that was printed in a work of Samuel Hartlib.[22] Some letters addressed to the Royal Society are buried in *The History of the Royal Society* by Thomas Birch, published in London in four volumes in 1757.

Leibniz is said to have been "obsessive about preserving his letters," of which about fifteen thousand are alleged to have survived. Some of his letters were published during his lifetime, such as a selection from correspondence with Jesuit missionaries in China, published in 1697. He intended to publish his correspondence and revise some of the letters he had sent for that purpose (Mason [1967] 1985:ix). Haller was also a prodigious letter writer. His correspondence, preserved in the city library of Bern, originally contained 13,202 letters involving 1,209 correspondents, according to an inventory made by his son (Boschung 1989:214). Many authors kept drafts or copies of their letters in the first place because of the insecurities of the delivery systems, as well as for reference in the future, and sometimes with a view to their being published (Dibon 1976:44). Some correspondents kept copies of those letters they received as well. In some instances, writers kept letter books in which they would keep copies of the letters they received as well as those they sent. In the preface to a facsimile of a seventeenth-century "letter book" in the Folger Library, we are advised that:

> The practice of drafting personal letters as well as official or business communications was much more common in the Elizabethan period than it is now; even educated writers tended to dictate letters to a professional scribe or secretary. . . . Copies were also made for the sender's information, or even by a third party who had some indirect interest in the sender, the address, or the subject of the letter. (*Seventeenth-Century Letter-Book* 1983:10)

Many collections of letters remain in manuscript form, scattered in various archives or private collections. Robert Schofield had to write over thirty letters to Austria, Italy, Sweden, England, Holland, the United States, and Belgium to request permission to publish Priestly letters that institutions and individuals had in their collections (Schofield 1966:ix–x). Although collections of the correspondence of some of the scientists of this era were deposited in institutions by their heirs or the executors of their estates along with their other papers, many others ended up being sold outright or at auction. Undiscovered or missing letters still appear from time to time in dealer catalogs or at auctions. The correspondence of some of the important scientists of this period has been published in collections of other writers. Michelangelo Ricci (1619–1682) made a number of contributions to mathematics, but he published only a sixteen-page booklet in 1666 (Campedello 1975). Many of his contributions are included in letters printed in various collections, such as that of Christiaan Huygens (Huygens 1888). Some of the letters of Adrien Auzout (1622–1691), an astronomer and early member of the Académie des Science, are also to be found in other collections. This is also true of the correspondence of Giovanni Alfonso Borelli (1608–1679), a member of the Accademia del Cimento and one of the outstanding scientists of the seventeenth century. Many of his letters are included in the correspondence of Malpighi, with whom he had a close relationship, and in other collections. The National Library in Florence is said to "contain a great deal of unpublished correspondence to, from, and relating to Borelli" (Settle 1970:313).

This also true in the case of Baron d'Holbach (1723–1789), an active participant in French chemistry and the Enlightenment. Although he was a prolific author, his correspondence is said to be "virtually non-existent" (Rappaport 1994:240). This does not always mean that his letters may not be discovered at some time in the future. Unpublished letters of key figures in early modern still science continue to appear from time to time. The editor of the correspondence of the French philosopher Claude Adrien Helvétius (1715–1771) echoed this sentiment when he said: "Nothing is more hazardous than to speak of an edition of correspondence as definitive" (Helvétius 1981, vol. 1:xv). The correspondence of Pierre Simon Laplace (1744–1829), the French scientist whose important contributions in astronomy and mathematics span the period before and after the Revolution (Singer 1959:311), has an interesting history. The collections were reported as having been destroyed in 1925 but surfaced later in the catalog of a Parisian book seller. It was purchased by the Bancroft Library of the University of California. It was calendared by Roger Hahn after consulting over ninety institutional sources in fourteen counties for the second edition (Hahn 1994:preface). All of this correspondence represents an irreplaceable and valuable resource for the history of science. Although much effective use has been made of it, it still offers opportunities for productive exploitation.

CORRESPONDENCE IN THE "INVISIBLE COLLEGE"

There is yet another metaphor to consider in discussing the commerce of letters in the early modern period, that of the "invisible college." The term has been attributed to Robert Boyle and it serves well to describe the relationships of a group of individuals who were interested in similar ideas, such as the new philosophy, and who communicate with each other by letters. Such groups may be considered "colleges" in that they are collegial and "invisible" in the sense that they are not formally organized. Sustained correspondence linkages were in place among similar groups long before scientific societies were organized and are apparent from studying some of the collections of letters that period. There has been some controversy about which of two groups the young Robert Boyle was referring to when he first used the term "invisible college." In three letters he wrote between 1646 and 1647, Thomas Birch in his *History of the Royal Society of London* refers to John Wallis who identified the group with the "worthy persons residing in London who were inquisitive into natural, and the new and experimental philosophy [who met weekly] to discourse upon such subjects" (Birch [1756] 1968:1–2). Robert Gunther associated the term with the group that met in London in the 1640s that included many of the founders of the Royal Society. It "was sometimes known," he says "by the name of the Invisible or Philosophical College." Boyle used both terms in his letters. The group established itself in Oxford as the Philosophical Society of Oxford when John Wilkins moved there in 1648 (Gunther [1925] 1968, vol. 4:1).

There are equally cogent arguments associating the term with the London group of followers of the reformer, Johann Amos Comenius (1586–1654), which included Samuel Hartlib and Theodore Haak, with both of whom Boyle had connections. Webster says: "Following Bacon Comenius encouraged the college to communicate with colleagues throughout the world. Hence the society might become an Invisible College [an institution that] did not exist until Boyle gave it a name." He also suggests that the term was "one of Boyle's characteristically cumbersome metaphors" (Webster 1975:47–60). On the other hand, Michael Hunter, suggests that the "stress on Boyle's links with Samuel Hartlib and his circle [is] perhaps excessive" (Hunter 1995:63). The issue has been discussed at length by other scholars (e.g., Webster 1974; Turnbull 1953). It is an interesting controversy because it reaches into the vigorous intellectual life of the period, but it is irrelevant for a purpose concerned more with the role of correspondence in maintaining the relationship among scholars of the period rather than those the constitution recognized.

Two of the metaphors discussed here that have been used to describe the structure of seventeenth- and eighteenth-century correspondence, "network" and "republic of letters," have also been employed in twentieth-century discussions of scientific communication. The "republic of letters" seems to have fallen by the wayside, primarily, because the "republic" has become too diffuse as a result of specialization and the lack of shared vocabularies among different disciplines. The term "network" probably has had more consistent use in the interval between the past and the present than the "invisible college." Derek de Solla Price used the term "networks" to describe how patterns of bibliographic references in research papers could be used to define specific research areas by plotting their linkages through citations of research papers (Price 1965). He expanded the concept in the following year, in conjunction with Donald De B. Beaver, in a paper entitled "Collaboration in an Invisible College" (Price and Beaver 1966). It described an experiment in 1961 in which the National Institutes of Health undertook to duplicate and distribute prepublication research papers to groups of scientists who had designated an interest in narrow research field.[23] The article also considered how clustering citations to scientific papers could be used to identify research fronts. Diana Crane used the term "invisible college" to characterize what she called a "group comprised of scientists who work on similar research problems," adding further that "its members are highly individualistic and widely separated geographically. Participation is voluntary" (Crane 1969:335). Some of the terms of the definition may have to be modified to apply it to the correspondents of erudite letters in early modern history, but the essence, the furtherance of collaboration among scientists, is similar in both cases.

It became feasible to study large clusters of citations to research papers with the introduction of the computer-based series of *Citation Indexes* produced by the Institute for Scientific Information. Eugene Garfield, who initiated the system of citation indexes as powerful bibliographic tools, early envisioned the potential of using these databases for the study of the history of science and the humanities, first in 1964 in a monograph titled *The Use of Citation Data in Writing the History of Sciences* (Garfield 1964), and then in a paper he presented to the Third International Congress of Medical Librarians in 1969 (Garfield 1970). He expanded on the concept in his book on citation indexing (Garfield 1979:81–97). The use of citations analysis has been considerably advanced by the development of the study of authors who tend to cite the same research papers ("co-citation" studies) by which shared research areas can be identified (Small 1973). Studies of this kind have appeared in the two decades following Small's paper (e.g., McCain 1990). It is not clear whether this kind of research is amenable to the study of the scientific correspondence of the seventeenth and eighteenth centuries. A sufficiently large volume of this correspondence, published or in inventories, is available to create a database that might be used for this purpose. Although we cannot claim that erudite letters are analogous to citations in research papers, they do represent relationships that can be investigated and studied. There do not seem to have been any formal or ethical standards for citing in the seventeenth and eighteenth centuries, although citations do occur in both explicit and implicit forms. As we have seen in the references to the Malpighi correspondence, not all (perhaps not much) of the erudite letters are concerned with erudite matters; some of them are concerned primarily with housekeeping or personal matters.

The protocols and algorithms that have been developed in modern times to study information networks may not be suitable for studying linkages represented by letters. However, it should be possible to develop software for the purpose. A modern project has been described that indicates not only that, even in the twentieth century, correspondence plays a part in scientific communication and that computers can be used as means to study it. The project, associated with the Office for History of Science and Technology at the University of California, has produced a study relating to the history of twentieth-century physics based on the correspondence of 5,700 physicists who contributed to the changes in the discipline since 1900. Pertinent letters or collections of letters were found, "to half a million letters in archives in thirty-five countries" (Wheaton 1984:155). Special computer software was developed to produce a microfiche record and printed guide to the collection (Wheaton 1993).

How feasible would it be to attempt to create a similar guide to the scientific correspondence of the seventeenth and eighteenth centuries? I have no answer, but it seems to me to be a question that is worth pursuing. Urs Boschung, in his paper on the history of the correspondence of Albrecht von Haller, provides us with some relevant data. The Haller correspondence in the City Library of Bern, according to a list made by his son (Gottlieb Emanuel, 1735–1786), consisted of 13,201 letters from 1,209 correspondents. An inventory made in 1935, however, was able to account for only 12,260 letters involving 1,114 correspondents. Much of this correspondence has already been published in various formats.

Haller was well aware of the importance of correspondence, having written the preface to the letters of his teacher, Thomas Bartholin (1616–1680),[24] and having published a collection of his own correspondence in six volumes during his lifetime. Boschung includes a list of over fifty published collections and individual letters of Haller (Boschung 1989:221–25). Boschung cites a study of the Haller correspondence by Franz Thormann that found that sixty-six correspondents (5.9 percent) who were responsible for thirty or more of the letters provide 68 percent of the total; that 60 percent were from Germany or Switzerland; and the rest were from twelve other countries. Not surprisingly, 62 percent of the letters were from physicians or various scientific disciplines (Thormann 1935). In an appendix, Boschung listed all the correspondents who were related to thirty or more letters. The first six on the list with the number of letters each contributed are: Paul Gottlieb Werlhof (1,585), Johannes Gessner (635), Gerlach Adolf Münchausen (563), Samuel Engel (558),

Samuel August Tissot (485), and Charles Bonnet (483) (Boschung 1989:220–21). Except for Werlhof and Engel, the other four represent important figures in eighteenth-century science and education. Werlhof is identified in the published Bonnet-Haller correspondence as "a court physician at Hannover since 1729 [and] a close friend of Haller and his most prolific correspondent." While this information is of some interest in itself, its importance can be validated only by the research that may be stimulated to determine its significance.

NOTES

1. There exists a large literature on this subject, such as Hall 1980; Feingold 1993; Westfall 1980; and Meli 1993.
2. Epistolary manuals continued to appear well into the nineteenth century. Between 1830 and 1900 in France alone, 195 were published (Chartier 1991:11).
3. Erasmus wrote the preface to the edition of Agricola's book printed by the Froben Press in Basle, but it was written in Latin and thus directed primarily to scholars rather than miners (Pacey 1976:155).
4. This distinction is seldom made in published correspondence where every letter can be significant.
5. The letter was not discovered until the end of the nineteenth century in the library of Trinity College, Cambridge.
6. See Merton (1993).
7. That is, he seemed to be motivated by his personal interests.
8. Dupuy served as a link in one of the important scientific information networks of the period.
9. Edmund Curll has been described as "one of the most colorful and unscrupulous of the Grub Street booksellers" (Pinkus 1968:57).
10. Hereafter, the correspondence will be cited as AC, and the history of embryology, which is paged consecutively through several volumes, will be cited as AE (e.g., AC:642 is equal to "letter no. 642 in the correspondence" and AE:1355 is equal to "page 1335 in the history of embryology." The quotations from the correspondence are from the brief English language abstracts by Adelmann. The letters are published in the original Italian or Latin in which they were written.
11. One of the original members of the group out of which the Royal Society evolved and which "was sometimes known by the name of the Invisible or Philosophical College" (Gunther [1925] 1968, vol. 4:1).
12. From an undated copy in the Royal Society Letterbook. Adelmann puts it between March 23 and 30.
13. The earliest reference to a Republic of Letters that Eisenstein was able to find was in a letter written by a Venetian humanist in 1417 (Eisenstein 1979:387, n. 287). Wacquet cites variants of the term in incunables in 1491, 1492, and 1497 (Wacquet 1989:476).
14. The literature on this issue is as large as the literature on the calculus controversy. See Hufbauer (1982) and Schneider (1989).

15. Translated in England as *A General Collection of Discourses of the Virtuosi of France*, London: 1664.

16. *La Presence du Absents* (Paris, 1642). Thorndike has called it the "earliest scientific society" (Thorndike 1923, vol. 8:35).

17. Thorndike called it "the earliest scientific society" (1923, vol. 8:35).

18. Its full title was *Société Patriotique de Hesse-Hombourg.pour l"Encouragement des Connaissances et des Moeurs.*

19. Its subtitle indicated that is function was "to make known in France what is produced in Germany which would be of interest and utility" (Voss 1980).

20. See Andrade 1965; Bluhm 1960; A. Hall 1962, 1968; M. Hall 1963, 1965, 2002; Hunter 1988, 1989; Oldenburg 1966; Van der Pas 1975.

21. *Epistolarum ab Eruditis vViris ad Albertum Hallerum Scriptarum*. Bern: Socitätis Typographicae, 1773–1775.

22. Samuel Hartlib, *Chymical, Medicinal, and Chyrurgical Addresses Made to Samuel Hartlib*. London: Calvert, 1655.

23. The *National Institutes of Health*'s network to serve the "Information Research Groups," as they were called, however, proved to be too cumbersome, and was abandoned in 1967.

24. See note 21.

REFERENCES

Aarsleff, Hans. "John Wilkins." In *Dictionary of Scientific Biography*. Edited by Charles C. Gillispie. Vol. 14. New York: Charles Scribner's Sons, 1976, pp. 361–81.

Adburgham, Alison. *Women in Print: Writing Women and Women's Magazines from the Restoration to the Accession of Victoria*. London: Allen and Unwin, 1972.

Adelmann, Howard B. *Marcello Malpighi and the Evolution of Embryology*. Ithaca, N.Y.: Cornell University Press, 1966.

Adelmann, Howard B. *The Correspondence of Marcello Malphigi*. 5 vols. Ithaca, N.Y.: Cornell University Press, 1975.

Altman, Janet G. "The Letter Book as a Literary Institution, 1539–1789: Toward a Cultural History of Published Correspondence in France." *Yale French Studies* 71 (1986): 17–62.

Andrade, E. N. da C. "The Birth and Early Days of the *Philosophical Transactions*." *Notes and Records of the Royal Society of London* 20 (1965): 9–27.

Banks, Joseph. *Catalogus Bibliothecae Historico Naturalis Josephi Banks . . . Auctore Jona Dryander*. 5 vols. London: Bulmer, [1866] 1965.

Banks, Joseph. *Catalogus Bibliothecae Historico-Naturalis*. 5 vols. London: Bulmer, 1798–1800.

Barnes, Annie. *Jean le Clerc (1657–1736) et le Republique des Lettres*. Paris: Drosz, 1938.

Baumgartner, F. J. "Galileo's French Correspondents." *Annals of Science* 45 (1988): 169–82.

Beretta, Marco. *A History of Non-Printed Science: A Select Catalogue of the Waller Collection*. Stockholm: Almgvist and Wiksell, 1993.

Bernhardt, Jean. "Les Activities Scientifiques de Nicolas-Claude Fabri de Peiresc (1580–1637)." *Annales Historique de la Revolution Française* (October/December 1957): 165–82.

Berry, Henry F. *A History of the Royal Dublin Society*. London: Longmans, Green, & Co., 1915.

Besterman, Theodore. *Voltaire*. New York: Harcourt, Brace & World, 1969.

Bietenholz, Peter G. "Erasmus and the German Public, 1518–15: The Authorized and Unauthorized Circulation of His Correspondence." *Sixteenth Century Journal* 8 (1977): 61–78.

Bigourdan, Guillaume. "Les Premières Sociétés Scientifique de Paris au 17e Siècle." *Comptes Rendus de l'Académie des Sciences* 163 (1916): 937–43; 164 (1917): 129–34.

Birch, Thomas. *The History of the Royal Society of London*. New York: Johnson, [1756–1757] 1968.

Bluhm, R. K. "Henry Oldenburg F.R.S. (c1615–1677)." *Notes and Records of the Royal Society of London* 15 (1960): 183–97.

Boschung, Urs. "Albrecht von Hallers Korrespondenz und Ihre Erschliessung." *Gesnerus* 46 (1989): 211–27.

Bossis, Mireille. "Methodological Journey through Correspondence." *Yale French Studies* 71 (1966): 63–75.

Boyer, Carl B. *A History of Mathematics*. Princeton: Princeton University Press, 1968.

Brown, Harcourt. *Scientific Organizations in Seventeenth-Century France (1620–1680)*. Baltimore, Md.: Williams and Wilkins, 1934.

Campedello, Luigi. "Rici, Michaelangelo." In *Dictionary of Scientific Biography*. Edited by Charles C. Gillispie. Vol. 11: New York: Charles Scribner's Sons, 1975, pp. 404–405.

Castiglioni, Arturo. *A History of Medicine: Translated from the Italian and Edited by E. B. Krumbhar*. New York: Knopf, 1941.

Catalogo dei Periodici Possedutti dell Università dalla Biblioteca Laurenziana, dalle Accademia a da altra Istitutizione de Firenze. Florence: Università degli Studi di firenze, 1963.

Caton, Hiram. *The Politics of Progress: The Origins and Development of the Commercial Republic, 1600–1835*. Gainesville: University of Florida Press, 1988.

Cavendish, J. M. *A Handbook of Copyright in British Publishing Practice*. 2d ed. London: Cassell, 1984.

Chartier, Roger. "Avant Propos." In *La Correspondence: Les Usages de la Lettre du XIX Siècle*. Edited by Alain Boureau, et al. Paris: Fayard, 1991.

Cilleuls, Jean des. "Un Precursor de la Presse Médicale en France." *Histoire des Sciences Médicales* 6 (1972): 153–59.

Clarke, J. A. "Sir Hans Sloane and Abbé Paul Bignon: Notes on Collection Building in the Eighteenth Century." *Library Quarterly* 50 (1980): 475–82.

Cohen, I. Bernard. *Benjamin Franklin's Experiments*. Cambridge, Mass.: Harvard University Press, 1941.

Constable, Giles. *Letters and Letter Collections*. Turnhout: Brepols, 1976.

Crane, Diana. "Social Structure in Group of Scientists: A Test of the 'Invisible College' Hypothesis." *American Sociological Review* 34 (1969):335–52.

Crombie, A. C. "Mersenne, Marin." In *Dictionary of Scientific Biography*. Edited by Charles C. Gillispie. Vol. 8. New York: Charles Scribner's Sons, 1973, pp. 316–22.

Darriceau, Raymond. "Erasme Éditeur de sa Correspondence." *Revue d'Histoire du Livre*. New series 46 (1985): 433–38.

Daston, Lorraine. "The Ideal and Reality of the Republic of Letters in the Enlightenment." *Science in Context* 42 (1991): 367–86.

Dawson, Warren R., ed. *The Banks Letters: A Calendar of the Manuscript Correspondence of Sir Joseph Banks Preserved in the British Museum, British Museum (Natural History) and Other Collections in Great Britain*. London: British Museum, 1958.

DeWaard, C. "A la Recherche de la Correspondance de Mersenne." *Revue d'Histoire de Sciences* 2 (1948): 13–28.

Dibon, Paul. "Les Échanges Épistolaires dans l'Europe Savante du XVIIe Siècle." *Revue de Synthèse* 97 (1976): 31–50.

Dolby, R. G. A. "The Transmission of Science." *History of Science* 15 (1977): 1–43.

Dooley, Brendan. *Science, Politics, and Society in Eighteenth-Century Italy*. New York: Garland, 1991.

Dougherty, F. W. P. "Correspondence of Lichtenberg." *History of Science* 26 (1988): 77–79.

Eamon, William. "From the Secrets of Nature to Public Knowledge." In *Reappraisals of the Scientific Revolution*. Edited by David C. Lindberg and Robert S. Westman. Cambridge: Cambridge University Press, 1990, pp. 333–65.

Eisenstein, Elizabeth. *The Printing Press as an Agent of Change*. 2 vols. London: Cambridge University Press, 1979.

Extrait des Registres du Conseil d'Administration de la Correspondance Générale et Gratuuit Pour le Sciences et les Arts. December 21, 1785. British Library 9346–349.

Feingold, Modechai. "Newton, Leibniz and Barrow Too." *Isis* 84 (1993): 310–38.

Ficino, Marsilio. *The Letters of Marsilio Ficino*. Translated from the Latin by Paul O. Kristeller. London: Shepheard-Wolwyn, 1975–1981.

Fletcher, John E. "Medical Men and Medicine in the Correspondence of Athanasius Kircher." *Janus* 56 (1969): 259–72.

Fletcher, John E. "Astronomy in the Life and Correspondence of Athanasius Kircher." *Isis* 61 (1970): 52–56.

Fontenelle, Bernard le Bovier de. *Histoire du Renouvellement de l'Académie en 1699*. Brussels: Culture et Civilization, [1709] 1969.

Frangsmyr, Tore, ed. *Science in Sweden: The Royal Swedish Academy of Science 1739–1989*. Canton, Mass.: Science History Publications, 1989.

Frick, George F. "Collinson, Peter." In *Dictionary of Scientific Biography*. Edited by Charles C. Gillispie. Vol. 3. New York: Charles Scribner's Sons, 1971, pp. 349–51.

Garfield, Eugene. *The Use of Citation Data in Writing the History of Science*. Philadelphia, Pa.: Institute for Scientific Information, 1964.

Garfield, Eugene. "Citation Indexing, Historico-Bibliography and the Sociology of Science." *Proceedings of the Third International Congress of Medical Librarianship*. Amsterdam 1970, pp. 187–204.

Garfield, Eugene. *Citation Indexing: Its Theory and Application in Science, Technology, and Humanities*. New York: Viking, 1979.

Gascoigne, John. *Joseph Banks and the English Enlightenment: Useful Knowledge and Political Culture*. Cambridge: Cambridge University Press, 1994.

Gillispie, Richard. "Ballooning in France and Britain, 1783–1786." *Isis* 75 (1984): 249–68.

Gillmor, C. Stewart. "Thévenot, Melchisédech." In *Dictionary of Scientific Biography*. Edited by Charles C. Gillisipie. Vol. 13. New York: Charles Scribner's Sons, 1976, pp. 334–37.

Gliozzi, Mario. "Torricelli." In *Dictionary of Scientific Biography*. Edited by Charles C. Gillispie. Vol. 13. New York: Charles Scribner's Sons, 1976, pp. 433–40.

Golinski, Jan. *Science as Public Culture: Chemistry and Enlightenment in Britain, 1760–1820*. Cambridge: Cambridge University Press, 1992.

Goodman, Dena. *The Republic of Letters: A Cultural History of the French Enlightenment*. Ithaca, N.Y.: Cornell University Press, 1994.

Goupil, Michelle. "Diffusion de la Chimie Française en Europe Occidentale, à Travers la Correspondance de Lavoisier, Celles de Ses Collaboateurs et Leurs Ouvrage Commune." In *Échanges d'Influences Scientifique et Technique Entre Pays Européens de 1780 à 1830*. Paris: Editions du CTHS, 1990.

Guenot, Hérve. "Le Salon de la Correspondance Générale Pour les Science et les Arts de Pahin de la Blancherie (1778–1788)." *Cahiers Hautmarnais* 3 (1985): 49–61.

Guenot, Hérve. "Musées et Lycées Parisiens (1780–1830)." *Dix-huitieme Siècle* 18 (1986): 249–67.

Gunther, Robert T. *Early Science in Oxford*. 14 vols. London: Dawson, [1923–1945] 1968.

Hahn, Roger. *The Anatomy of a Scientific Institution: The Paris Academy of Sciences, 1666–1803*. Berkeley: University of California Press, 1971.

Hahn, Roger. *The New Calendar of the Correspondence of Pierre Simon Laplace*. Berkeley: Office for History of Science and Technology, University of California at Berkeley, 1994.

Hall, A. Rupert. "English Medicine in the Royal Society's Correspondence, 1660–1677." *Medical History* 15 (1971): 111–25.

Hall, A. Rupert. *Philosophers at War: The Quarrel between Newton and Leibniz*. Cambridge: Cambridge University Press, 1980.

Hall, A. Rupert, and Marie Boas Hall. "Why Blame Oldenburg." *Isis* 53 (1962): 482–91.

Hall, A. Rupert, and Marie Boas Hall. "Further Notes on Henry Oldenburg." *Notes and Records of the Royal Society of London* 23 (1968): 33–42.

Hall, Marie Boas. "Some Hitherto Unknown Facts about the Private Career of Oldenburg." *Notes and Records of the Royal Society of London* 18 (1963): 94–103.

Hall, Marie Boas. "Oldenburg and the Art of Scientific Communication." *British Journal for the History of Science* 2 (1965): 277–90.

Hall, Marie Boas. "The Royal Society's Role in the Diffusion of Information in the Seventeenth Century." *Notes and Records of the Royal Society of London* 29 (1974/1975): 173–92.

Hall, Marie Boas. "Leibniz and the Royal Society 1670–1676." *Studia Leibnitiana* suppl.17 (1978): 171–82.

Hall, Marie Boas. "Salomon's House Emergent: The Royal Society and Cooperative Research." In *The Analytic Spirit*. Edited by Harry Woolf. Ithaca, N.Y.: Cornell University Press, 1981, pp. 177–94.

Hall, Marie Boas. "Oldenburg, the *Philosophical Transactions*, and Technology." In *The Uses of Science in the Age of Newton*. Edited by John E. Burke. Berkeley: University of California Press, 1983, pp. 21–47.

Hall, Marie Boas. *Promoting Experimental Learning: Experiment and the Royal Society 1660–1727*. Cambridge: Cambridge University Press, 1991.

Hall, Marie Boas. *Henry Oldenburg: Shaping the Royal Society*. Oxford: Oxford University Press, 2002.

Hannaway, Caroline. *Medicine, Public Welfare, and the State in Eighteenth-Century France: The Société Royale de Médecine of Paris*. Ph.D. dissertation. Johns Hopkins University, 1974.

Hatin, Eugène. *Bibliograpie Historique et Critique de la Pressse Periodique Française*. Paris: Éditions Anthropos, [1866] 1965.

Heilbron, John L. *Electricity in the Seventeenth and Eighteenth Centuries: A Study of Early Modern Physics*. Berkeley: University of California Press, 1979.

Helvétius, Claude Adrien. *Correspondance Générale d'Helvétius, 1715–1771*. 3 vols. Oxford: Voltaire Foundation, 1981–1991.

Henderson, J. R. "Defining the Genre of the Letter." *Renaissance and Reformation*. New series 7 (1983): 89–105.

Herrlinger, R. "Das *Collegium Naturae Curiosorum* und Niels Stensen Beziehungen zu den deutschen Akademikern." *In Steno and Brain Research in the Seventeenth Century*. Edited by G. Scherz. Oxford: Pergamon Press, 1968, pp. 201–72.

Hindle, Brooke. *The Pursuit of Science in Revolutionary America, 1735–1789*. Chapel Hill: University of North Carolina Press, 1956.

Howland, John W. *The Letter Form and the French Enlightenment: The Epistolary Paradox*. New York: Peter Lang, 1991.

Hubrig, Hans. *Die patriotischen Gesellschaften des 18. Jahrhunderts*. Weinheim: Velag Julius Beltz, 1957.

Hufbauer, Karl. *The Formation of the German Chemical Community (1720–1795)*. Berkeley: University of California Press, 1982.

Hunter, Michael. *Science and Society in Restoration England*. Cambridge: Cambridge University Press, 1981.

Hunter, Michael. "Promoting the New Science: Henry Oldenburg and the Early Royal Society." *History of Science* 26 (1988): 165–81.

Hunter, Michael. *Establishing the New Science: The Experience of the Early Royal Society*. Woodbridge: Boydell Press, 1989.

Hunter, Michael. "How Boyle Became a Scientist" *History of Science* 33 (1995): 59–103.

Huygens, Christiaan. *Oevers Complete*. 21 vols. La Haye: Nyhoff, 1888–1950.

Im Hof, Ullrich. *Das gesellige Gesellschaften in Zeitalter der Aufklärung*. Munich: G. H. Beck, 1982.

Jarcho, Saul. "Seventeenth-Century Journalism as Exemplified by the *Ephemerides Naturae Curiosorum*." *Journal of the American Medical Association* 220 (1972): 64–68.

Jones, Richard F. *Ancients and Moderns: A Study of the Rise of the Scientific Movement in Seventeenth-Century England*. New York: Dover Publications, [1961] 1982.

Kidwell, Clara S. *The Accademia dei Lincei and the Apiarium*. Ph.D. dissertation. University of Oklahoma, 1970.

Kirchner, Joachim. *Die Zeitschriften des deutschen Sprachgebietes von den Anfängen bis 1830*. 4 vols. Stuttgart: Hiersemann, 1969–1977.

Koumantos, Georges. Les Correspondances Inédits en Droit Comparée." In *Les Correspondeances Inédits*. Edited by André Françon and Claude Goyard. Paris: Economica, 1984, pp. 201–209.

Koyré, Alexander. "An Unpublished Letter of Robert Hooke to Isaac Newton." *Isis* 43 (1952): 312–37.

Kronick, David A. "The Fielding H. Garrison List of Medical and Scientific Periodicals of the 17th and 18th Centuries: Addenda et Corrigenda." *Bulletin of the History of Medicine* 32 (1958): 456–74.

Kronick, David A. "Nicolas de Blegny, Medical Journalist." *Bulletin of the Cleveland Medical Library* 7 (1960): 47–56.
Kronick, David A. *History of Scientific and Technical Periodicals: The Origins and Development of the Scientific and Technical Press, 1665–1790.* 2d ed. Metuchen, N.J.: Scarecrow Press, 1976.
Kronick, David A. *The Literature of the Life Sciences*. Philadelphia, Pa.: ISI Press, 1985.
Laeven, Hubertus. *The Acta Eruditorum under the Editorship of Otto Mencke (1644–1707).* Translated from the Dutch by Lynne Richards. Amsterdam: APA Holland University Press, 1990.
Lamoine, Georges. "L'Europe de l'Esprit ou la Royal Society de Londres." *Dix-huitième Siècle* 25 (1993): 167–98.
Lindeboom, Herman, ed. *Boerhaave's Correspondence*. Leiden: Brill, 1962.
Löfstedt, Bengt. "Notizen zu Boerhaave's Briefwechsel." *Orpheus* 15 (1995): 161–64.
McCain, Katherine W. "Mapping Authors in Intellectual Space: A Technical Overview." *Journal of the American Society for Information Science* 41 (1990): 433–43.
McClellan, James E., III. "The Scientific Press in Transition: Rozier's Journal and the Scientific Societies in the 1770s." *Annals of Science* 36 (1979): 425–49.
McClellan, James E., III. "The Académie Royale des Science, 1699–1793: A Statistical Portrait." *Isis* 72 (1981): 541–67.
McClellan, James E., III. *Science Reorganized: Scientific Societies in the Eighteenth Century*. New York: Columbia University Press, 1985.
Mason, H. T., ed. *The Leibniz-Arnaud Correspondence, With an Introduction by G. H. R. Parkinson*. New York: Garland, [1967] 1985.
Massarella, Derek. "Philip Henry Zollman, the Royal Society's First Assistant Secretary for Foreign Correspondence." *Notes and Records of the Royal Society of London* 46 (1992): 219–34.
Meli, Bertolini. *Equivalence and Priority: Newton Versus Leibniz: Including Leibniz's Unpublished Manuscript on the Principia*. Oxford: Clarendon Press, 1993.
Merton, Robert K. *On the Shoulders of Giants: A Shandian Postscript*. Chicago: University of Chicago Press, 1993.
Metoyer-Duran, Cheryl. "Information Gatekeepers." *Annual Review of Information Science and Technology* 28 (1993): 111–50.
More, Louis T. *Isaac Newton: A Biography*. New York: Charles Scribner's Sons, 1934.
Musson, A. E., and Eric Robinson. *Science and Technology in the Industrial Revolution*. Manchester: Manchester University Press, 1969.
Nellen, H. J. M. "La Correspondance Savante au XVIIe Siècle." *XVII Siècle* 45 (1993): 87–98.
Oellers, von Norbert. "Der Brief als Mittel privater und öffentlicher Kommunikation in Deutschland im 18. Jahrhundert." In *Brief und Briefwechsel in Mittel-und Osteuropa im 18. und 19. Jahrhnderts*. Edited by Wolfgang Kessler. Essen: Hobbing, 1989, pp. 9–36.
Oldenburg, Henry. *Correspondence*. Edited and translated by A. Rupert Hall and Marie Boas Hall. Madison: University of Wisconsin Press, vols. 1–9, 1966–1973; London: Mansell, vols. 10–11, 1975–1977; London: Taylor and Francis, vols. 12–13, 1986.
O'Malley, C. D. "Bartholin, Thomas." In *Dictionary of Scientific Biography*. Edited by Charles C. Gillispie. Vol. 1. New York: Charles Scribner's Sons, 1970, pp. 482–83.

Ornstein, Martha. *The Role of Scientific Societies in the Seventeenth Century*. Chicago: University of Chicago Press, 1929.
Pacey, Arnold. *The Maze of Ingenuity*. Cambridge: MIT Press, 1976.
Palm, L. C. "Leeuwenhoek and Other Dutch Correspondents of the Royal Society." *Notes and Records of the Royal Society of London* 43 (1983): 191–207.
Perrin, C. F. "Lavoisier, Monge, and the Synthesis of Water, a Case of Pure Coincidence." *British Journal of the History of Science* 6 (1973): 424–28.
Pinkus, Philip. *Grub St. Stripped Bare*. London: Constable, 1968.
Pomian, Krzystof. "De la Lettre au Periodique: La Circulation des Information dans la Mileux des Historiens au XVII Siècle." *Organon* 10 (1974): 25–43.
Price, Derek J. de Solla. "Networks of Scientific Papers." *Science* 149 (1965): 510–14.
Price, Derek J. de Solla, and Donald de B. Beaver. "Collaboration in an Invisible College." *American Psychologist* 21 (1966): 1011–18.
Rappaport, Rhoda. "Baron d'Holbach's Campaign for German (and Swedish) Science." *Studies on Voltaire and the Eighteenth Century* 323 (1994): 225–47.
Reglement Pour une Académie du Chirurgie. British Library, 557d2(13).
Ribier, L. de. "La Fondaation de l'Académie de Chirurgie et le Province." *La France Médicale* 52 (1905): 386–89.
Rigaud, S. J., and S. P. Rigaud. *Correspondence of Scientific Men in the Eighteenth Century*. Hildesheim: Olms, [1841] 1965.
Robertson, Jean. *The Art of Letter-Writing*. London: Folcroft Library Editions, [1942] 1973.
Rochot, Bernard. "Le P. Mersenne et les Relations Intellectuelles dans l'Europe de XVII Siècle." *Cahiers d'Histoire Mondiale* 10 (1966): 55–73.
Roger, Jacques. "Correspandances de Savants et Naturalists aux XVIIIe Siècle." *Revue de Synthèse* 97 (1976): 144–46.
Rose, Mark. *Authors and Owners: The Invention of Copyright*. Cambridge: Harvard University Press, 1993.
Rozier, Jean Baptiste François. *Observations periodique sur la physique*. Preface, vol. 1, 1756.
Rusnock, Andrae. "Correspondence Networks and the Royal Society, 1700–1750." *British Journal for the History of Science* 32 (1999): 159–69.
Sarasohn, Lisa T. "Nicolas-Claude de Peiresc and the Patronage of the New Science in the Seventeenth Century." *Isis* 84 (1993): 70–90.
Scherz, Gustav. "Niels Stensen und Antonio Magliabecchi." *Sudhof's Archiv* 45 (1961): 223–33.
Schneider, Georg. *Theory and History of Bibliography*. Translated by Ralph Shaw. New York: Columbia University Press, 1934.
Schneider, Hans-Georg. "The Fatherland of Chemistry: Early Nationalistic Currents in Late Eighteenth-Century German Chemistry." *Ambix* 36 (1989): 14–21.
Schofield, Robert E. *A Scientific Autobiography of Joseph Priestly (1733–1800)*. Cambridge: MIT Press, 1966.
Schouten, J. "Johannes Walaeus (1604–1649) and His Experiments on the Circulation of the Blood." *Journal of the History of Medicine* 29 (1974): 259–79.
Scriba, Christopher I. "Mercator's Kinckhuysen Translations in the Bodleian Library at Oxford." *British Journal of the History of Science* 2 (1964): 45–58.

Settle, Thomas B. "Borelli, Giovanni Alfonso (1608–1679). In *Dictionary of Scientific Biography*. Edited by Charles C. Gillispie. Vol 2. New York: Charles Scribner's Sons, 1970, pp. 303–14.

Seventeenth-Century Letter-Book. Newark: University of Delaware Press, 1983.

Shapin, Steven. "O Henry. Essay Review: The Correspondence of Henry Oldenburg." *Isis* 78 (1987): 417–24.

Sigerist, Henry E. *A History of Medicine*. 2 vols. New York: Oxford University Press, 1951.

Singer, Charles. *A Short History of Scientific Ideas to 1900*. New York: Oxford University Press, 1959.

Sinnerstad, Ulf. "Astronomy and the First Observatory." In *Science in Sweden: The Royal Swedish Academy of Science, 1739–1900*. Edited by Tore Frängsmyr. Canton, Mass.: Science History Publications 1989, pp. 45–71.

Small, H. "Co-citation in the Scientific Literature: A New Measure of the Relationships between Two Documents." *Journal of the American Society for Information Science* 24 (1973): 265–68.

Smith, James E., ed. *A Selection of the Correspondence of Linnaeus and Other Naturalists, from Original Manuscripts*. 2 vols. London: Longmans, 1821.

Snorrasen, Egill. "Danish Physicians and Periodicals of the Seventeenth and Eighteenth Centuries." *Danish Medical Bulletin* 5 (1958): 206–209.

Sobel, Dava. *Longitude: The True Story of a Lone Genius Who Solved the Greatest Scientific Problem of His Time*. New York: Walker and Co., 1995.

Solomon, Howard M. *Public Welfare, Science, and Propaganda in the Seventeenth Century: The Innovations of Theophraste Renaudot*. Princeton: Princeton University Press, 1972.

Sonntag, Otto, ed. *The Correspondence between Albrecht von Haller and Charles Bonnet*. Bern: Hans Huber, 1983.

Stegman, Saskia. "How to Set Up a Scholarly Correspondence." *LIAS* 26 (1993): 227–42.

Stroup, Alice. "Royal Funding of the Parisian *Académie Royale des Sciences* during the 1690s." *Transactions of the American Philosophical Society* 77 (1987), part 4.

Tamizey de Larroque, Philippe, ed. *Lettres de Peiresc*. 10 vols. Paris: Imprimerie Nationale, 1888–1897.

Taylor, Archer. *The Bibliographic History of Anonyms and Pseudonyms*. Chicago: University of Chicago Press, 1951.

Thormann, Franz. *Register der Briefwechsel von Albrecht von Haller in die Berner Stadtbibliothek*. Bern, 1935.

Thorndike, Lynn. *A History of Magic and Experimental Science*. 8 vols. New York: Macmillan, 1923–1958.

Tolbert, Jane Thornton. *A Study of a Seventeenth-Century Gatekeeper: The Role of Nicolas-Claude Fabri de Peiresc in the Dissemination of Science through the Correspondence Network*. Ph.D. dissertation. University of Florida, 1992.

Torricelli, Evangelista. *Opera*. 4 vols. Edited by Gino Loria and Giuesepe Faenza. Fenza: 1919–1944.

Turnbull, G. H. "Samuel Hartlib's Influence on the Early History of the Royal Society." *Notes and Records of the Royal Society of London* 10 (1953): 101–30.

Turnbull, H. W., ed. *The Correspondence of Isaac Newton*. 7 vols. Cambridge: Cambridge University Press, 1959–1977.

Ultee, Maarten. "The Republic of Letters: Learned Correspondence, 1680–1720." *Seventeenth Century* 2 (1987): 95–112.

Van der Pas, Peter W. "Leeuwenhoek's Correspondence at the Royal Society." *Aere Perenniius* 18 (1975): 3–11.

Voss, Jürgen. "Die *Société Patriotique* de Hessse-Homburg (1775–1781)." In *Deutsche patriotische und gemeinnützige Gemeinschaften*. Edited by Rudolf Vierhaus. Munich: Kraus International Publications, 1980, pp. 195–221.

Voss, Jürgen. "Bibliothèque du Nord (1778–1780)." In *Dictionnaiare des Journaux*. Edited by Jean Sgard. Paris: Universitas, 1991, pp. 184–85.

Wacquet, Françoise. "Qu'est-ce que la Republique des Lettres? Essai de Sémantique Historique." *Bibliothèque de l'École des Chartes* 147 (1989): 473–502.

Waquet, Françoise. "Les Éditions de Correspondances Savantes et le Idéaux de la Republique des Lettres." *XIIe Siècle* 45 (1993): 99–118.

Webster, Charles. "New Light on the Invisible College: The Social Relations of English Science in the Mid-Eighteenth Century." *Transactions of the Royal Historical Society* 24 (1974): 19–42.

Webster, Charles. *The Great Instauration: Science, Medicine, and Reform, 1626–1660*. London: Duckworth, 1975.

Webster, Charles. "Benjamin Worsley: Engineering for Universal Reform from Invisible College to the Navigation Act." In *Samuel Hartlib and Universal Reformation: Studies in Intellectual Communication*. Edited by Mark Greengrass, Michael Leslie, and Timothy Raylor. Cambridge: Cambridge University Press, 1995, pp. 213–35.

Westfall, Richard S. "The Priority Dispute." In *Never At Rest: A Biography of Isaac Newton*. Cambridge: Cambridge University Press, 1980, pp. 698–780.

Westfall, Richard S. "Science and Patronage: Galileo and the Telescope." *Isis* 76 (1985): 11–30.

Westfall, Richard S. *The Life of Isaac Newton*. Cambridge: Cambridge University Press, 1993.

Wheaton, Bruce R. "Inventory of Sources for History of Twentieth-Century Physics." *Isis* (1984): 153–57.

Wheaton, Bruce R. *Inventory of Sources for the History of Twentieth-Century Physics: Report and Microfilm Index to 100,000 Letters*. Stuttgart: Velag für Geschhichte der Naturwissenschaften und der Technik, 1993.

Widmaier, Rita, ed. *Leibniz Korrespondent mit China: Der Briefwechsel mit den Jesuitenmission (1659–1714)*. Frankfurt am Main: Vittorio Klostermann, 1990.

Index

Page references in italics refer to tables.

About the Author

David A. Kronick has been involved in studying the history of scientific communication since he wrote his dissertation at the University of Chicago in 1956. He has spent the major part of his career working in university medical libraries and at the National Library of Medicine. He has had many opportunities to travel abroad to visit other libraries in the United Kingdom and Europe.

Kronick is a fellow of the Medical Library Association. He was also recognized as a "Pioneer in Information Science" at the Conference on the History and Heritage of Information Science in Pittsburgh in 1992. His *History of Scientific and Technical Periodicals*, which has been widely cited, was published by Scarecrow in 1962 and revised in 1976. He is also the author of *The Literature of the Life Sciences* (1985) and *Scientific and Technical Periodicals of the Seventeenth and Eighteenth Centuries: A Guide* (Scarecrow, 1991).